Higher Engineering Science

Higher Engineering Science

Second Edition

W. Bolton

AMSTERDAM • BOSTON • HEIDELBERG • LONDON • NEW YORK • OXFORD
PARIS • SAN DIEGO • SAN FRANCISCO • SINGAPORE • SYDNEY • TOKYO

Newnes is an imprint of Elsevier

Newnes
An imprint of Elsevier
Linacre House, Jordan Hill, Oxford OX2 8DP
200 Wheeler Road, Burlington, MA 01803

First published 1999
Second edition 2004

British Library Cataloguing in Publication Data
A catalogue record for this book is available from the British Library

Library of Congress Cataloguing in Publication Data
A catalogue record for this book is available from the Library of Congress

ISBN 0 7506 6253 0

For information on all Newness publications
visit our website at www.newnespress.com

Printed and bound in The Netherlands

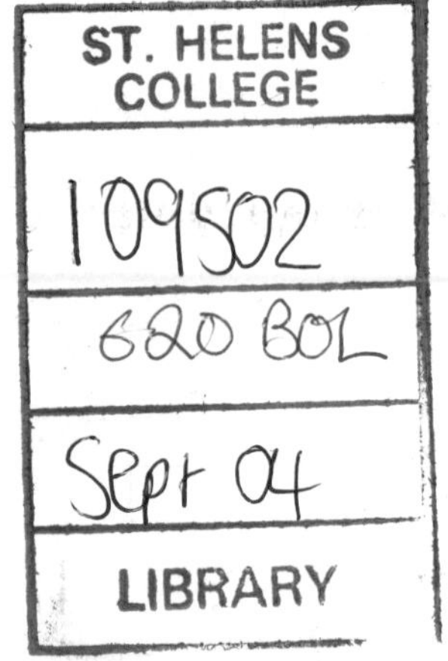

Contents

Preface ix

1 **Structural analysis**
1.1 Introduction 1
1.2 Axial loading 1
1.3 Axially loaded members 6
1.4 Poisson's ratio 10
Problems 11

2 **Beams and columns**
2.1 Introduction 13
2.2 Beams 13
2.3 Shear force and bending moment 17
2.4 Bending stresses 25
2.5 Columns 35
Problems 42

3 **Torsion**
3.1 Introduction 45
3.2 Torsion of circular shafts 47
3.3 Transmission of power 52
Problems 53

4 **Linear and angular motion**
4.1 Introduction 56
4.2 Linear motion 56
4.3 Two dimensional linear motion 63
4.4 Angular motion 66
4.5 Force and linear motion 73
4.6 Torque and angular motion 80
Problems 88

5 **Energy transfer**
5.1 Introduction 94
5.2 Work 94
5.3 Potential energy 100
5.4 Linear and angular kinetic energy 102
5.5 Conservation of mechanical energy 104
5.6 Mechanical power transmission 108

			Problems	110
6	**Mechanical oscillations**	6.1	Introduction	115
		6.2	Simple harmonic motion	116
		6.3	Undamped oscillations	121
		6.4	Damped oscillations	128
		6.5	Forced oscillations	130
			Problems	132
7	**D.c. theory**	7.1	Introduction	135
		7.2	Resistors in d.c. circuits	137
		7.3	Kirchhoff's laws	141
		7.4	Capacitors	149
		7.5	Inductors	160
		7.6	D.c. motor	167
			Problems	170
8	**A.c. theory**	8.1	Introduction	176
		8.2	Phasors	179
		8.3	Reactance and susceptance	181
		8.4	Phasor relationships for pure components	182
		8.5	Impedance and admittance	184
		8.6	Series a.c. circuits	184
		8.7	Parallel circuits	193
		8.8	Power	199
			Problems	206
9	**Complex waveforms**	9.1	Introduction	210
		9.2	The Fourier series	210
		9.3	Circuit analysis with complex waveforms	214
		9.4	Production of harmonics	217
			Problems	220
10	**Transformers**	10.1	Introduction	222
		10.2	Basic principles	222
		10.3	Transformer construction	225
		10.4	Impedance matching	228
			Problems	230
11	**Systems**	11.1	Introduction	232
		11.2	Basic principles	232
		11.3	Information and signals	235
		11.4	Signal processing	239
		11.5	Measurement systems	247

	11.6	Feedback systems	253
	11.7	Noise	258
		Problems	259
12 Control systems	12.1	Introduction	262
	12.2	Basic principles	262
	12.3	Electrical switching	274
	12.4	Speed control of motors	279
	12.5	Problems	282
		Answers	285
		Index	301

Preface

Aims

This book is a comprehensive revision of the first edition of Higher Engineering Science with the aim of covering the revised specification of the Edexcel Level 4 BTEC mandatory unit Engineering Science for the Higher National Certificates and Diplomas in Engineering. It thus differs from the first edition in that some chapters have been removed and replaced by others to enable the new BTEC unit content to be covered. This unit is a broad based unit covering the mechanical principles, electrical principles and engineering system principles underpinning the design and operation of engineering systems and provide the basis for further study in specialist areas of engineering.

The book is likely to contain more material than might appear in any one interpretation of the unit, the aim having been to supply all the material that is likely to appear in any interpretation of the unit.

Structure of the book

The book has been designed to give a clear exposition and guide readers through the scientific principles, reviewing background principles where necessary. Each chapter includes numerous worked examples, self-check revision questions and problems. Answers are supplied to all revision questions and problems. A change from the first edition is the inclusion in the second edition of application notes and activities.

The book can be considered to consist of four main sections, mirroring those in the BTEC unit:

Static engineering systems

Chapter 1: Structural analysis
Chapter 2: Beams and columns
Chapter 3: Torsion

Dynamic engineering systems

Chapter 4: Linear and angular motion
Chapter 5: Energy transfer
Chapter 6: Mechanical oscillations

D.c. and a.c. theory

Chapter 7: D.c. theory
Chapter 8: A.c. theory

Chapter 9: Complex waveforms
Chapter 10: Transformers

Systems

Chapter 11: Systems
Chapter 12: Control systems

Performance outcomes

The following indicate the outcomes for which each chapter has been planned. At the end of the chapters the reader should be able to:

Chapter 1: Structural analysis

Solve engineering problems involving the axial loading of structures.

Chapter 2: Beams and columns

Draw shear force and bending moment diagrams for beams subject to bending.
Determine the stresses arising from bending.
Determine critical buckling loads for columns.
Determine the effects of eccentric loading of columns.
Select standard universal sections for beams and columns.

Chapter 3: Torsion

Determine shear stresses and angular deflections arising from the torsion of circular shafts.
Determine the power transmitted by shafts.

Chapter 4: Linear and angular motion

Determine the behaviour of dynamic mechanical systems in which uniform acceleration is present.
Solve problems involving linear motion with constant acceleration.
Solve problems involving projectiles.
Solve problems involving angular motion with constant acceleration.
Solve problems involving Newton's laws.
Solve problems involving torque and angular motion.

Chapter 5: Energy transfer

Determine the effects of energy transfer in mechanical systems.
Solve problems involving work, potential energy, linear and angular kinetic energy.
Solve problems involving mechanical power transmission by gears and belts.

Chapter 6: Mechanical oscillations

Explain the principles of simple harmonic motion and solve problems involving such motion.
Describe the effects of damping and the forcing of oscillations.

Chapter 7: D.c theory

Solve d.c. electrical circuit problems involving series and parallel resistors.
Determine the transient effects on currents and voltages in circuits involving capacitors and resistors, inductors and resistors.
Explain the basic principles of d.c. motors.

Chapter 8: A.c. theory

Solve single-phase a.c. series and parallel circuit problems by the use of phasors and the use of drawings and simple trigonometry.
Explain and use the terms reactance, susceptance, impedance and admittance.
Determine the conditions for series and parallel resonance and the Q-factors of such circuits.
Determine the power developed in a.c. circuits.
Explain the term power factor, its significance in electrical power transmission, and how it can be improved.

Chapter 9: Complex waveforms

Describe the nature of complex waveforms in terms of harmonics and synthesise graphically such waveforms.
Describe how electrical and electronic circuit elements can produce such waveforms.
Solve problems involving circuits supplied by a constant complex waveform voltage.

Chapter 10: Transformers

Explain the basic principles of transformers.
Explain how transformers can be used for impedance matching.

Chapter 11: Systems

Represent systems by block models and analyse such systems.
Describe the methods by which electrical signals convey information.
Describe the characteristics of electronic systems used for the signal processing operations of analogue-to-digital conversion, digital-to-analogue conversion and amplification.
Describe the basic elements of measurement systems and select and interface components to enable chosen measurements to be made
Explain the uses of negative and positive feedback with systems.
Explain the use of feedback with amplifiers and oscillators.

Chapter 12: Control systems

Describe and analyse open-loop and closed-loop control systems in terms of block diagram models.
Describe methods used for electrical and electronic switching of actuators.
Describe methods used for the speed control of motors.

W. Bolton

1 Structural analysis

1.1 Introduction

Load-bearing structures can take many forms. For example, for the building column shown in Figure 1.1(a) the load of the floors and structure above it are applying forces which tend to axially squash the column. For the simple beam bridge in Figure 1.1(b) the load arising from a car crossing it will tend to bend the beam. For the aeroplane in Figure 1.1(c), the lift forces on the wings will tend to bend the wings. For the electric motor in Figure 1.1(d), the load will cause the shaft to become twisted as the motor rotates it, the loading being said to be torsional. To analyse structures so that we can predict their behaviour when loaded it is usual to consider certain basic forms of loading, namely axial tension or compression, bending and torsion.

In this chapter, axial loading is considered, in chapter 2 bending and in chapter 3 torsional loading. Such analysis is necessary for the safe design of structures, whether they be buildings, bridges, aeroplanes, or motors rotating loads.

1.2 Axial loading

Application
Tensile members of structures can be steel cables or ropes which have no compression resistance. Examples of compressive members are concrete and brickwork columns which are weak in tension but strong in compression

Consider a straight bar of constant cross-sectional area when external axial forces are applied at its ends. If the forces stretch the bar (Figure 1.2(a)) then the bar is said to be in *tension*, if they compress it (Figure 1.2(b)) in *compression*. In structures, a member that is in tension is called a *tie* while a member in compression is called a *strut*.

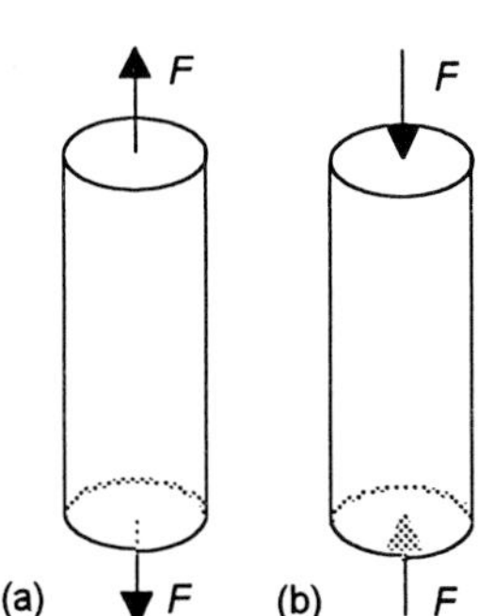

Figure 1.2 *Bar (a) in tension, (b) in compression*

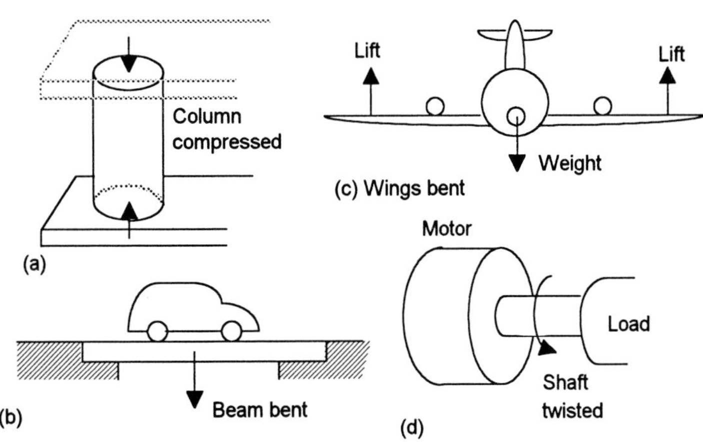

Figure 1.1 *Examples of loading*

1.2.1 Direct stress

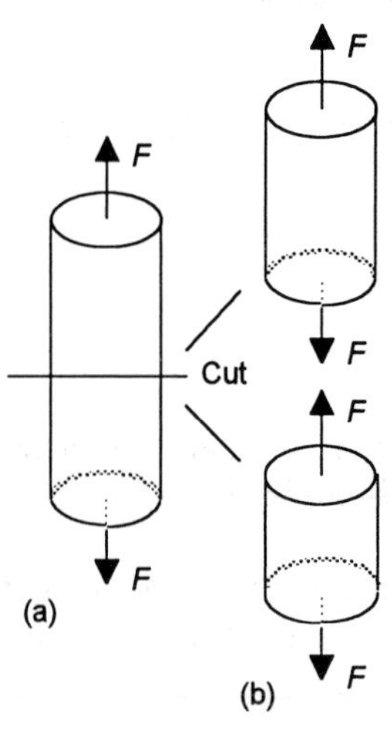

Figure 1.3 *(a) Bar in tension, and (b) with an imaginary sectional cut*

It is necessary in analysing structures to be able to distinguish between external forces, such as applied loads, and internal forces which are produced in structural members as a result of applying the loads. Consider an axially loaded bar (Figure 1.3(a)). We can think of the bar as being like a spring which, when stretched by external forces F, sets up internal forces which resist the external forces extending it. Consider a plane in the bar which is at right angles to its axis and suppose we make an imaginary cut along that plane (Figure 1.3(b)). Equal forces F are required at the break to maintain equilibrium of the two lengths of the bar. This is true for any section across the bar and hence there is a force F acting on any imaginary section perpendicular to the axis of the bar. Thus we can consider, in this case, that there are internal forces F across any section at right angles to the axis. Internal forces are responsible for what is termed *stress.*

The term *direct stress* is used for the value of this internal force per unit area of the plane, with the stresses being termed *tensile stresses* if the forces are in such directions as to stretch the bar and *compressive stresses* if they compress the bar. It is customary to denote tensile stresses as positive and compressive stresses as negative. The symbol used for the stress is σ and, it is defined by:

$$\sigma = \frac{F}{A} \quad [1]$$

where A is the cross-sectional area. It has the units of force per unit area. The unit of N/m^2 is termed the pascal (Pa). We are often concerned in engineering with very large values of stress and so the unit the mega-pascal MPa, one million pascals, i.e. 10^6 Pa, is frequently used.

Example

A bar with a uniform rectangular cross-section of 20 mm by 40 mm is subjected to an axial force of 50 kN. Determine the tensile stress in the bar.

Using equation [1]:

$$\sigma = \frac{F}{A} = \frac{50 \times 10^3}{0.020 \times 0.040} = 62.5 \times 10^6 \text{ Pa} = 62.5 \text{ MP}$$

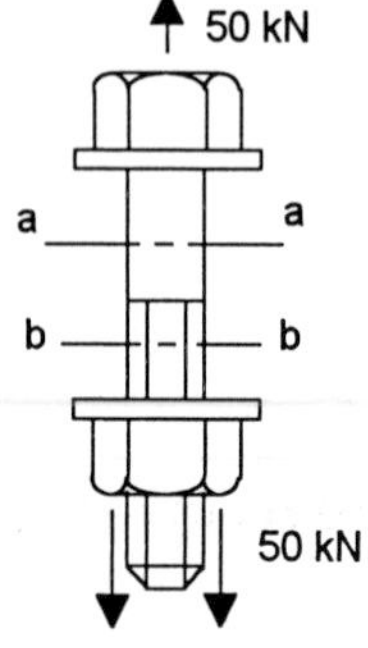

Figure 1.4 *Example*

Example

A steel bolt (Figure 1.4) has a diameter of 25 mm and carries an axial tensile load of 50 kN. Determine the average tensile stress at the shaft section *aa* and the screwed section *bb* if the diameter at the root of the thread is 21 mm.

The average tensile stress at section *aa* is given by equation [1] as:

$$\sigma_{aa} = \frac{50 \times 10^3}{\frac{1}{4}\pi 0.025^2} = 101.9 \times 10^6 \text{ Pa} = 101.9 \text{ MPa}$$

The average tensile stress at section *bb* is given by equation [1] as:

$$\sigma_{bb} = \frac{50 \times 10^3}{\frac{1}{4}\pi 0.021^2} = 144.4 \times 10^6 \text{ Pa} = 144.4 \text{ MPa}$$

Revision

1 A circular cross-section rod with a diameter of 20 mm is stretched by axial forces of 10 kN. What is the stress in the bar?

2 A circular cross-section column with a diameter of 100 mm is compressed by axial forces of 70 kN. What is the stress in the column?

1.2.2 Direct strain

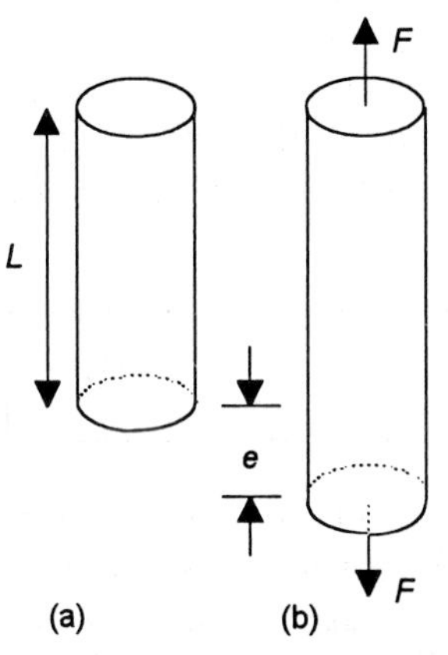

Figure 1.5 *(a) Unstretched bar, (b) stretched bar*

An axial loaded bar undergoes a change in length, increasing in length when in tension (Figure 1.5) and decreasing in length when in compression. The change in length e for an original length L is termed the *direct strain* ε.

$$\varepsilon = \frac{e}{L} \qquad [2]$$

Since strain is a ratio of two lengths it is a dimensionless number, i.e. it has no units. Engineering strains are usually quite small.

When the change in length is an increase in length then the strain is termed *tensile strain* and is positive. When the change in length is a decrease in length then the strain is termed *compressive strain* and is negative.

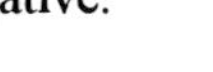

Example

Determine the strain experienced by a rod of length 100.0 cm when it is stretched by 0.2 cm.

Using equation [2]:

$$\text{strain} = \frac{e}{L} = \frac{0.2}{100} = 0.0002$$

Revision

3 A column of height 40 cm contracts axially by 0.02 cm when loaded. What is the strain experienced by the column?

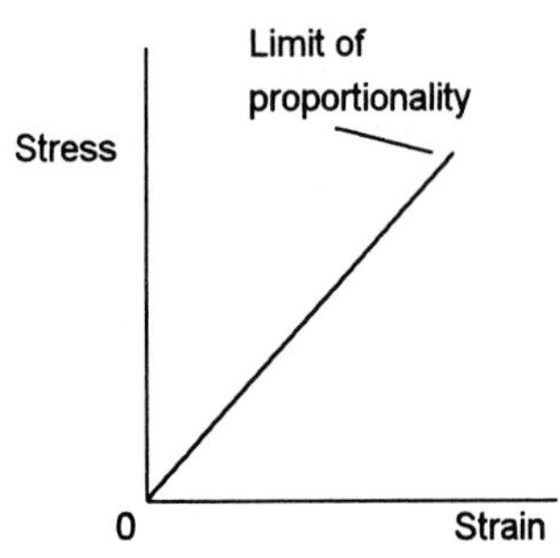

Figure 1.6 *Hooke's law*

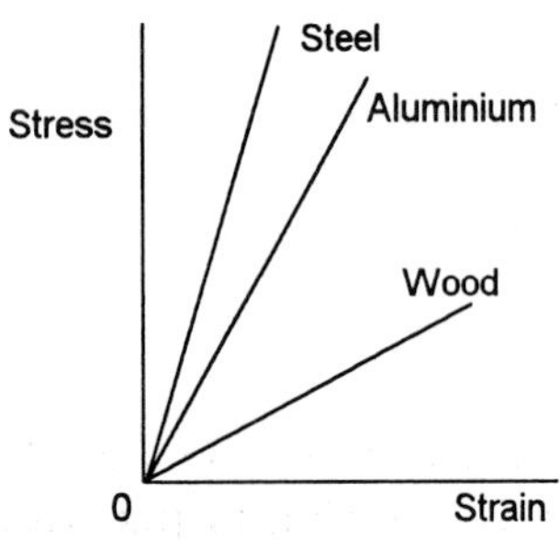

Figure 1.7 *Graphs for some materials*

Application
Strawberry jelly is not stiff and has a low modulus of elasticity. A digestive biscuit is stiff and has a much higher modulus of elasticity. However, both the jelly and the biscuit cannot withstand a high stress before breaking.

In the material used for, say, a bridge a high modulus of elasticity material is required but also one which can withstand a higher stress without breaking. Steel has a high modulus of elasticity and a high strength.

1.2.3 Hooke's law

Hooke's law can be stated as: strain is proportional to the stress producing it (Figure 1.6). This law can generally be assumed to be obeyed within certain limits of stress by most of the metals used in engineering. Within the limits to which Hooke's law is obeyed, the ratio of the direct stress to the strain produced is called the *modulus of elasticity E*:

$$E = \frac{\sigma}{\varepsilon} \qquad [3]$$

For a bar of uniform cross-sectional area A and length L, subject to axial force F and extending by e, equation [3] becomes:

$$E = \frac{FL}{Ae} \qquad [4]$$

Since the modulus of elasticity is stress divided by strain and strain has no units, then the modulus has the same unit as stress. Since, in engineering, strain is generally rather small we tend to have very large values for a modulus of elasticity and so units of giga-pascals (GPa), are generally used. 1 GPa = 1000 000 000 Pa, i.e. 10^9 Pa.

The gradient of the stress–strain graph, i.e. the value of the modulus of elasticity, depends on the material concerned. Figure 1.7 illustrates this. Typically, steels have a modulus of elasticity of about 210 GPa, aluminium alloys about 70 GPa and wood, along the grain, about 14 GPa.

The modulus of elasticity is a measure of how much stress has to be applied to a material to strain it and is sometimes referred to as a measure of the elastic stiffness of a material. A stiff material has a high modulus of elasticity.

Example

A circular cross-section steel bar of uniform diameter 12 mm and length 1 m is subject to tensile forces of 10 kN. If the steel has a modulus of elasticity of 200 GPa, what will be the stress and strain in the bar?

Using equation [1]:

$$\text{stress} = \frac{F}{A} = \frac{10 \times 10^3}{\frac{1}{4}\pi 0.012^2} = 88.4 \times 10^6 \text{ Pa} = 88.4 \text{ MP}$$

Using equation [3]:

$$\text{strain} = \frac{\sigma}{E} = \frac{88.4 \times 10^6}{200 \times 10^9} = 4.42 \times 10^{-4}$$

Example

Loads are suspended from a vertical wire of length 2.0 m and diameter 0.08 mm. If each added load of 50 g causes an extension of 1 mm, what is the modulus of elasticity of the wire?

Using equation [1], for a 50 g load:

$$\text{stress} = \frac{F}{A} = \frac{0.050 \times 9.8}{\frac{1}{4}\pi(0.08/1000)^2} = 97.5 \text{ MPa}$$

Using equation [2], for an extension of 1 mm:

$$\text{strain} = \frac{e}{L} = \frac{0.001}{2} = 0.0005$$

Using equation [3], the modulus of elasticity is:

$$E = \frac{\text{stress}}{\text{strain}} = \frac{97.5 \times 10^6}{0.0005} = 195 \text{ GPa}$$

Revision

4 A rod has a uniform cross-sectional area of 400 mm^2 and a length of 1.6 m. Determine the elongation of the rod when it is subject to axial tensile forces of 28 kN if the material has a modulus of elasticity of 200 GPa.

5 A mass of 6 kg is suspended by a vertical steel wire, modulus of elasticity 200 GPa, from a beam. If the wire has a uniform diameter of 2.5 mm and a length of 5 m, by how much will the wire stretch? Ignore the weight of the wire.

Activity

Carry out the following simple experiments to obtain information about the tensile properties of materials when commercially made tensile testing equipment is not available. *Safety note*: when doing experiments involving the stretching of wires, or other materials, the specimen may fly up into your face when it breaks. When a taut wire snaps, a lot of stored elastic energy is suddenly released. *Safety spectacles should be worn.*

Obtain a force–extension graph for rubber by hanging a rubber band (e.g. 74 mm by 3 mm by 1 mm band) over a clamp or other fixture, adding masses to a hanger suspended from it and measuring the extension with a ruler (Figure 1.8). A force–extension graph for a nylon fishing line can be obtained in a similar way, the fishing line being tied to form a loop (e.g. about 75 cm long).

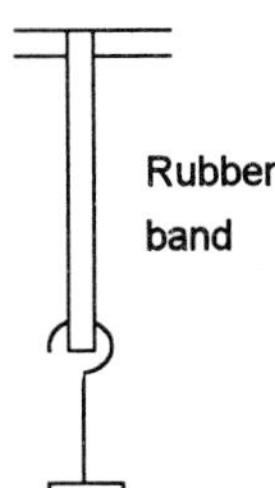

Figure 1.8 *Obtaining a force–extension graph for a rubber band*

Carry out a tensile testing of a metal wire. Figure 1.9 shows how the force–extension graph, and hence the tensile modulus of elasticity, can be determined for a metal wire (e.g. iron wire with a diameter of about

0.2 mm, copper wire about 0.3 mm diameter, steel wire about 0.08 mm diameter, all having lengths of about 2.0 m). The initial diameter d of the wire is measured using a micrometer screw gauge. The length L of the wire from the clamped end to the marker (a strip of paper attached by Sellotape) is measured by using a rule, a small load being used to give a taut wire. Masses are then added to the hanger and the change in length e from the initial position recorded. Hence data can be obtained to plot a graph of force (F) against extension (e).

We have, for a wire of length L and diameter d:

$$E = \frac{\text{stress}}{\text{strain}} = \frac{\left(F/\frac{1}{4}\pi d^2\right)}{e/L} = \frac{F}{e} \times \frac{4L}{\pi d^2}$$

This can be rearranged as:

$$F = \left(\frac{\pi d^2 E}{4L}\right) e$$

Thus the graph of F against e is a straight line with a gradient of $(\pi d^2 E/4L)$. Hence, determine the modulus of elasticity E from the gradient of the graph.

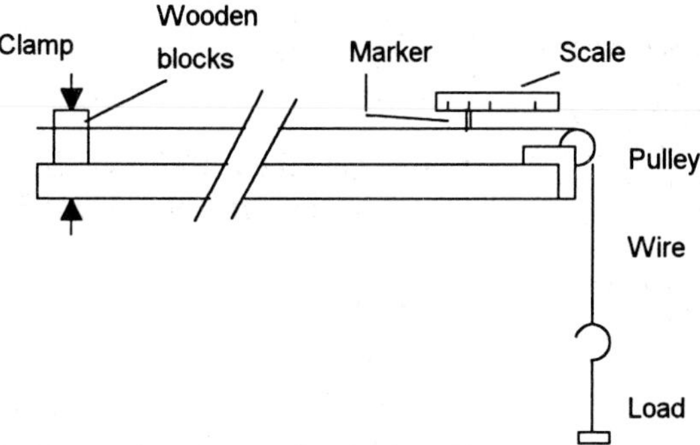

Figure 1.9 *Obtaining a force–extension graph for a wire*

1.3 Axially loaded members

Consider two forms of axially loaded members, one in which a bar is formed by combining two members in parallel and one in which they are combined in series.

1.3.1 Members in parallel

Tensile or compressive members which consist of two or more bars or tubes in parallel are termed *compound bars*. Figure 1.10 shows such an arrangement involving a central rod A of one material in a tube B of another material, the load being applied to plates fixed across the tube ends so that the load is applied to both A and B.

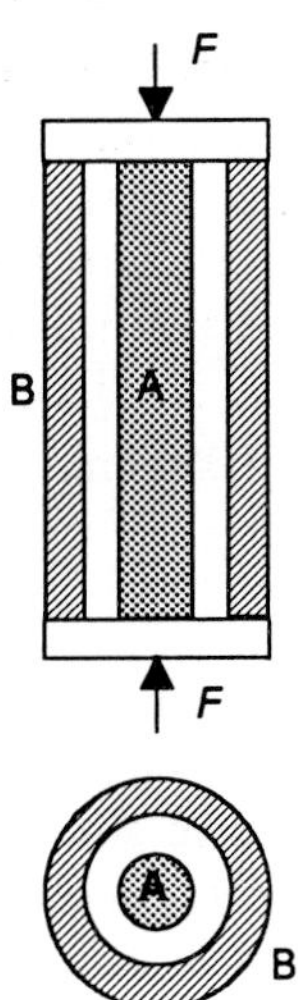

Figure 1.10 *Example of a compound bar*

Application

An example of a parallel-connected compound material is a reinforced concrete beam, this consisting of axial steel rods in a matrix of concrete. The reason for this is that though concrete is strong in compression it is weak in tension. The addition of the steel rods gives a composite material with a higher tensile strength than is possible with the concrete alone.

With such a compound bar, the load F applied is shared by the members. Thus if F_A is the force acting on member A and F_B is the force acting on member B, for *equilibrium* we must have:

$$F_A + F_B = F \qquad [5]$$

If σ_A is the resulting stress in element A and A_A is its cross-sectional area, $\sigma_A = F_A/A_A$. Likewise, if σ_B is the stress in element B and A_B is its cross-sectional area, $\sigma_B = F_B/A_B$. Thus the equilibrium equation [5] can be written as:

$$\sigma_A A_A + \sigma_B A_B = F \qquad [6]$$

Since the elements A and B are the same initial length and must remain together when loaded, the strain in A of ε_A must be the same as that in B of ε_B. Thus, for *compatibility*, we have:

$$\varepsilon_A = \varepsilon_B \qquad [7]$$

Thus, using Hooke's law, we must have:

$$\frac{\sigma_A}{E_A} = \frac{\sigma_B}{E_B} \qquad [8]$$

where E_A is the modulus of elasticity of the material of element A and E_B that of the material of element B.

Example

A compound bar consists of a brass rod of diameter 30 mm inside a cylindrical steel tube of internal diameter 35 mm. What should be the external diameter of the steel tube if the stress in the brass rod is not to exceed 80 MPa when the compound bar is subject to an axial load of 200 kN. The modulus of elasticity for the steel is 200 GPa and that for the brass 120 GPa.

Using equation [8]:

$$\frac{\sigma_S}{E_S} = \frac{\sigma_B}{E_B}$$

$$\sigma_S = \frac{\sigma_B E_S}{E_B} = \frac{80 \times 10^6 \times 200 \times 10^9}{120 \times 10^9} = 133.3 \times 10^6 \text{ P}$$

Using equation [6]:

$$F = \sigma_S A_S + \sigma_B A_B = \sigma_S A_S + 80 \times 10^6 \times \tfrac{1}{4}\pi 0.030^2 = 200 \times 10^3 \text{ N}$$

Hence substituting the value of stress obtained above:

$$A_S = \frac{200 \times 10^3 - 80 \times 10^6 \times \frac{1}{4}\pi 0.030^2}{133.3 \times 10^6} = 1.076 \times 10^{-3}\ \text{m}$$

Thus:

$$1.076 \times 10^{-3} = \tfrac{1}{4}\pi D^2 - \tfrac{1}{4}\pi 0.035^2$$

and D, the external diameter of the steel tube is 0.0509 m.

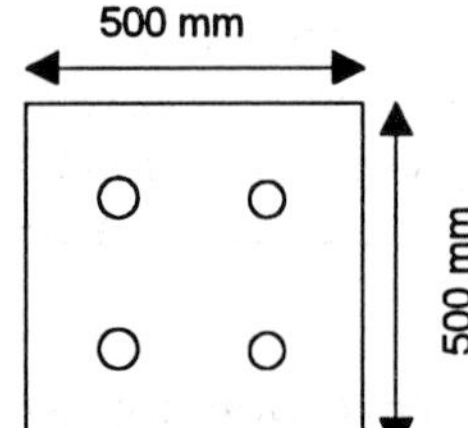

Figure 1.11 *Revision problem 6*

Revision

6 A reinforced concrete column is uniformly 500 mm square and consists of four steel rods, each of diameter 25 mm, embedded in the concrete (Figure 1.11). Determine the compressive stresses in the concrete and the steel when the column is subject to a compressive load of 1 MN, the modulus of elasticity of the steel being 200 GPa and that of the concrete 14 GPa.

7 A compound bar of length 500 mm consists of a steel rod of diameter 20 mm in a brass tube of internal diameter 20 mm and external diameter 30 mm. Determine the stress in each material and the extension of the bar when axial tensile forces of 30 kN are applied. The modulus of elasticity for the steel is 200 GPa and that for the brass 90 GPa.

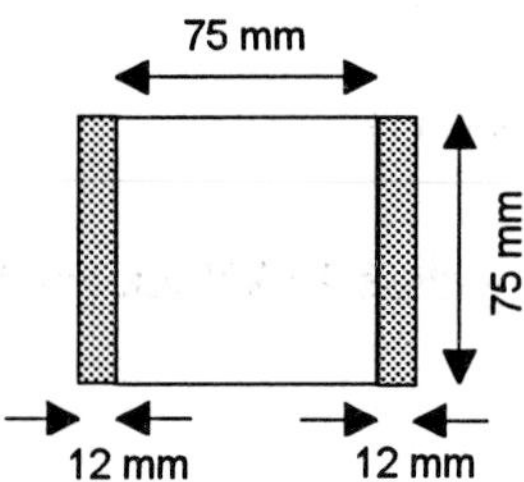

Figure 1.12 *Revision problem 8*

8 A compound beam consists of a square cross-section timber core 75 mm by 75 mm with steel plates 75 mm by 12 mm bolted along its entire length to opposite faces (Figure 1.12). Determine the maximum permissible axial tensile load if the maximum permissible stress in the timber is 6.3 MPa and in the steel 140 MPa, the modulus of elasticity of the timber being 8 GPa and for the steel 200 GPa.

9 A copper rod of diameter 25 mm is inserted into a steel tube of internal diameter 35 mm and external diameter 40 mm, the rod and tube being attached at each end. Determine the stresses in the rod and tube when the compound arrangement is subject to an axial tensile load of 40 kN, the modulus of elasticity of the steel being 200 GPa and that of the copper 95 GPa.

1.3.2 Members in series

Consider a composite bar consisting of two, or more, members in series. These may be of different materials and/or different cross-sections. Figure 1.13 illustrates the type of situation that might occur. Here we have three rods connected end-to-end, the rods perhaps being of different cross-sections and perhaps materials. The composite is subject to a tensile load.

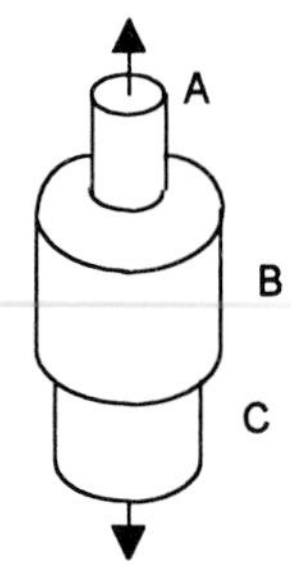

Figure 1.13 *Members in series*

With just a single load we must have, for equilibrium, the same forces acting on each of the series members. Thus the forces stretching member

A are the same as those stretching member B and the same as those stretching member C. The extensions of the members may, however, differ. The total extension of the composite bar will be the sum of the extensions arising for each series element.

Example

Steel rods of diameter 10 mm and 15 mm are connected to either end of a copper bar of diameter 20 mm (a similar situation to that shown in Figure 1.13). The 10 mm diameter steel rod has a length of 600 mm, the 15 mm rod a length of 400 mm and the copper rod a length of 800 mm. Determine the stresses in each of the rods and the total elongation if the composite is subject to axial tensile forces of 12 kN. The steel has a modulus of elasticity of 200 GPa and the copper one of 100 GPa.

The forces acting axially on the 10 mm steel rod are 12 kN and thus the stress in that member is:

$$\sigma = \frac{12 \times 10^3}{\frac{1}{4}\pi 0.010^2} = 152.8 \text{ MPa}$$

Assuming Hooke's law, this rod will extend by:

$$e = L \times \frac{\sigma}{E} = \frac{0.600 \times 152.8 \times 10^6}{200 \times 10^9} = 0.000\,458$$

Likewise the forces acting on the copper rod are 12 kN and so the stress in it is:

$$\sigma = \frac{12 \times 10^3}{\frac{1}{4}\pi 0.020^2} = 38.2 \text{ MPa}$$

and it extends by:

$$e = \frac{0.800 \times 38.2 \times 10^6}{100 \times 10^9} = 0.000\,306$$

The forces acting on the other steel rod are also 12 kN. Thus the stress in it is:

$$\sigma = \frac{12 \times 10^3}{\frac{1}{4}\pi 0.015^2} = 67.9 \text{ MPa}$$

and its extension is:

$$e = \frac{0.400 \times 67.9 \times 10^6}{200 \times 10^9} = 0.000\,136$$

The total extension is the sum of the three extensions and so is 0.458 + 0.306 + 0.136 = 0.900 mm.

Revision

10 A steel bar with a total length of 240 mm has a diameter of 40 mm for a length of 100 mm, a diameter of 30 mm for a length of 60 mm and a diameter of 20 mm for the remaining 80 mm of its length. Determine the tensile load required to produce a total elongation of 0.177 mm for the rod. The modulus of elasticity is 200 GPa.

11 A steel rod with a diameter of 12 mm and a length of 3 m is joined to the end of an aluminium rod of diameter 12 mm and length 2 m. Determine the overall extension of the rod when it is subject to an axial tensile load of 18 kN. The modulus of elasticity of the steel is 200 GPa and that of the aluminium 70 GPa.

1.4 Poisson's ratio

When a material is longitudinally stretched it contracts in a transverse direction (Figure 1.14). The ratio of the transverse strain to the longitudinal strain is called *Poisson's ratio.*

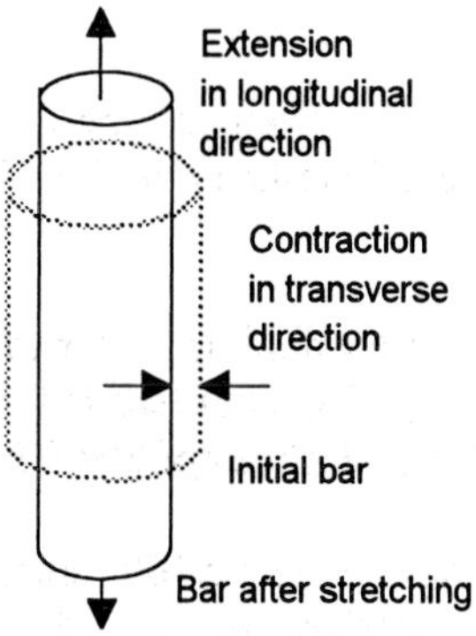

Figure 1.14 *Transverse contraction as a result of longitudinal stretching*

$$\text{Poisson's ratio} = -\frac{\text{transverse strain}}{\text{longitudinal strain}} \qquad [9]$$

The minus sign is because when one of the strains is tensile the other is compressive and by including the minus sign we end up with a positive value for Poisson's ratio. For most engineering metals, Poisson's ratio is about 0.3.

Example

A steel bar of length 1 m is extended by 0.1 mm. By how much will the width of the bar contract if initially the width was 100 mm? Poisson's ratio is 0.31.

The longitudinal strain is 0.1/1000 = 0.000 1. Thus, using equation [9], the transverse strain is:

$$\text{transverse strain} = -0.31 \times 0.000\ 1 = -3.1 \times 10^{-5}$$

Thus:

$$\text{change in width} = \text{original width} \times \text{transverse strain}$$

$$= 100 \times (-3.1 \times 10^{-5}) = -3.1 \times 10^{-3}\ \text{mm}$$

The minus sign indicates that the width is reduced by this amount.

Revision

12 A steel bar with a rectangular cross-section 75 mm by 25 mm is subject to a tensile longitudinal load of 200 kN. Determine the

decrease in the lengths of the sides of the resulting cross-section. The material has an elastic modulus of 200 GPa and Poisson's ratio of 0.3.

Problems

1 A circular cross-section rod with a diameter of 25 mm and initial length 500 mm is subject to axial forces of 50 kN which cause it to extend by 0.25 mm. Determine the stress and strain in the rod.

2 A circular cross-section rod with a diameter of 25 mm and initial length 15 m is subject to axial forces of 80 kN. The material has a modulus of elasticity of 200 GPa. Determine the stress and strain in the rod.

3 A hollow circular column of length 1.5 m has an outside diameter of 300 mm and a wall thickness of 25 mm. The column material has a modulus of elasticity of 200 GPa. Determine the compressive stress in the column and its change in length when it carries a compressive load of 700 kN.

4 An aluminium wire of length 3 m is subject to a tensile stress of 70 MPa. If the aluminium has a modulus of elasticity of 70 GPa, determine the elongation of the wire.

5 A bar has a uniform cross-sectional area of 50 mm^2 and a length of 5 m. What will be its elongation when subject to axial tensile forces of 40 kN if the bar material has a modulus of elasticity of 200 GPa?

6 A hollow circular cross-section column has a length of 600 mm, an external diameter of 75 mm and a wall thickness of 7.5 mm. Determine the change in length of the cylinder when it is subject to axial compressive forces of 50 kN, the material having a modulus of elasticity of 100 GPa.

7 An electrical conductor consists of a 5 mm diameter steel wire coated with copper to give an external diameter of 7 mm. Determine the stresses in the two materials when a length of the conductor is subject to tensile forces of 2 kN. The modulus of elasticity of the steel is 200 GPa and that of the copper 120 GPa.

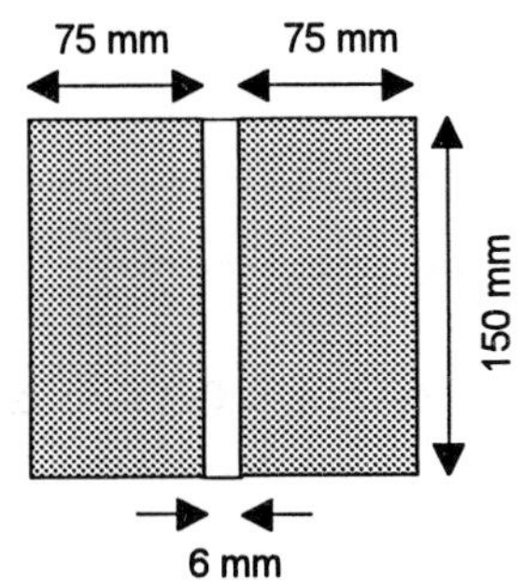

Figure 1.15 *Problem 9*

8 A reinforced concrete column has a 450 mm square uniform cross-section and contains four steel bars, each of diameter 25 mm. Determine the stresses in the steel and the concrete when the column is subject to an axial compressive load of 1.5 MN, the modulus of elasticity of the steel being 200 GPa and that of the concrete 14 GPa.

9 A compound beam is made by sandwiching a steel plate 150 mm by 6 mm by timber members 150 mm by 75 mm (Figure 1.15), the three members being bolted together. The maximum permissible stress for the timber is 6 MPa and for the steel 130 MPa, the

modulus of elasticity for the timber being 8.2 GPa and for the steel 205 GPa. Determine the maximum permissible tensile load for the compound beam.

10 A component consists of a steel bar of diameter 40 mm and length 200 mm with a copper rod of length 400 mm joined to its end. Determine the diameter necessary for the copper rod if the extension of each of the constituent rods is to be the same when the component is subject to an axial load. The steel has a modulus of elasticity of 200 GPa and the copper 100 GPa.

11 A bar of length 3 m and with a circular cross-section of diameter 30 mm is stretched by tensile forces of 85 kN. Determine the elongation of the bar and the decrease in diameter. The material has an elastic modulus of 200 GPa and Poisson's ratio of 0.3.

2 Beams and columns

2.1 Introduction

As discussed in chapter 1 in relation to possible forms of loading structures, one basic form involves bending. Thus, for the simple beam bridge in Figure 2.1(a) the load arising from a car crossing it will tend to bend the beam.

As also discussed in chapter 1, another form of loading involves the loading in compression. Such members might be concrete or brick columns supporting the floors or roof of a building (Figure 2.2(b)), the applied loads of the floors or roof above applying compressive loads.

This chapter is a discussion of loading due to bending, the various forms it can take, and the stresses that can arise from such bending and also the loading of columns.

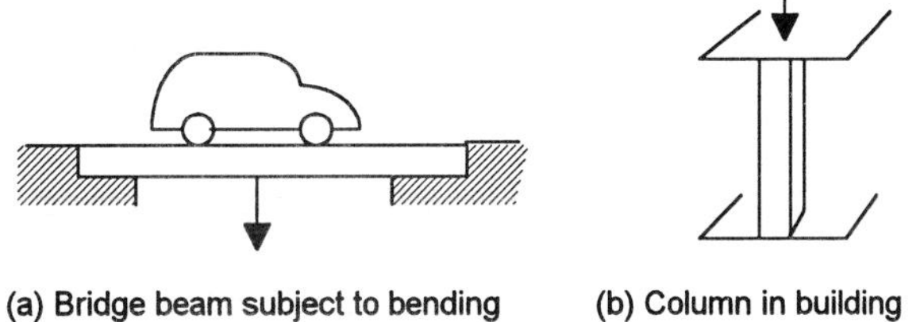

Figure 2.1 *Examples of structures where (a) bending (b) compressive loading occurs*

2.2 Beams

A *beam* can be defined as a structural member, generally horizontal, to which loads are applied and which cause it to bend. As a result of loads causing one surface of a beam to become longer and the opposite surface shorter, when beams bend they become curved.

2.2.1 Types of beams

The following are some examples of types of beams:

1 *Cantilever* (Figure 2.2(a))
This is a beam which is rigidly fixed at just one end, the other end being free.

2 *Simply supported beam* (Figure 2.2(b))
This is a beam which is supported at its ends on rollers or smooth surfaces or one of these combined with a pin at the other end.

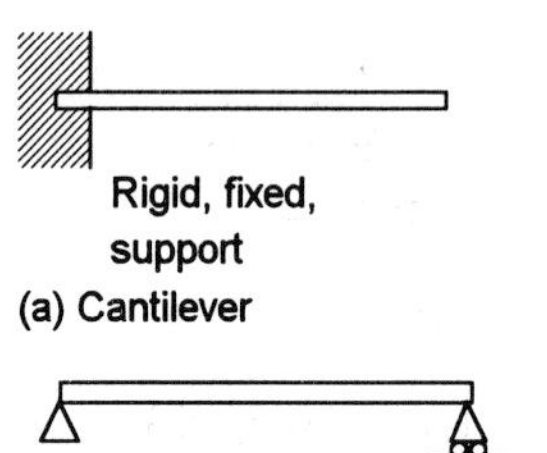

(a) Cantilever

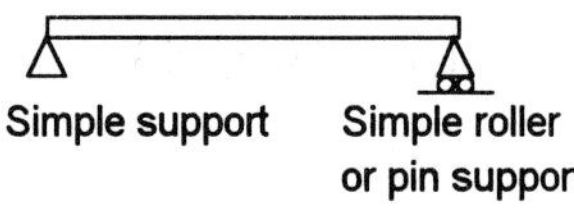

(b) Simply supported beam

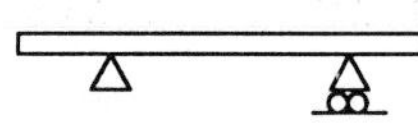
(c) Overhanging beam

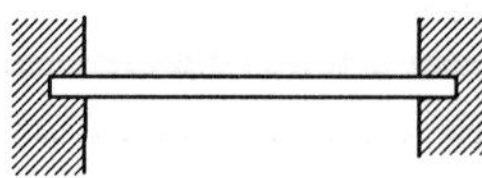
(d) Built-in beam, both ends being rigidly fixed

Figure 2.2 *Examples of beams*

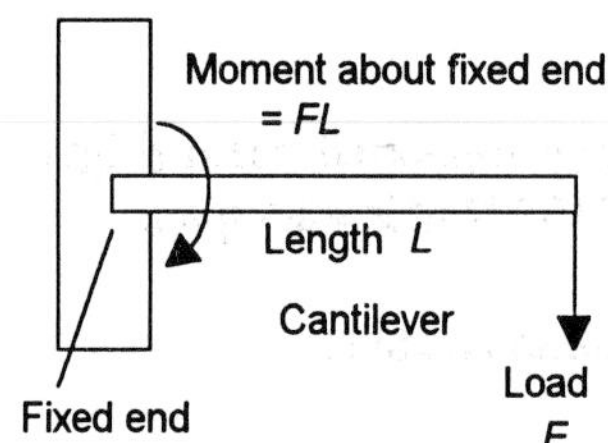

Figure 2.3 *Moment*

Application
An example of a fully fixed support is where a beam is built into a wall. A simple support would be where a beam end rests on a wall. An example of a roller type of support is where the end of a bridge rests on a low resistance bearing plate which allows horizontal movement of the end of the bridge. This is necessary since changes in temperature cause bridges to change in length.

3 *Simply supported beam with overhanging ends* (Figure 12.2(c))
This is a simple supported beam with the supports set in some distance from the ends.

4 *Built-in beam* (Figure 2.2(d))
This is a beam which is built-in at both ends and so both ends are rigidly fixed.

Where an end is rigidly fixed there is a reaction force and a resisting moment. At a supported end or point there are reactions but no resisting moments. At a free end there are no reactions and no resisting moments. Thus, with a cantilever (Figure 2.3), the rigid fixing at one end prevents rotation of the beam at that end when a load is applied to the cantilever. Thus, if you apply a vertical force some distance from the fixed end there will be a moment which, for equilibrium, has to be balanced by a resisting moment at the fixed end. Without the balancing moment there would be nothing to stop the beam rotating.

2.2.2 Supports

In Figure 2.2 a number of forms of support were illustrated. Basically there are three types of support:

1 *Fully fixed* (Figure 2.4(a))
With this type of support, the end of the beam is clamped in a fixed position by the member supporting it. An example of this is where the end of the beam is built into a wall.

2 *Simple support* (Figure 2.4 (b))
This type of support offers no resistance to bending, the end of the beam resting on the supporting structure or held with a single pin or bolt. Such a support provides resistance to vertical or horizontal movement of the end of the beam but the beam is free to rotate at the support.

3 *Roller support* (Figure 2.4(c))
This is just another version of a simple support and allows horizontal movement, but no vertical movement, of the end of the beam as well as allowing the beam to freely rotate at the support.

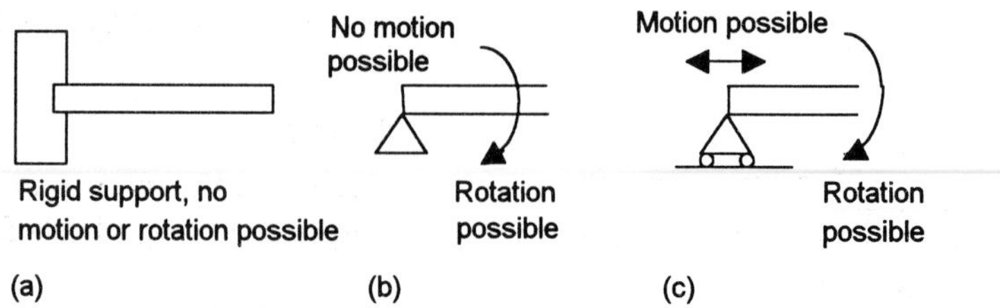

Figure 2.4 *Forms of support: (a) fully fixed, (b) simple support, (c) roller support*

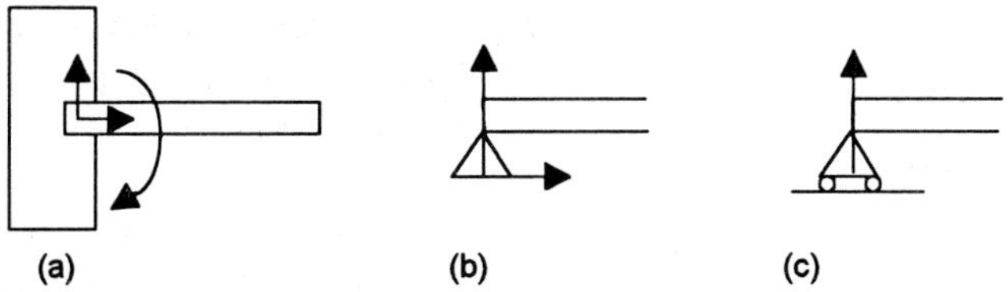

(a) (b) (c)

Figure 2.5 *Reactive forces at supports: (a) fully fixed, (b) simple support, (c) roller support*

Figure 2.5 shows the reactive forces that arise at the three types of supports. Beam reactive forces can be determined by considering equilibrium conditions, namely we must have:

1 The sum of the vertical forces must be zero.

2 The sum of the horizontal forces must be zero.

3 The sum of the moments about any axis must be zero.

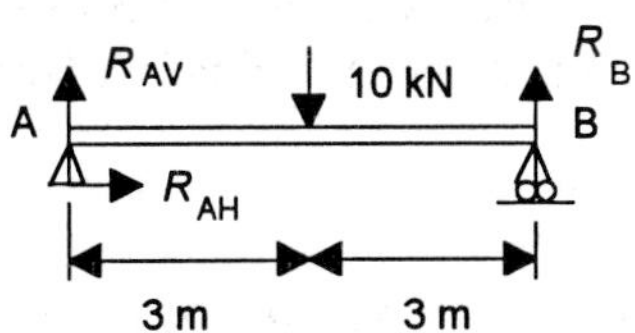

Figure 2.6 *Example*

Example

Determine the reactive forces for a simply supported beam (Figure 2.6) of length 6 m and carrying a load of 10 kN at is mid-span.

For equilibrium with the vertical forces we must have:

$$R_{AV} + R_B = 10$$

For equilibrium of the horizontal forces we must have:

$$R_{AH} = 0$$

Taking moments about support A, for equilibrium we must have:

$$R_B \times 6 = 10 \times 3$$

Hence $R_B = 5$ kN, $R_{AV} = 5$ kN and $R_{AH} = 0$.

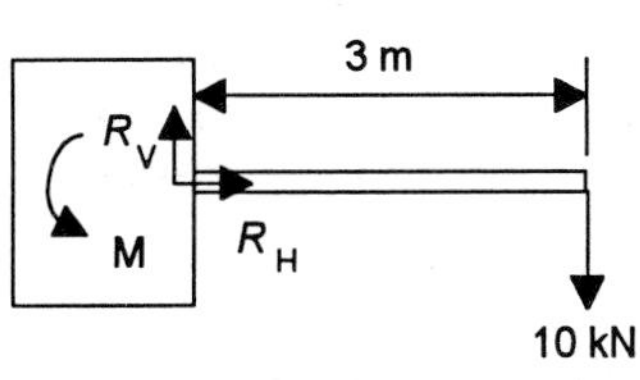

Figure 2.7 *Example*

Example

Determine the reactive forces for a cantilever (Figure 2.7) of length 3 m and carrying a load of 10 kN at its free end.

For equilibrium with the vertical forces we must have:

$$R_V = 10 \text{ kN}$$

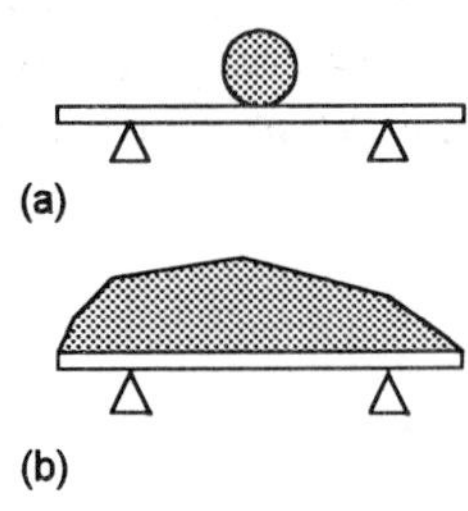

Figure 2.8 *Loads: (a) concentrated, (b) distributed*

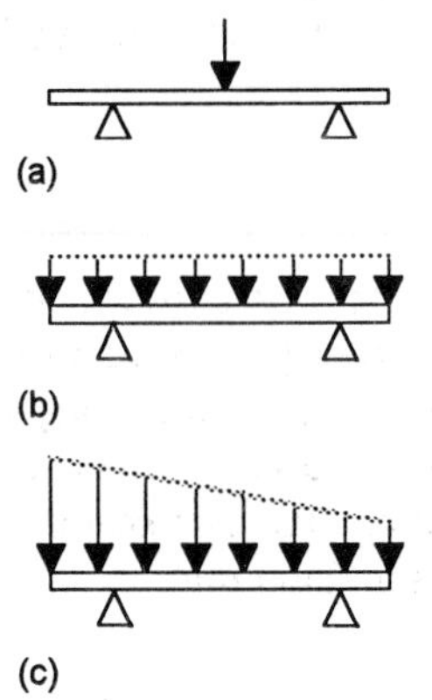

Figure 2.9 *Loads: (a) concentrated, (b) uniformly distributed, (c) non-uniformly distributed*

Application
An example of a point load applied to a beam might be the load applied to a beam bridge by a car through its centre of gravity. An example of a distributed load could be the actual weight of a beam acting over its entire length.

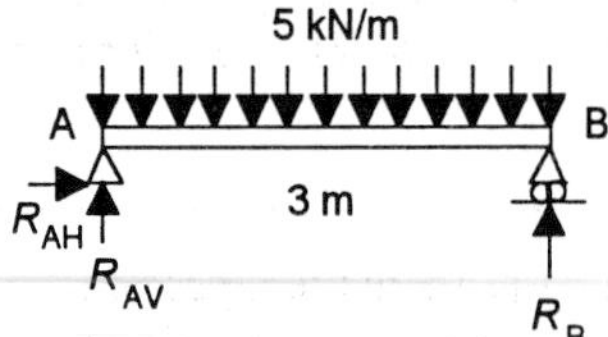

Figure 2.10 *Example*

For equilibrium with the horizontal forces we must have for equilibrium:

$$R_H = 0$$

Taking moments about the fixed end gives a clockwise moment of 10 × 3 kN m. For equilibrium we must have this moment balanced by the resistive moment. Thus. $M = 10 \times 3 = 30$ kN m in an anti-clockwise direction.

Revision

1 Determine the reactive forces for a simply supported beam of length 8 m with a mid-span point load of 6 kN.

2 Determine the reactive forces for a simply supported beam of length 8 m with point loads of 3 kN at 2 m from one end and 5 kN at 5 m from the same end.

2.2.3 Loads

The loads that can be carried by beams may be concentrated or distributed. A concentrated load is one which can be considered to be applied at a point (Figure 2.8(a)); a distributed load is one that is applied over a significant length of the beam (Figure 2.8(b)). An obvious example of a distributed load is the weight of the beam, there being a weight force for each unit length of the beam. On figures, concentrated loads are represented by single arrows acting along the line concerned (Figure 2.9(a)); distributed loads are represented by a series of arrows along the length of beam over which the load is distributed (Figure 2.9(b)). With a uniformly distributed load the arrows are all the same length; if the distributed load is not uniform then the lengths of the arrows are varied to indicate how the distributed load is varying (Figure 2.9(c)). The loading on a beam may be a combination of fixed loads and distributed loads.

Example

Determine the reactive forces for a simply supported beam (Figure 2.10) of length 3 m carrying a uniformly distributed load of 5 kN/m along its entire length.

The total load on the beam is 5 × 3 = 15 kN. For vertical equilibrium we must have:

$$R_{AV} + R_B = 15$$

For horizontal equilibrium we must have:

$$R_{AH} = 0$$

The entire distributed load can be considered to act at the centroid of the beam and thus, since it is uniformly distributed along the length, will be at mid-span. Hence, taking moments about A gives, for equilibrium:

$$3 \times R_B = 15 \times 1.5$$

and so $R_B = 7.5$ kN. Consequently, $R_{AV} = 7.5$ kN.

Revision

3 Determine the reactive forces for a simply supported beam of length 8 m carrying a uniformly distributed load of 10 kN/m along its entire length.

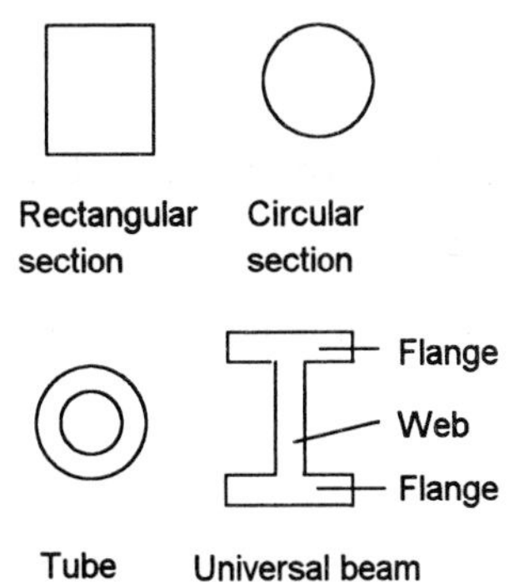

Figure 2.11 *Forms of beams*

2.2.4 Types of beams

Beams can take a number of forms (Figure 2.11). For example, they might be simple rectangular sections, e.g. the timber joists used in the floor construction of houses, circular sections, tubes, e.g. tubes carrying liquids and supported at a number of points, and the very widely used *universal beam*. The universal beam is an I-section and such beams are widely available from stockists in a range of sizes and weights.

2.3 Shear force and bending moment

Shear force and bending moment diagrams are graphs which plot how shear forces and bending moments change along the length of a beam and are useful assets in the design of structures involving beams. In this section we look at what these terms mean and plot such diagrams for commonly encountered forms of beams.

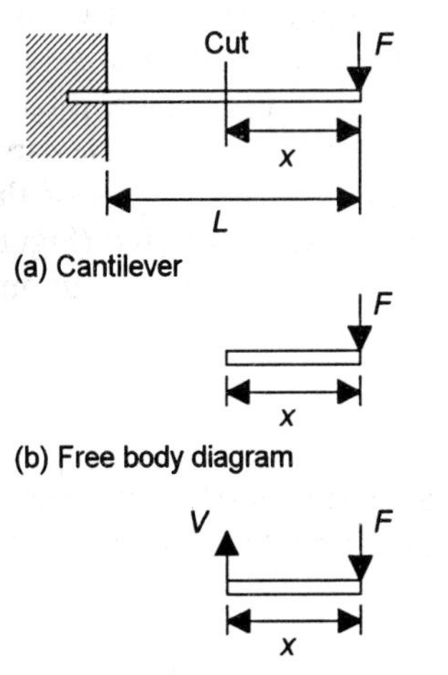

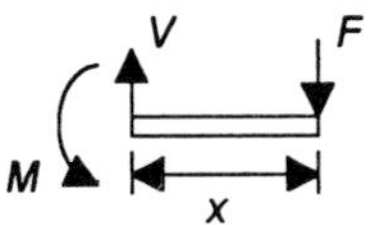

(d) Vertical and moment equilibrium

Figure 2.12 *Cantilever*

2.3.1 Shear force

Consider a cantilever (Figure 2.12(a) which has a concentrated load F applied at the free end. Suppose we now make an imaginary cut through the beam at a distance x from the free end. We will think of the cut section of beam as a *free body*, isolated from the rest of the beam and effectively floating in space. Now consider the conditions for the equilibrium of the section of beam to the right of the cut (Figure 2.12(b)).

For the section of beam to be in vertical equilibrium, we must have a vertical force V acting on it such that $V = F$ (Figure 2.12(c)). This force V is called the *shear force* because the combined action of V and F on the section is to shear it (Figure 2.13). In general:

> The shear force at a transverse section of a beam is the algebraic sum of the external forces acting at right angles to the axis of the beam on one side of the section concerned.

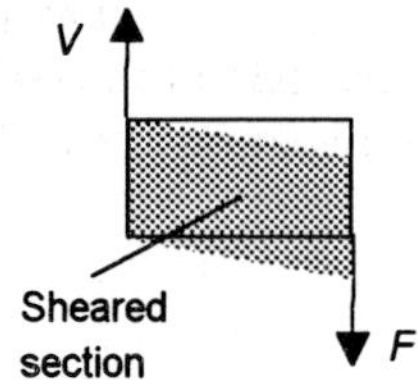

Figure 2.13 *Shear*

2.3.2 Bending moment

In addition to vertical equilibrium we must also have the section of beam in rotational equilibrium. For the section of the beam to be in moment equilibrium and not rotate, we must have a moment M applied (Figure 2.12(d)) at the cut so that $M = Fx$. This moment is termed the *bending moment.*

> The bending moment at a transverse section of a beam is the algebraic sum of the moments about the section of all the forces acting on one (either) side of the section concerned.

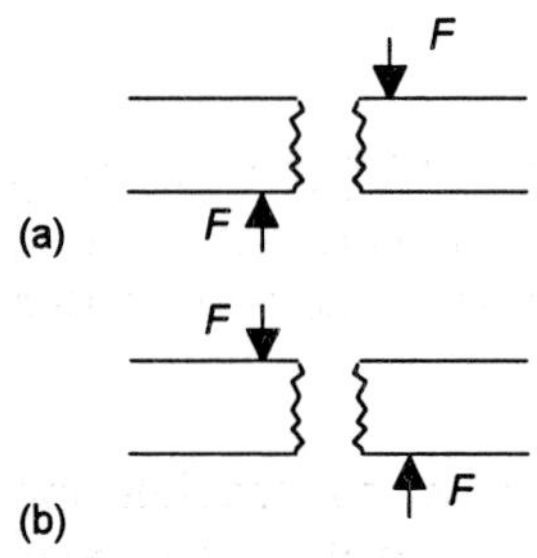

Figure 2.14 *Shear force: (a) positive, (b) negative*

2.3.3 Conventions for shear and bending moments

The conventions most often used for the signs of shear forces and bending moments are:

1 *Shear force*
When the shear forces on either side of a section are clockwise (Figure 2.14(a)), i.e. the left-hand side of the beam is being pushed upwards and the right-hand side downwards, the shear force is taken as being positive. When the shear forces on either side of a section are anticlockwise (Figure 2.14(b)), i.e. the left-hand side of the beam is being pushed downwards and the right-hand side upwards, the shear force is taken as being negative.

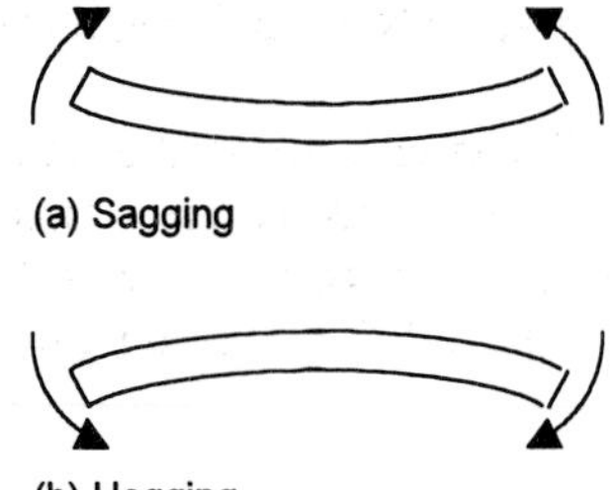

Figure 2.15 *Bending moment: (a) positive, (b) negative*

2 *Bending moment*
Bending moments are positive if they give rise to sagging (Figure 2.15(a) and negative if they give rise to hogging (Figure 2.15(b)).

Example

Determine the shear force and bending moment at points 1 m and 4 m from the right-hand end of the beam shown in Figure 2.16. Neglect the weight of the beam.

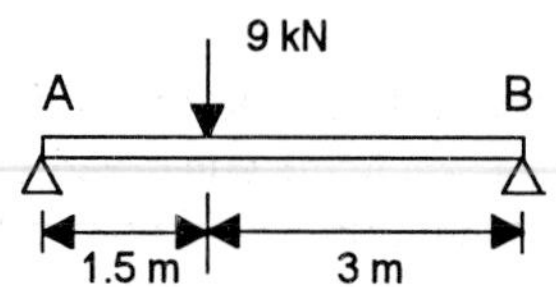

Figure 2.16 *Example*

The reactions at the ends A and B can be found by taking moments about A:

$$R_B \times 4.5 = 9 \times 1.5$$

to give $R_B = 3$ kN and then considering the vertical equilibrium which gives:

$$R_A + R_B = 9$$

and thus $R_A = 6$ kN. Figure 2.17 shows the forces acting on the beam.

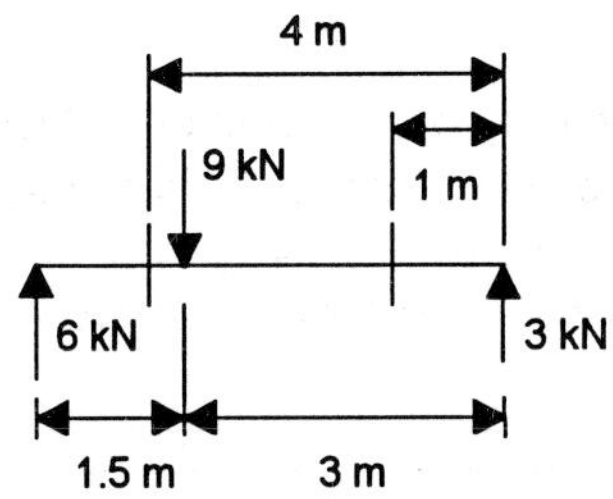

Figure 2.17 *Example*

If we make an imaginary cut in the beam at 1 m from the right-hand end, then the force on the beam to the right of the cut is 3 kN upwards and that to the left is 9 – 6 = 3 kN downwards. The shear force is thus negative and –3 kN.

If we make an imaginary cut in the beam at 4 m from the right-hand end, then the force on the beam to the right of the cut is 9 – 3 = 6 kN downwards and that to the left is 6 kN upwards. The shear force is thus positive and +6 kN.

The bending moment at a distance of 1 m from the right-hand end of the beam, when we consider that part of the beam to the right, is 3 × 1 kN m. Since the beam is sagging the bending moment is +3 kN m. At a distance of 4 m from the right-hand end of the beam, the bending moment is 3 × 4 – 9 × 0.5 = +7.5 kN m.

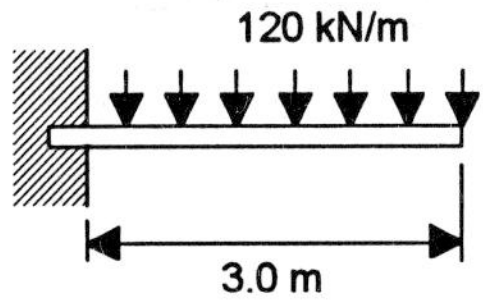

Figure 2.18 *Example*

Example

A uniform cantilever of length 3.0 m (Figure 2.18) has a weight per metre of 120 kN. Determine the shear force and bending moment at distances of 1.0 m and 3.0 m from the free end if no other loads are carried by the beam.

At 1.0 m from the free end, there is 1.0m of beam to the right and it has a weight of 120 kN (Figure 2.19(a)). Thus the shear force is +120 kN; it is positive because the forces are clockwise. The weight of this section can be considered to act at its centre of gravity which, because the beam is uniform, is at its midpoint. Thus the 120 kN weight force can be considered to be 0.5 m from the 1.0 m point and so the bending moment is –120 × 0.5 = –60 kN m; it is negative because there is hogging.

At 3.0 m from the free end, there is 3.0 m of beam to the right and it has a weight of 360 kN (Figure 2.19(b)). Thus the shear force is +360 kN. The weight of this section can be considered to act at its midpoint, a distance of 1.5 m from the free end. Thus the bending moment is –360 × 1.5 = –540 kN m.

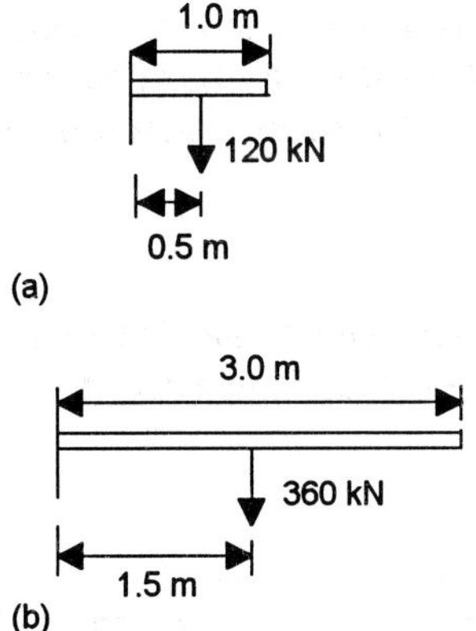

Figure 2.19 *Example*

Revision

4 A beam of length 4.0 m rests on supports at each end and a concentrated load of 500 N is applied at its midpoint. Determine the shear force and bending moment at distances of (a) 1.0 m, (b) 2.5 m from the right-hand end of the beam. Neglect the weight of the beam.

5 A cantilever has a length of 2 m and a concentrated load of 8 kN is applied to its free end. Determine the shear force and bending moment at distances of (a) 0.5 m, (b) 1.0 m from the fixed end. Neglect the weight of the beam.

6 A uniform cantilever of length 4.0 m has a weight per metre of 10 kN. Determine the shear force and bending moment at 2.0 m from the free end if no other loads are carried by the beam.

2.3.4 Shear force and bending moment diagrams

Figures which graphically show how the variations of the shear forces and bending moments along the length of a beam are termed *shear force diagrams* and *bending moment diagrams.*

> In shear force and bending moment diagrams, the convention adopted is of drawing positive shear forces above the centre line of the beam if positive and below it if negative; bending moments being drawn above the centre line if negative, i.e. hogging, and below the centre line if positive, i.e. sagging (it then gives some indication of the deflected shape of the beam).

The following show how such diagrams can be developed for commonly occurring situations.

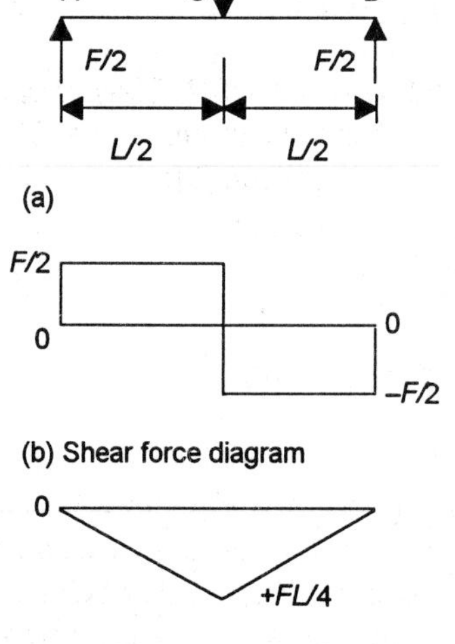

Figure 2.20 *Simply supported beam with point load*

1 *Simply supported beam with point load at mid-span*

Figure 2.20(a) shows the beam and the forces concerned, the weight of the beam being neglected. For a central load F, the reactions at each end will be $F/2$.

Consider the shear forces. At point A, the forces to the right are $F - F/2$ and so the shear force at A is $+F/2$; it is positive because the forces are clockwise about A. This shear force value will not change as we move along the beam from A until point C is reached. To the right of C we have just a force of $F/2$ and this gives a shear force of $-F/2$; it is negative because the forces are anticlockwise about it. To the left of C we have just a force of $F/2$ and this gives a shear force of $+F/2$; it is positive because the forces are clockwise about it. Thus at point C, the shear force takes on two values. For points between C and B, the forces to the left are constant at $F/2$ and so the shear force is constant at $-F/2$. Figure 2.20(b) shows the shear force diagram.

Consider the bending moments. At point A, the moments to the right are $F \times L/2 - F/2 \times L = 0$. The bending moment is thus 0. At point C the moment to the right is $F/2 \times L/2$ and so the bending moment is $+FL/4$; it is positive because sagging is occurring. At point B the moment to the right is zero, likewise that to the left $F \times L/2 - F/2 \times L = 0$. Between A and C the bending moment will vary, e.g. at one-quarter the way along the beam it is $FL/8$. In general, between A and C the bending moment a distance x from A is $Fx/2$ and between C and B is $Fx/2 - F(x - L/2) = F/2(L - x)$. Figure 2.20(c) shows the bending moment diagram. The maximum bending moment occurs under the load and is:

$$\text{maximum bending moment} = \frac{FL}{4} \qquad [1]$$

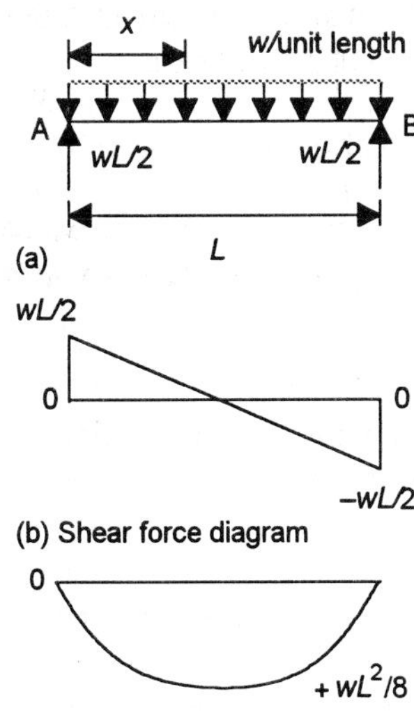

Figure 2.21 *Simply supported beam with distributed load*

2 *Simple supported beam with uniformly distributed load*
Consider a simple supported beam which carries just a uniformly distributed load of w/unit length (Figure 2.21). The reactions at each end are $wL/2$.

Consider the shear force a distance x from the left-hand end of the beam. The load acting on the left-hand section of beam is wx. Thus the shear force is:

$$V = wL/2 - wx = w(\tfrac{1}{2}L - x) \qquad [2]$$

When $x = \tfrac{1}{2}L$, the shear force is zero. When $x < \tfrac{1}{2}L$ the shear force is positive and when $x > \tfrac{1}{2}L$ it is negative. Figure 2.21(b) shows the shear force diagram.

Consider the bending moment. At A the moment due to the beam to the right is $-wL \times L/2 + wL/2 \times L = 0$. At the midpoint of the beam the moment is $-wL/2 \times L/4 + wL/2 \times L/2 = wL^2/8$; the bending moment is thus $+wL^2/8$. At the quarter-point along the beam, the moment due to the beam to the right is $-3L/4 \times 3L/8 + wL/2 \times 3L/4 = 3wL^2/32$. In general, the bending moment due to the beam at distance x is:

$$M = -wx \times x/2 + wL/2 \times x = -wx^2/2 + wLx/2 \qquad [3]$$

Differentiating equation [3] gives $\mathrm{d}M/\mathrm{d}x = -wx + wL/2$. Thus $\mathrm{d}M/\mathrm{d}x = 0$ at $x = L/2$. The bending moment is thus a maximum at $x = L/2$, the value given by substituting this value in equation [3]:

$$\text{maximum bending moment} = \frac{wL^2}{8} \qquad [4]$$

Figure 2.21(c) shows the bending moment diagram.

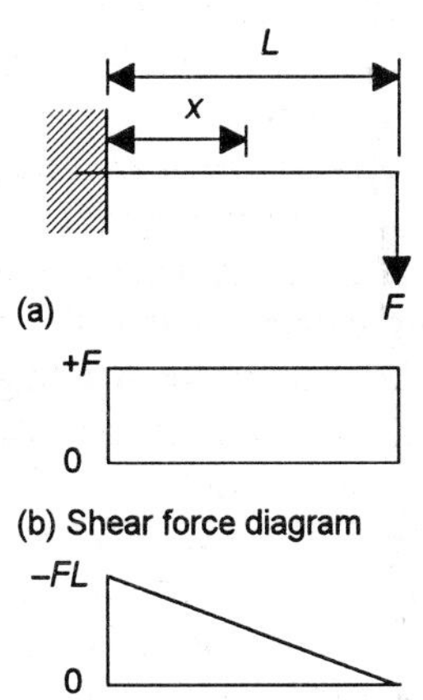

Figure 2.22 *Cantilever with point load at free end*

3 *Cantilever with point load at free end*
Consider a cantilever which carries a point load F at its free end (Figure 2.22(a)), the weight of the beam being neglected. The shear force at any section will be $+F$, the shear force diagram thus being as shown in Figure 2.22(b). The bending moment at a distance x from the fixed end is:

$$M = -F(L - x) \qquad [5]$$

The minus sign is because the beam shows hogging. We have $\mathrm{d}M/\mathrm{d}x = F$ and thus the bending moment diagram is a line of constant slope F. At the fixed end, when $x = 0$, the bending moment is FL; at the free end it is 0.

4 *Cantilever with uniformly distributed load*
Consider a cantilever which has just a uniformly distributed load of w per unit length (Figure 2.23(a)). The shear force a distance x from the fixed end is:

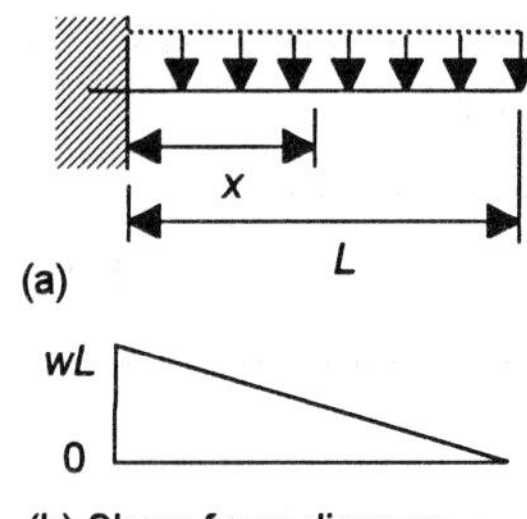

(b) Shear force diagram

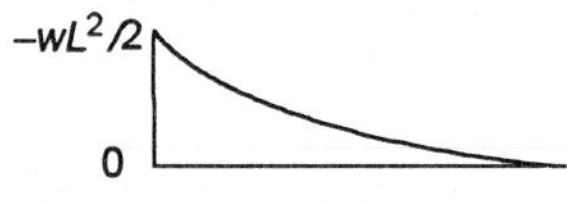

(c) Bending moment diagram

Figure 2.23 *Cantilever with uniformly distributed load*

Application
In the design of reinforced concrete beams a designer can utilise shear force and bending moment diagrams to determine the layout of the reinforcing bars. The reinforcement needs to resist the forces and moments applied. Thus most bars will be placed where the maximum bending moment occurs and the numbers reduced in the regions where the bending moment is less. Steel links are included in the reinforcement to improve a beams resistance to shear, being placed closer together where the shear forces are high.

For example, for a simply supported beam (Figure 2.24), the maximum bending moment is mid-span and so there will be more reinforcing bars in that region. The maximum shear forces will occur near the supports and so more steel links are inserted there.

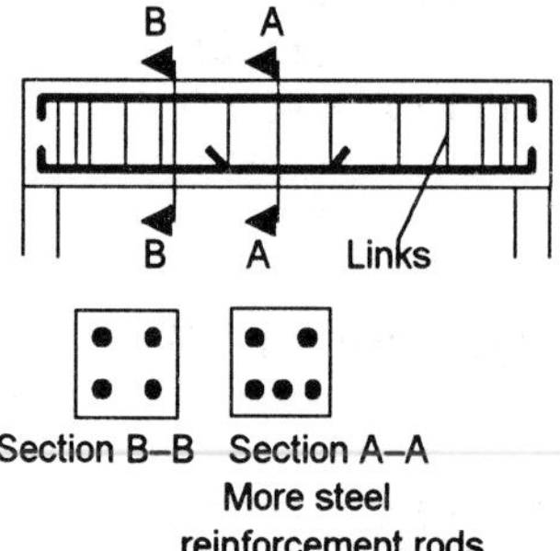

Figure 2.24 *Reinforced concrete*

$$V = +w(L - x) \qquad [6]$$

Thus at the fixed end the shear force is $+wL$ and at the free end it is 0. Figure 2.23(b) shows the shear force diagram. The bending moment at a distance x from the fixed end is, for the beam to the right of the point, given by:

$$M = -w(L - x) \times (L - x)/2 = -\tfrac{1}{2}w(L - x)^2 \qquad [7]$$

This is a parabolic function. At the fixed end, where $x = 0$, the bending moment is $-\tfrac{1}{2}wL^2$. At the free end the bending moment is 0. Figure 2.23(c) shows the bending moment diagram.

In general, when drawing shear force and bending moment diagrams:

1 Between point loads, the shear force is constant and the bending moment gives a straight line.

2 Throughout a length of beam with a uniformly distributed load, the shear force varies linearly and the bending moment is parabolic.

3 The bending moment is a maximum when the shear force is zero. The proof of this follows.

4 The shear force is a maximum when the slope of the bending moment diagram is a maximum and zero when the slope is zero.

5 For point loads, the shear force changes abruptly at the point of application of the load by an amount equal to the size of the load.

As proof that the maximum value of the bending moment occurs at a point of zero shear force, consider a very small segment of beam (Figure 2.25) of length δx and which is supporting a uniformly distributed load of w per unit length. The load on the segment is $w\delta x$ and can be considered to act through its centre. The values of the shear force V and bending moment M increase by δV and δM from one end of the segment to the other. If we take moments about the left-hand edge of the segment then:

$$M + V\delta x = w\delta x \times \delta x/2 + M + \delta M$$

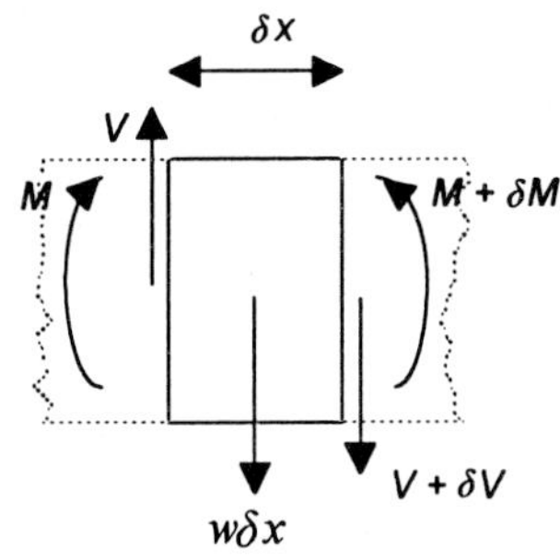

Figure 2.25 *Small segment of a beam*

Neglecting multiples of small quantities gives $V\delta x = \delta M$ and hence, as δx tends to infinitesimally small values, we can write:

$$V = \frac{dM}{dx} \quad [8]$$

Thus $V = 0$ when $dM/dx = 0$.

Tables 2.1 and 2.2 give a summary showing the shear force and bending moment diagrams and values for the commonly encountered standard cases of simple supported and cantilever beams.

Table 2.1 *Shear force and bending moment diagrams for simply supported beams*

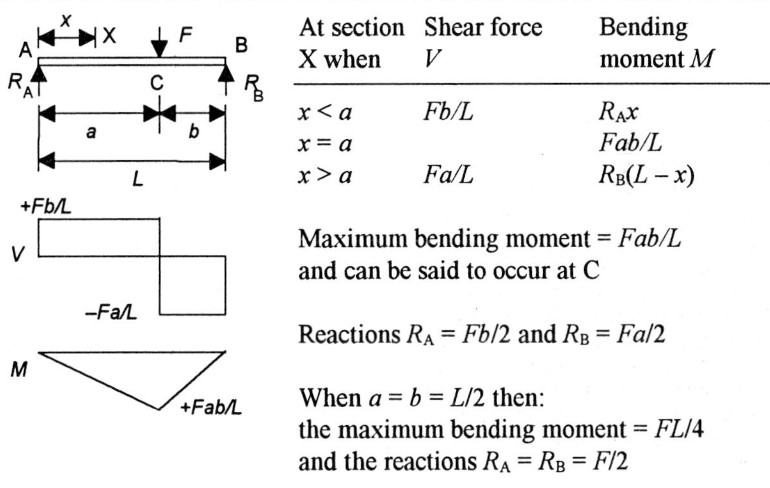

At section X when	Shear force V	Bending moment M
$x < a$	Fb/L	$R_A x$
$x = a$		Fab/L
$x > a$	Fa/L	$R_B(L - x)$

Maximum bending moment = Fab/L and can be said to occur at C

Reactions $R_A = Fb/2$ and $R_B = Fa/2$

When $a = b = L/2$ then:
the maximum bending moment = $FL/4$
and the reactions $R_A = R_B = F/2$

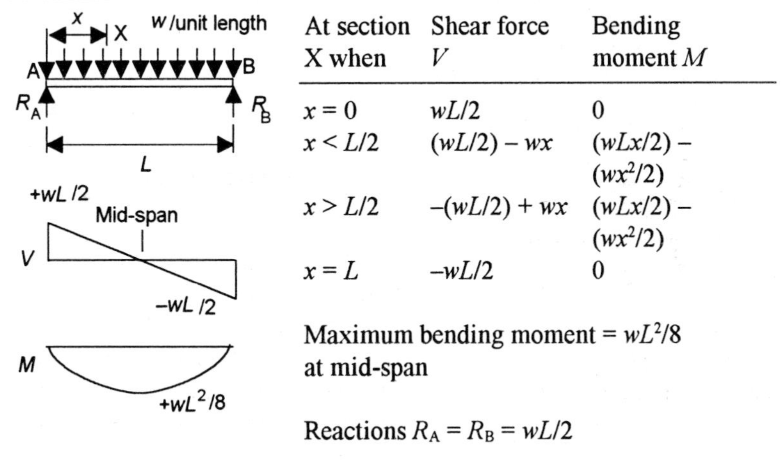

At section X when	Shear force V	Bending moment M
$x = 0$	$wL/2$	0
$x < L/2$	$(wL/2) - wx$	$(wLx/2) - (wx^2/2)$
$x > L/2$	$-(wL/2) + wx$	$(wLx/2) - (wx^2/2)$
$x = L$	$-wL/2$	0

Maximum bending moment = $wL^2/8$ at mid-span

Reactions $R_A = R_B = wL/2$

Table 2.2 *Shear force and bending moment diagrams for cantilevers*

At section X when	Shear force V	Bending moment M
$x < a$	F	$-F(a - w)$
$x \geq a$	0	0
$x = 0$	F	$-Fa$

Maximum bending moment = $M_A = -Fa$ and occurs at the fixed end

When $a = L$ then:
$M_X = -F(L - x)$ and $M_A = -FL$

At section X when	Shear force V	Bending moment M
$x < a$	$w(a - x)$	$-(w/2)(a - x)^2$
$x > a$	0	0
$x = 0$	wa	$-wa^2/2$

Maximum bending moment = $M_A = -wa^2/2$ at fixed end

When $a = L$ then:
$M_X = -(w/2)(L - x)^2$ and $M_A = -wL^2/2$

Example

A horizontal beam has a length of 6 m and is supported at its ends. A point load of 20 kN is applied at 1.5 m from the left-hand end and another point load of 5 kN at 3.0 m from the left-hand end. A uniformly distributed load of 10 kN/m is applied over the 3.0 m length between the 50 kN load and the right-hand end. Draw the shear force and bending moment diagrams and determine the position and size of the maximum bending moment and maximum shear force.

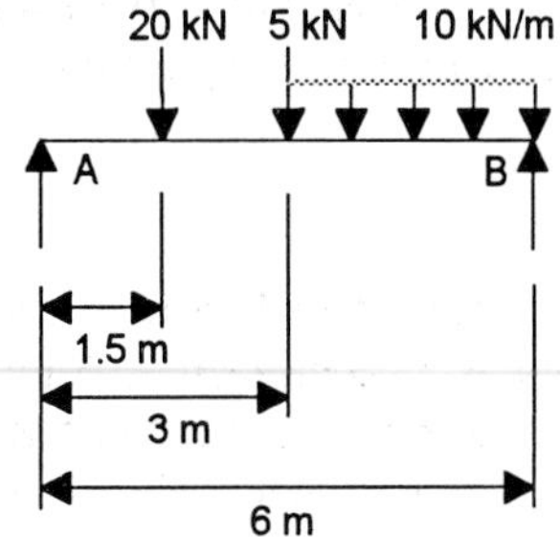

Figure 2.26 *Example*

Figure 2.26 described the arrangement. First the reactions at A and B are calculated. Taking moments about A:

$$6R_B = 20 \times 1.5 + 5 \times 3 + 30 \times 4.5$$

Hence $R_B = 30$ kN. For vertical equilibrium we have:

$$R_A + R_B = 20 + 5 + 30$$

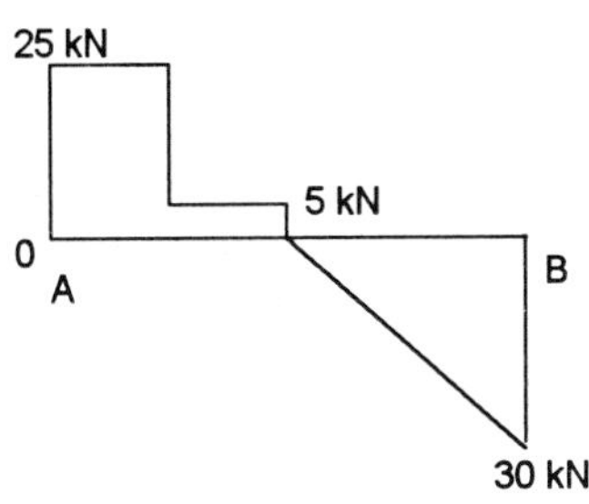

(a) Shear force diagram

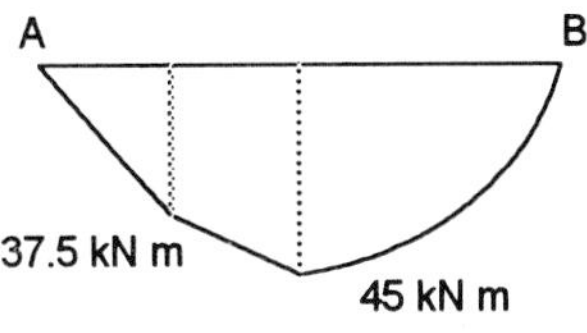

(b) Bending moment diagram

Figure 2.27 *Example*

and so $R_A = 25$ kN.

The shear force diagram will have constant values between the points where there are just point loads. Between A and the 20 kN point force, the shear force will have the value 25 kN. Between the 20 kN and 5 kN point forces the shear force will have the value 25 – 20 = 5 kN. At the 5 kN load point, the shear force will drop to 0. Over the region where there is a distributed load the shear force will vary linearly from 0 to –30 kN. Figure 2.27(a) shows the shear force diagram.

For the region of the beam where we have just point loads, the bending moment will vary linearly between these points. It is zero at A, 37.5 kN m at the 20 kN point and 45 kN m at the 5 kN point. Over the distributed load portion of the beam the bending moment will vary parabolically to become 0 at B. Figure 2.27(b) shows the bending moment diagram.

The maximum bending moment is 45 kN m and occurs at the midpoint of the beam where the 5 kN point load is applied. The maximum shear force is 30 kN and occurs at B.

Revision

7 A beam of length 6 m is simply supported at points 1 m from each end. Draw the shear force and bending moment diagrams if it carries a uniformly distributed load of 10 kN per metre and determine the position and size of the maximum bending moment.

8 A beam of length 4 m is supported at points 1 m from each end. It carries point loads of 15 kN at one end, 10 kN at the other end and 80 kN at its midpoint. Draw the shear force and bending moment diagrams and determine the position and size of the maximum bending moment.

9 A simply supported beam of length 6 m has supports at its ends and carries a point load of 18 kN at a distance of 2 m from one end. Draw the shear force and bending moment diagrams and determine the position and size of the maximum bending moment.

10 A simply supported beam of length 6 m has supports at its ends and carries a uniformly distributed load of 12 kN/m. Draw the shear force and bending moment diagrams and determine the position and size of the maximum bending moment.

2.4 Bending stresses

When a beam bends, one surface becomes extended and so in tension and the other surface becomes compressed and so in compression (Figure 2.28). This implies that between the upper and lower surface there is a plane which is unchanged in length when the beam is bent. This plane is called the *neutral plane* and the line where the plane cuts the cross-section of the beam is the *neutral axis*.

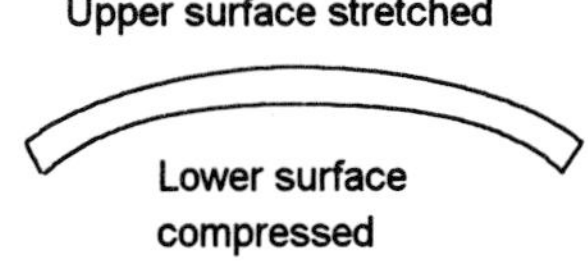

Figure 2.28 *Bending*

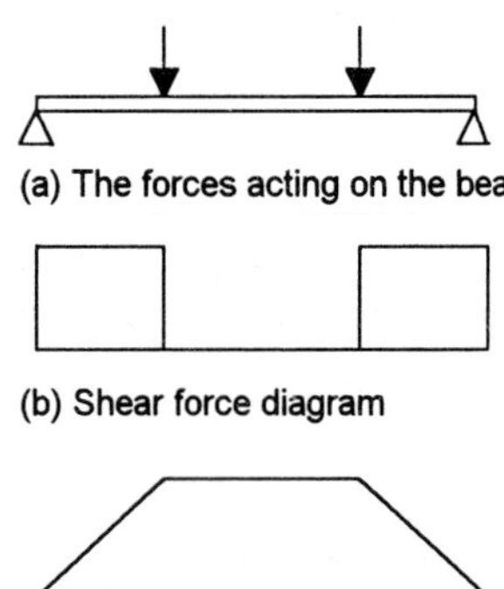

Figure 2.29 *Beam with central region in pure bending*

Consider a beam, or part of a beam, where it can be assumed that it is bent to form the arc of a circle. This is termed *pure bending* and occurs when there is a constant bending moment, Figure 2.29 showing one way this can be realised. Consider the section through the beam aa which is a distance y from the neutral axis (Figure 2.30). It has increased in length as a consequence of the beam being bent. The strain it experiences will be its change in length ΔL divided by its initial unstrained length L. But for circular arcs, the arc length is the radius of the arc multiplied by the angle it subtends. Thus, since aa is of radius $R + y$, we have:

$$L + \Delta L = (R + y)\theta$$

The neutral axis NA will, by definition, be unstrained. Thus:

$$L = R\theta$$

Hence, the strain on aa is:

$$\text{strain} = \frac{\Delta L}{L} = \frac{(R + y)\theta - R\theta}{R\theta}$$

and so:

$$\text{strain} = \frac{y}{R} \qquad [9]$$

The strain thus varies linearly through the thickness of the beam being larger the greater the distance y from the neutral axis. For a uniform rectangular cross-section beam the neutral axis is located symmetrically between the two surfaces and thus the maximum strain occurs on the surfaces of the beam.

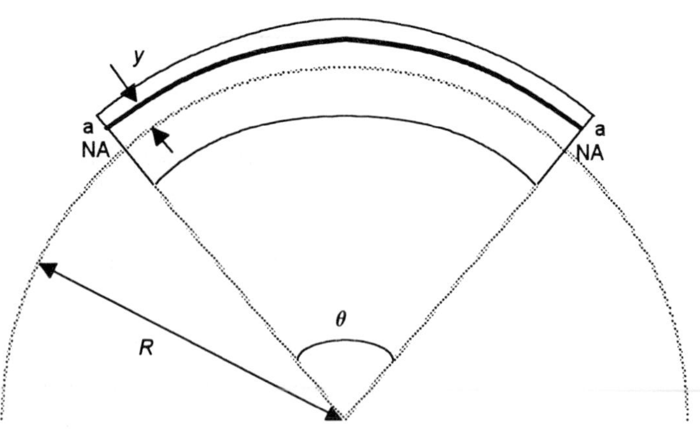

Figure 2.30 *Bending into the form of an arc of a circle*

Provided we can use Hooke's law the stress due to bending which is acting on aa is:

$$\text{stress} = E \times \text{strain} = \frac{Ey}{R} \qquad [10]$$

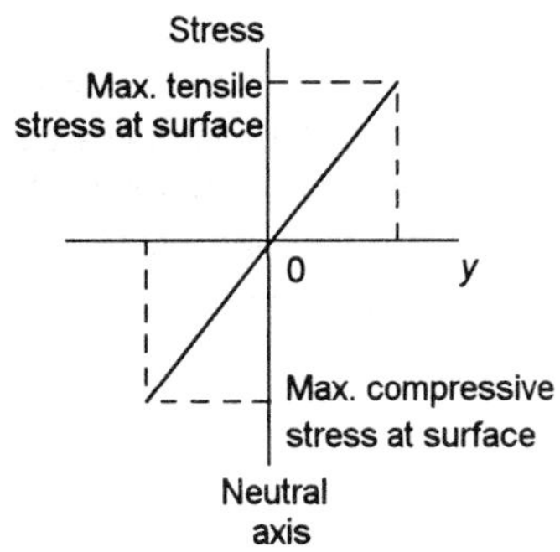

Figure 2.31 *Stress variation across beam section*

With a uniform rectangular cross-section beam, the maximum bending stresses will be on the surfaces. Figure 2.31 shows how the stress will vary across the section of the beam.

Example

A uniform square cross-section steel strip of side 4 mm is bent into a circular arc by bending it round a drum of radius 4 m. Determine the maximum strain and stress produced in the strip. Take the modulus of elasticity of the steel to be 210 GPa.

The neutral axis of the strip will be central and so the surfaces will be 2 mm from it. The radius of the neutral axis will be 4 + 0.002 m. Thus, using equation [9]:

$$\text{maximum strain} = \frac{y}{R} = \frac{2 \times 10^{-3}}{4.002} = 0.5 \times 10^{-3}$$

This will be the value of the compressive strain on the inner surface of the strip and the value of the tensile stress on the outer surface. The maximum stress will be $E \times$ max. strain and thus:

$$\text{maximum stress} = 210 \times 10^{9} \times 0.5 \times 10^{-3} = 105 \text{ MPa}$$

This will be tensile on the outer surface of the strip and compressive on the inner.

Revision

11 Steel strip is to be bent round a drum of radius 1 m. What is the maximum thickness of strip that can be bent in this way if the stress in the strip is not to exceed 100 MPa. The steel has a modulus of elasticity of 210 GPa.

2.4.1 The general bending equations

Consider a beam which has been bent into the arc of a circle so that its neutral axis has a radius R and an element of area dA in the cross-section of the beam at a distance y from the neutral axis (Figure 2.32). The element will be stretched as a result of the bending. The stress σ due to the bending acting on this element is given by equation [10] as Ey/R, where E is the modulus of elasticity of the material. The forces stretching this element aa are:

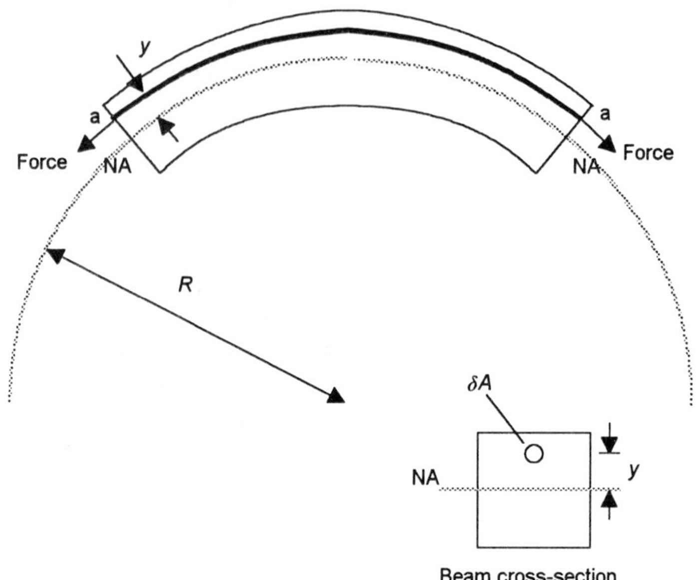

Figure 2.32 *Beam bent into the arc of a circle*

$$\text{force} = \text{stress} \times \text{area} = \sigma\delta A = \frac{Ey}{R}\delta A \qquad [11]$$

The moment of the force acting on this element about the neutral axis is:

$$\text{moment} = \text{force} \times y = \frac{Ey}{R}\delta A \times y = \frac{E}{R}y^2\delta A$$

The total moment M produced over the entire cross-section is the sum of all the moments produced by all the elements of area in the cross-section. Thus, if we consider each such element of area to be infinitesimally small, we can write:

$$M = \int \frac{E}{R}y^2\,\mathrm{d}A = \frac{E}{R}\int y^2\mathrm{d}A \qquad [12]$$

The integral is termed the *second moment of area I* of the section:

$$I = \int y^2\,\mathrm{d}A \qquad [13]$$

Thus equation [12] can be written as:

$$M = \frac{EI}{R} \qquad [14]$$

Since the stress σ on a layer a distance y from the neutral axis is yE/R then we can also write equation [14] as:

$$M = \frac{\sigma I}{y} \qquad [15]$$

Equations [14] and [15] are generally combined and written as the *general bending formula*:

$$\frac{M}{I} = \frac{\sigma}{y} = \frac{E}{R} \qquad [16]$$

This formula is only an exact solution for the case of pure bending, i.e. where the beam is bent into the arc of a circle and the bending moment is constant. However, many beam problems involve bending moments which vary along the beam. In these cases, equation [16] is generally still used since it provides answers which are usually accurate enough for engineering design purposes.

2.4.2 First moment of area

Consider the beam bent into the arc of a circle. The forces acting on a segment a distance y from the neutral axis is given by equation [11] as:

$$\text{force} = \text{stress} \times \text{area} = \sigma \delta A = \frac{Ey}{R}\delta A$$

The total longitudinal force will be the sum of all the forces acting on such segments and thus, when we consider infinitesimally small areas, is given by:

$$\text{total force} = \int \frac{Ey}{R}\,\mathrm{d}A = \frac{E}{R}\int y\,\mathrm{d}A$$

But the beam is only bent and so only acted on by a bending moment, there is no longitudinal force stretching the beam. Thus, since E and R are not zero, we must have:

$$\int y\,\mathrm{d}A = 0 \qquad [17]$$

The integral $\int y\,\mathrm{d}A$ is called the *first moment of area* of the section. The only axis about which we can take such a moment and obtain 0 is an axis through the centre of the area of the cross-section, i.e. the centroid of the beam. Thus the neutral axis must pass through the centroid of the section when the beam is subject to just bending.

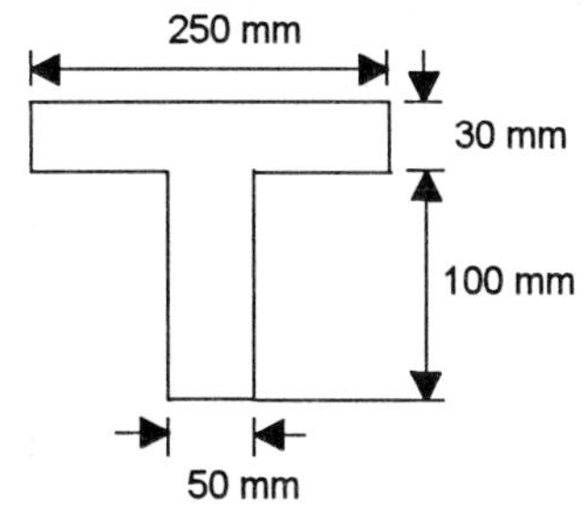

Figure 2.33 *Example*

Example

Determine the position of the neutral axis for the T-section beam shown in Figure 2.33.

The neutral axis will pass through the centroid. We can consider the T-section to be composed of two rectangular sections. The centroid of each will be at its centre. Hence, taking moments about the base of the T-section:

$$\text{total moment} = 250 \times 30 \times 115 + 100 \times 50 \times 50$$

$$= 1.11 \times 10^6 \text{ mm}^4$$

Hence the distance of the centroid from the base is (total moment)/(total area):

$$\text{distance from base} = \frac{1.11 \times 10^6}{250 \times 30 + 100 \times 50} = 89 \text{ mm}$$

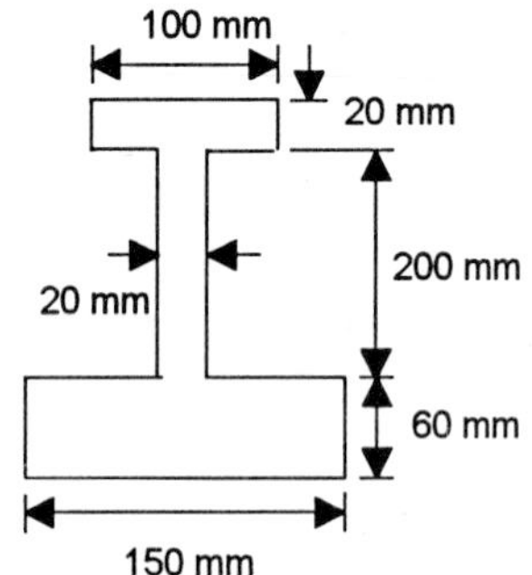

Figure 2.34 *Revision problem 12*

Revision

12 Determine the position of the neutral axis from the base for the non-symmetrical I-section shown in Figure 2.34.

2.4.3 Second moments of area

The integral $\int y^2 \, dA$ defines the *second moment of area I* about an axis. Consider a rectangular cross-section of breadth b and depth d (Figure 2.35). For a layer of thickness δy a distance y from the neutral axis, which passes through the centroid, the second moment of area for the layer is:

second moment of area of strip = $y^2 \delta A = y^2 b \delta y$

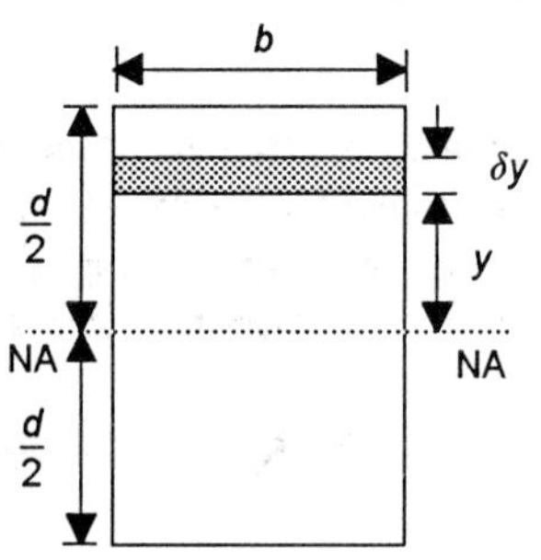

Figure 2.35 *Second moment of area*

The total second moment of area for the section is thus:

$$\text{second moment of area} = \int_{-d/2}^{d/2} y^2 b \, dy = \frac{bd^3}{12} \qquad [18]$$

If we had a second moment of area $I = \int y^2 \, dA$ of an area about an axis and then considered a situation where the area was moved by a distance h from the axis, the new second moment of area I_h would be:

$$I_h = \int (y+h)^2 \, dA = \int y^2 \, dA + 2h \int y \, dA + h^2 \int dA$$

But $\int y \, dA = 0$ and $\int dA = A$. Hence:

$$I_h = I + Ah^2 \qquad [19]$$

This is called the *theorem of parallel axes* and is used to determine the second moment of area about a parallel axis.

Figure 2.36 shows some values of second moments of area about the neutral axes for commonly encountered beam sections.

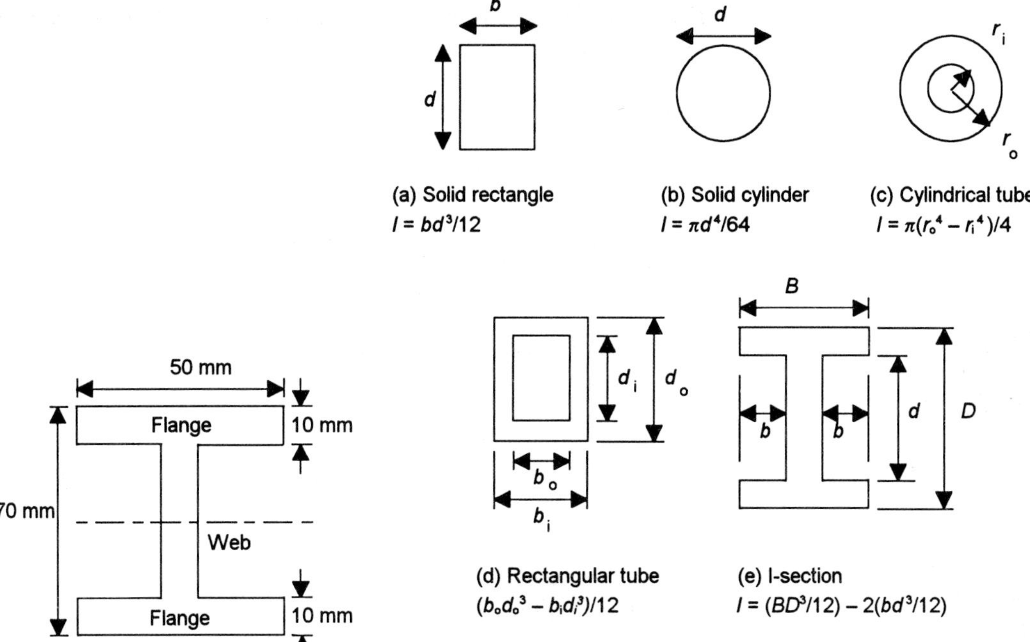

Figure 2.36 *Second moments of area*

Figure 2.37 *Example*

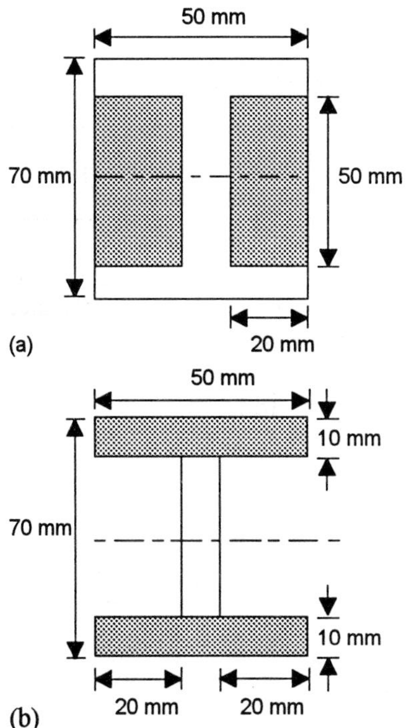

Figure 2.38 *Example*

Example

Determine the second moment of area about the neutral axis of the I-section shown in Figure 2.37.

One way of determining the second moment of area for such a section involves determining the second moment of area for the entire rectangle containing the section and then subtracting the second moments of area for the rectangular pieces 'missing' (Figure 2.38(a)).

Thus for the rectangle containing the entire section, the second moment of area is given by $I = bd^3/12$ as:

$$\frac{50 \times 70^3}{12} = 1.43 \times 10^6 \text{ mm}^4$$

For the two 'missing' rectangles, each will have a second moment of area of:

$$\frac{20 \times 50^3}{12} = 0.21 \times 10^6 \text{ mm}^4$$

Thus the second moment of area of the I-section is:

$$1.43 \times 10^6 - 2 \times 0.21 \times 10^6 = 1.01 \times 10^6 \text{ mm}^4$$

Another way of determining the second moment of area of the I-section is to consider it as three rectangular sections, one being the central rectangular section, the web, and the others a pair of rectangular sections, the flanges, with their neutral axes displaced from the neutral axis of the I-section by 30 mm (Figure 2.38(b)). The central rectangular section has a second moment of area of:

$$\frac{10 \times 50^3}{12} = 0.104 \times 10^6 \text{ mm}^4$$

Each of the outer rectangular areas will have a second moment of area given by the theorem of parallel axes as:

$$\frac{50 \times 10^3}{12} + 50 \times 10 \times 30^2 = 0.454 \times 10^6 \text{ mm}$$

Thus the second moment of area of the I-section is:

$$0.104 \times 10^6 + 2 \times 0.454 \times 10^6 = 1.01 \times 10^6 \text{ mm}^4$$

Application
Consider the problem of bridging a gap using stone beams. Stone is strong in compression and weak in tension. Thus the beam must not develop large tensile forces. The maximum stress on a rectangular section beam will be My_{max}/I where $I = bd^3/3$ with b being the breadth and d the depth. Thus, since $y_{max} = d/2$ the maximum stress is $3M/2bd^2$. Thus to keep the stress low means having a large value of bd, i.e. cross-sectional area. Also we need to keep M low and so small length beams are required. Thus the characteristic appearance of ancient stone buildings as many supporting pillars, relative close together with large cross-section beams bridging between them (Figure 2.39).

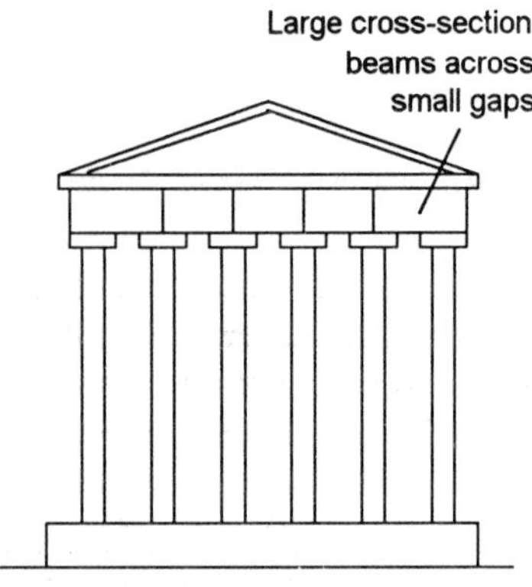

Figure 2.39 *Bridging gaps with stone beams*

Example

A horizontal beam with a uniform rectangular cross-section of breadth 100 mm and depth 150 mm is 4 m long and rests on supports at its ends. It supports a concentrated load of 10 kN at its midpoint. Determine the maximum tensile and compressive stresses in the beam.

The second moment of area is:

$$I = bd^3/12 = 0.100 \times 0.150^3/12 = 2.8 \times 10^{-5} \text{ m}^4$$

The reactions at each support will be 5 kN and so the maximum bending moment, which will occur at the midpoint, is 10 kN m. The maximum bending stress will occur at the cross-section where the bending moment is a maximum and on the outer surfaces of the beam, i.e. $y = \pm 75$ mm. Using equation [22]:

$$\sigma = \frac{My}{I} = \pm \frac{10 \times 10^3 \times 0.075}{2.8 \times 10^{-5}} = \pm 26.8 \text{ MPa}$$

Revision

13 Determine the second moment of area of an I-section, about its horizontal neutral axis when the web is vertical, if it has rectangular flanges each 120 mm by 10 mm, a web of thickness 12 mm and an overall depth of 150 mm.

14 For the I-section in the above problem, determine the maximum bending moment that can be applied if the maximum bending stress is not to exceed 80 MPa.

2.4.4 Section modulus

For a beam which has been bent, the maximum stress σ_{max} will occur at the maximum distance y_{max} from the neutral axis. Thus, using equation [15], we can write:

$$M = \frac{I}{y_{max}} \sigma_{max}$$

The quantity I/y_{max} is a purely geometric function and is termed the *section modulus* Z. Thus:

$$M = Z\sigma_{max} \quad [20]$$

with:

$$Z = \frac{I}{y_{max}} \quad [21]$$

For a rectangular cross-section beam, the second moment of area $I = bd^3/12$ and the maximum stress occurs at the surfaces which are $d/2$ from the neutral axis. Thus $Z = (bd^3/12)/(d/2) = bd^2/6$. Standard section handbooks give values of section modulus for different section beams. Table 2.3 shows an example. BS5950 Part 1: 2000 gives more comprehensive and detailed tables.

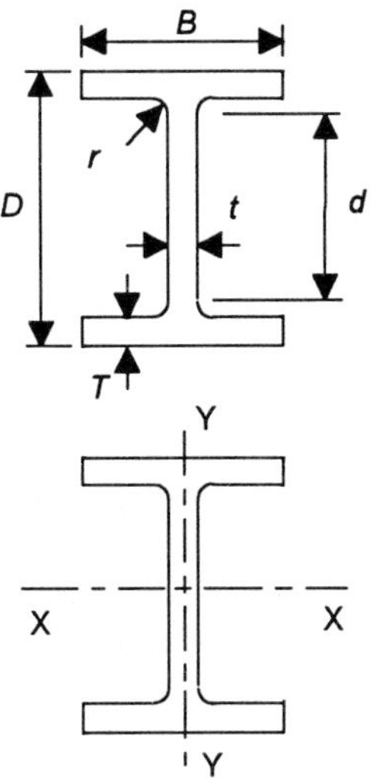

Figure 2.40 *Universal beam*

Table 2.3 *Some of the properties for a few universal beams (Figure 2.40) found in standard tables (British Standards)*

Serial size mm × mm	Mass per unit length kg/m	Depth D mm	Width B mm	Web thickness t mm	Flange thickness T mm	Root radius r mm	Depth between fillets d mm	Area of section cm^2	Second moment of area cm^4 X–X axis	Second moment of area cm^4 Y–Y axis
914 × 419	388	920.5	420.5	21.5	36.6	24.1	799.0	494.5	718 742	45 407
	343	911.4	418.5	19.4	32.0	24.1	799.0	437.5	625 282	39 150
914 × 305	289	926.6	307.8	19.6	32.0	19.1	824.4	368.8	504 594	15 610
	253	918.5	305.5	17.3	27.9	19.1	824.4	322.8	436 610	13 318
	224	910.3	304.1	15.9	23.9	19.1	824.4	285.3	375 924	11 223
	201	903.0	303.4	15.2	20.2	19.1	824.4	256.4	325 529	9 427
838 × 292	226	850.9	293.8	16.1	26.8	17.8	761.7	288.7	339 747	11 353
	194	840.7	292.4	14.7	21.7	17.8	761.7	247.2	279 450	9 069
	176	834.9	291.6	14.0	18.8	17.8	761.7	224.1	246 029	7 792

Example

A beam has a section modulus of 2×10^6 mm^3, what will be the maximum bending moment that can be used if the stress must not exceed 6 MPa?

Using equation [20]:

$$M = Z\sigma_{max} = 2 \times 10^6 \times 10^{-9} \times 6 \times 10^6 = 12 \text{ kN m}$$

Example

An I-section beam has a section modulus of 25×10^{-5} m^3. What will be the maximum bending stress produced when the beam is subject to a bending moment of 30 kN m?

Using equation [20]:

$$\sigma_{max} = \frac{M}{Z} = \frac{30 \times 10^3}{25 \times 10^{-5}} = 120 \text{ MPa}$$

Example

A rectangular cross-section timber beam of length 4 m rests on supports at each end and carries a uniformly distributed load of 10 kN/m. If the stress must not exceed 8 MPa, what will be a suitable depth for the beam if its width is to be 100 mm?

For a simply supported beam with a uniform distributed load over its full length, the maximum bending moment is $wL^2/8$ (equation [4]) and thus the maximum bending moment for this beam is $10 \times 4^2/8 =$ 20 kN m. Using equation [14]:

$$Z = \frac{M}{\sigma_{max}} = \frac{20 \times 10^3}{8 \times 10^6} = 2.5 \times 10^{-3} \text{ m}^3$$

For a rectangular cross-section $Z = bd^2/6$ and thus:

$$d = \sqrt{\frac{6Z}{b}} = \sqrt{\frac{6 \times 2.5 \times 10^{-3}}{0.100}} = 0.387$$

A suitable beam might thus be one with a depth of 400 mm.

Revision

15 A steel scaffold tube has a section modulus of 7.2×10^{-6} m^3. What will be the maximum allowable bending moment on the tube if the bending stresses must not exceed 100 MPa?

16 An I-section beam has a section modulus of 3.2×10^{-5} m^3. What will be the maximum allowable bending moment on the beam if the bending stresses must not exceed 150 MPa?

17 Determine the section modulus required of a steel beam which is to span a gap of 6 m between two supports and support a uniformly distributed load over its entire length, the total distributed load amounting to 65 kN. The maximum bending stress permissible is 165 MPa.

2.5 Columns

The term *column* is used for a compression member such as a main supporting in a building and might be made of timber, steel or reinforced concrete, the term *strut* being used for a smaller compression member in a framed structure such as a roof truss. When the line of action of the load on such a member is coincident with the axis through the centre of gravity, the member is said to be *axially loaded* and the stress produced is said to be a *direct compressive stress.* When the load is not axial, it is said to be *eccentric*, i.e. off-centre, loading and bending stress, as well as direct compressive stress, occurs (Figure 2.41).

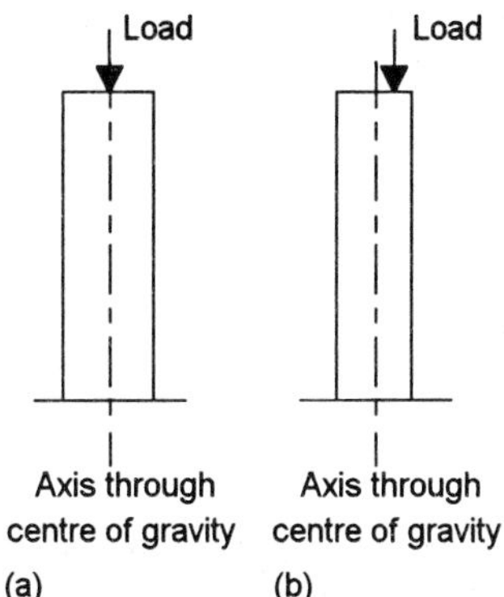

Figure 2.41 *(a) Axial loading, (b) eccentric loading*

2.5.1 Short columns

The term *short* is applied to a column if the result of applying a direct axial load is to just squash it. Such a column will fail internally by yielding if a ductile material or by shearing if a brittle material. The load such a column can withstand depends only on its cross-sectional area and the strength of the material used. A rectangular column is described as short if the height to minimum width ratio is less than 15 when the top is restrained against sideways movement and less than 10 if unrestrained against sideways movement.

Application
If you stand on the end of an aluminium drinks' can, the can collapses as a result of the walls buckling.

2.5.2 Slender columns

When a slender column is axially loaded, it buckles, i.e. the vertical axial load causes a sideways deflection of the column (Figure 2.42). At low axial loads the column springs back to its original straight vertical position when the load is removed. However, when a particular load is applied, the column no longer springs back when the load is removed and retains a sideways buckle. This load at which a slender column buckles is called the *critical buckling load.* A slender column fails by buckling.

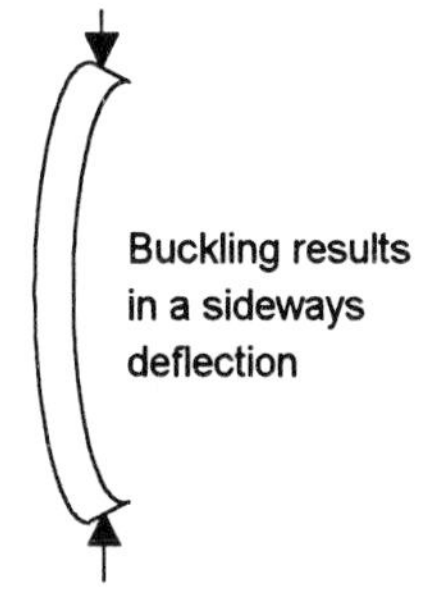

Figure 2.42 *Buckling*

The critical buckling stress depends on the material concerned and how slender a column is. The measure of slenderness used is the *slenderness ratio*.

$$\text{Slenderness ratio} = \frac{\text{effective length}}{\text{radius of gyration}} \qquad [22]$$

The effective length is essentially the length between the points at which the member bows out (Figure 2.43). Fixing an end inhibits bowing out. For a pin-ended member of length L the *effective length* L_e is the distance between the pins, i.e. L. When both ends are fixed, the effective

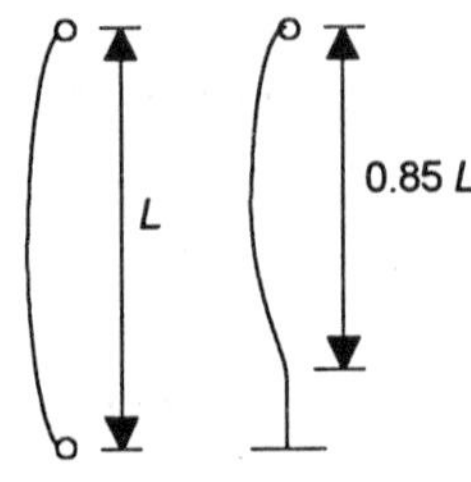

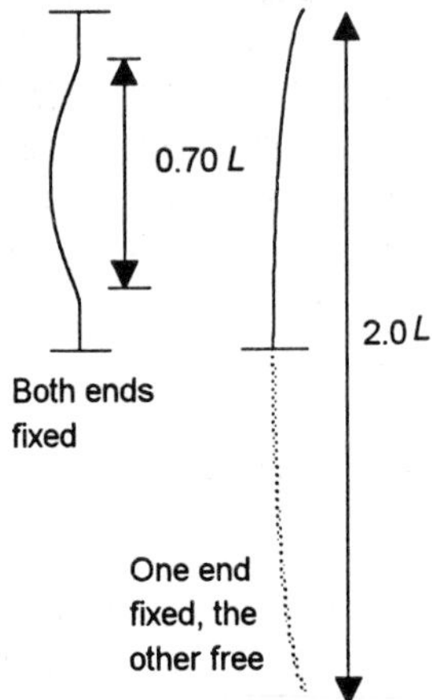

Figure 2.43 *Effective lengths*

Application
For a given amount of material, the greater the distance it is from the buckling axis of a column the greater is the radius of gyration and hence the stronger the column. Thus hollow sections are more efficient than solid sections. Thus, round tubes are more efficient than solid sections.

A tube, however, has two possible modes of buckling. It can buckle over its full length in the fashion described here in the text, or it can buckle in a more local fashion with a crease occurring in the tube wall. It is this which sets a limit to the use of tubes. Local buckling in a thin walled tube occurs when the stress in the wall reaches the value $Et/4r$, where E is the modulus of elasticity, t is the wall thickness and r the radius of the tube. To guard against local buckling, stiffeners can be added to a tube. These may be small ribs which run longitudinally along or circumferentially round the tube.

length is $0.70L$. When one end is pinned and the other fixed, the effective length is $0.85L$. When one end is fixed and the other free (a vertical cantilever) the effective length is $2.0L$.

The *radius of gyration* is defined by reference to the second moment of area of a section (see Section 2.4.3). The second moment of area I for a section is given by equation [13] as $\int y^2 \, dA$, where y is the distance of an element of the area δA from the axis. If, however, we consider all the area A of a section to be concentrated into a single dimensionless point, then the distance r of this point from the axis to give the second moment of area I of the section is termed the radius of gyration and must be such that:

$$I = Ar^2$$

$$r = \sqrt{\frac{I}{A}} \qquad [23]$$

British Standards 5950: Part 1: 2000 gives tables from which the radius of gyration can be obtained for steel sections. Table 2.4 shows the type of information supplied.

Table 2.4 *Some of the properties for a few universal columns (Figure 2.40) found in standard tables (British Standards)*

Serial size mm	Mass per unit length kg/m	Area of section cm^2	Minimum radius of gyration cm
305 × 305	283	360.4	8.25
	198	252.3	8.02
	137	174.6	7.82
	97	123.3	7.68
254 × 254	167	212.4	6.79
	107	136.6	6.57
	89	114.0	6.52
	73	92.9	6.46
203 × 203	86	110.1	5.32
	71	91.1	5.28
	60	75.8	5.19
	52	66.4	5.16
	46	58.8	5.11

In selecting compression members, tables a graph or a formula can be used to determine the compressive strength for particular slenderness ratios. Tables in British Standards 5950: Part 1: 2000, can be used which give values of compressive strength for different values of slenderness ratio for steels. Table 2.5 shows the type of data available. Figure 2.44

shows a graph for mild steel of compressive strength against slenderness ratio.

Table 2.5 *Compressive strength for struts*

Slenderness ratio	Compressive strength MPa							
	Axis of buckling X–X				Axis of buckling Y–Y			
	Yield stress MPa				Yield stress MPa			
	265	275	340	355	265	275	340	355
25	261	270	333	347	258	267	328	342
50	237	245	267	309	221	228	275	285
75	208	214	246	252	187	192	221	226
100	155	157	169	171	138	141	153	155
150	80	80	82	83	74	74	77	78

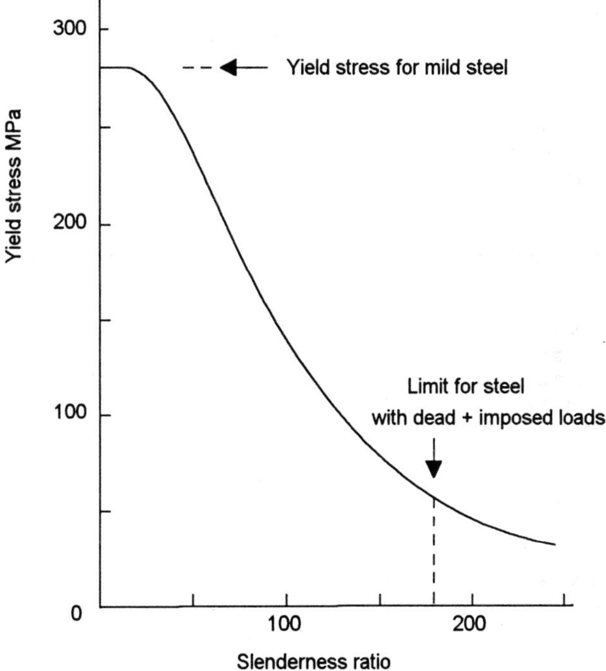

Figure 2.44 *Compressive yield stress for mild steel I sections*

The maximum slenderness ratio for steel columns carrying dead and imposed loads is limited to 180; as the table and graph indicates, when the slenderness ratio approaches this value the compressive strength drops to quite a low value. If, in the design of a structure, a larger slenderness ratio is indicated then it indicates the need to use a larger section size in order to decrease the slenderness ratio.

The general procedure for designing axially loaded columns is:

1. Determine the effective length of the required column.
2. Select a trial section.
3. Using the radius of gyration value for the trial section, calculate the slenderness ratio.
4. If the slenderness ratio is greater than 180, try a larger cross-section trial section.
5. Using the slenderness ratio, obtain the compression strength from tables.
6. Compare the compression strength with that which is likely to be required in practice. If the compression strength is adequate then the trial section is suitable. If the compression strength is much greater than is required then it would be more economical to choose a section with a smaller radius of gyration.

Example

What is the slenderness ratio for a column, with a radius of gyration of 60 mm, which is fixed at both ends and has a length of 5 m?

The effective length for a column fixed at both ends is 0.70 times the actual length and so is $0.7 \times 5 = 3.5$ m. The slenderness ratio is thus:

$$\text{slenderness ratio} = \frac{3.5}{0.060} = 58.3$$

Example

Use tables to select a universal column to be a pin-ended column of length 4 m for an axial load of 1000 kN. The steel has a yield stress of about 265 MPa.

Suppose we consider a $203 \times 203 \times 52$ kg/m universal column. Tables indicate that such a column has a cross-sectional area of 66.4 cm^2 and a minimum radius of gyration of 5.16 cm. A pin-ended column will have an effective length equal to its actual length. Thus, such a column will have a slenderness ratio of:

$$\text{slenderness ratio} = \frac{4}{0.0516} = 77.5$$

As Table 2.5 indicates, such a member is likely to have a compressive strength of about 180 MPa, for buckling in the Y–Y direction. The actual stress on the member is:

$$\text{stress} = \frac{1000 \times 10^3}{66.4 \times 10^{-4}} = 150.6 \text{ MPa}$$

Thus, such a column would be suitable.

Revision

18 What is the slenderness ratio for a column, with a radius of gyration of 50 mm, which is pinned at both ends and has a length of 4 m?

19 What will be (a) the slenderness ratio and (b) the compressive strength of a 203 × 203 × 60 kg/m universal column if it is pinned at both ends and has a length of 4.0 m? The steel has a yield stress of about 265 MPa.

Application
An example of eccentric loading of a column is where a beam is connected to the face of the column (Figure 2.45).

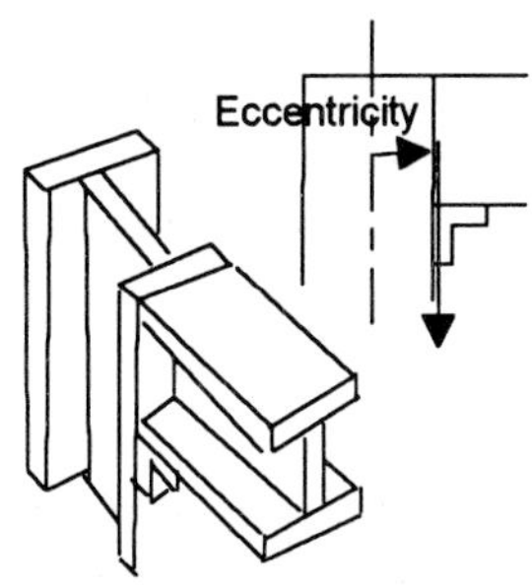

Figure 2.45 *An example of eccentric loading*

Activity

Investigate the effect of length on the buckling of slender columns by taking some drinking straws and applying axial loads to the ends and observing whether they buckle. Try a full length straw, three-quarter length straw, half-length straw and quarter-length straw.

2.5.3 Eccentric loading of columns

An eccentric load as well as causing a direct compressive stress as a result of its downward action on a column, also sets up a bending moment and so bending stresses. Thus we are concerned with the addition of a direct and a bending stress.

The bending moment M_e set up is equal to the load F times the eccentricity e:

$$M_e = Fe \quad [24]$$

This bending moment will result in bending such that one face of the column is in tension and the opposite face in compression. The stress σ_e due to this eccentric loading is given by equation [16] as:

$$\sigma_e = \frac{M_e y}{I}$$

where y is the distance from the neutral axis of the column and I is the second moment of area of the column. The stress distribution arising from this bending is shown in Figure 2.46(a). Figure 2.46(b) shows the stress distribution σ_d arising from the direct loading.

$$\sigma_d = \frac{F}{A}$$

where F is the load and A the cross-sectional area. Figure 2.46(c) is the resultant stress distribution as the sum of those in Figure 2.46(a) and (b). Thus we have the stress as:

$$\sigma = \frac{F}{A} \pm \frac{M_e y}{I} \quad [25]$$

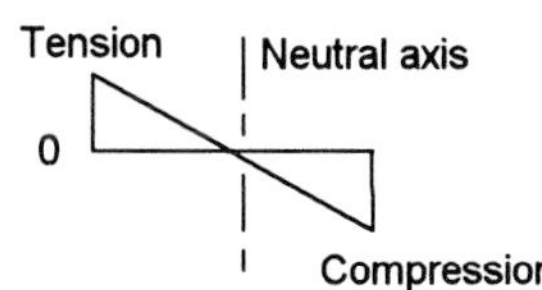

(a) Stress distribution from the bending arising from eccentric load

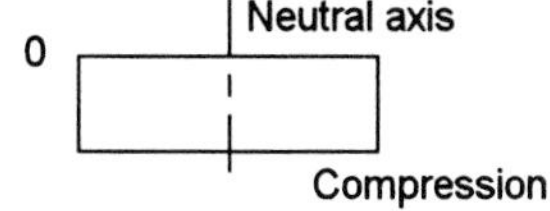

(b) Stress distribution from the direct compressive loading

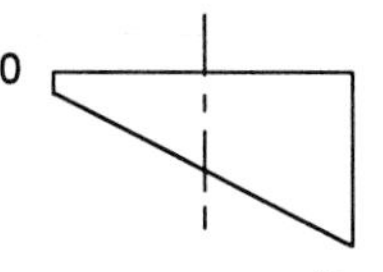

(c) The resultant stress as the sum of (a) and (b)

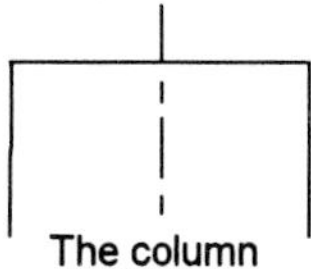

Figure 2.46 *Stresses due to eccentric loading*

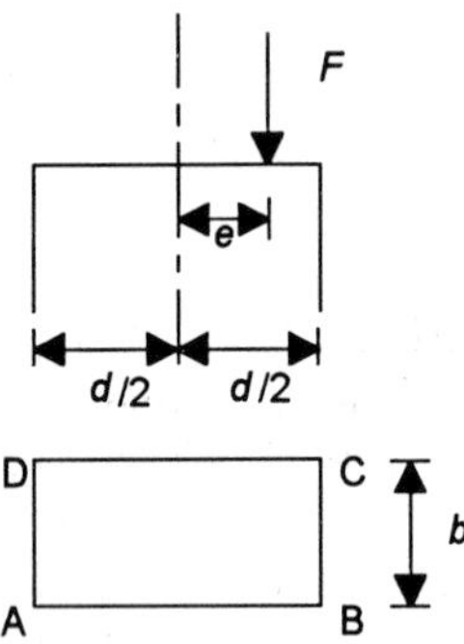

Figure 2.47 *Eccentric force applied to a rectangular section*

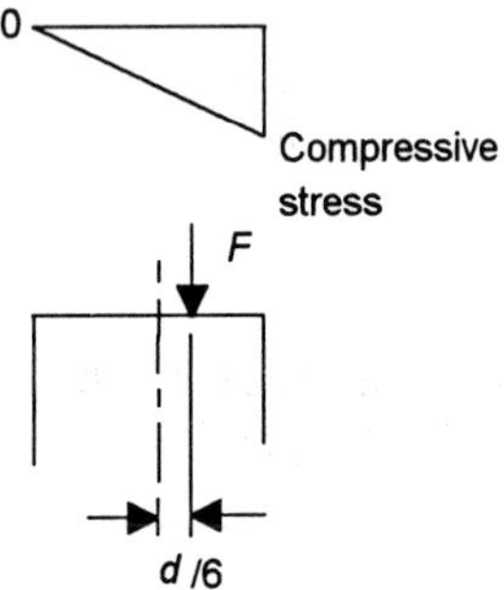

Figure 2.48 *The arrangement for no tensile stress*

Application
Whenever a force is not applied through the centroidal axis of a member it results in a moment in addition to an axial force. Prestressed concrete beams are an example (Figure 2.50) where eccentric forces are deliberately introduced in order that the concrete beam when subject to bending is not subject to tensile stresses. Concrete is weak in tension but strong in compression.

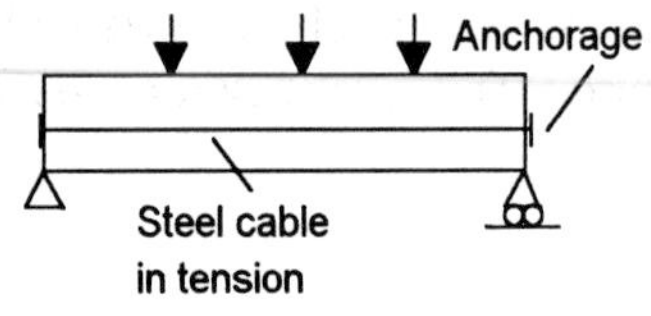

Figure 2.50 *Prestressed beam*

For a force F applied at an eccentricity e (Figure 2.47), the moment $M_e = Fe$.

The stress at the outermost fibres of a rectangular section, i.e. at $y = d/2$, of cross-section $d \times b$ is:

$$\sigma = \frac{F}{bd} \pm \frac{Fe(d/2)}{I}$$

and so, since $I = bd^3/12$:

$$\sigma = \frac{F}{bd} \pm \frac{6Fe}{bd^2}$$

We can obtain no tensile stress along the edge DA if:

$$\frac{F}{bd} = \frac{6Fe}{bd^2}$$

i.e. if we have the eccentricity e such that:

$$e = \frac{d}{6} \quad [26]$$

Figure 2.48 shows the resulting stress variation across the section. If e is less than $d/6$ there will be no tensile stress. Only if e is greater than $d/6$ will there be any tensile stress acting on the column. Thus if the applied load has an eccentricity within $+d/6$ or $-d/6$ there is no tensile stress (Figure 2.49). This condition for obtaining no tensile stress is known as the *middle-third rule*.

> Middle-third rule: tensile stresses will not occur within an eccentrically loaded rectangular cross-section column if the applied load is applied within the middle-third of the section.

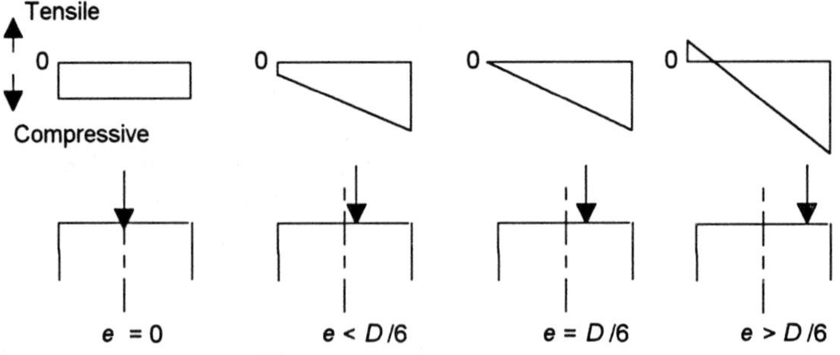

Figure 2.49 *Stress distribution with different eccentricity*

If we have loads applied eccentrically to both axes of a column (Figure 2.51), then there will be moments about both the axes and so the resulting stress in the column is:

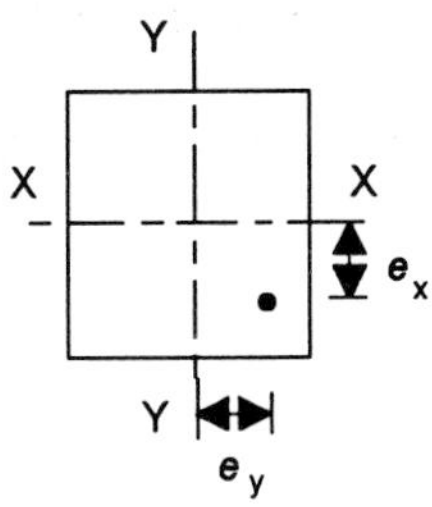

Figure 2.51 *Eccentricity about both axes*

$$\sigma = \frac{F}{A} \pm \frac{M_{XX}y}{I_{XX}} \pm \frac{M_{YY}x}{I_{YY}} \qquad [27]$$

Example

A short 203 × 203 × 46 kg/m universal column carries an axial load of 300 kN and a further load of 120 kN from an incoming beam supported on a bracket (as in Figure 2.45). The 120 kN load can be considered to act with an eccentricity of 100 mm from the X–X axis of the column. Determine the stress distribution across the column. For the column I_{XX} = 4560 cm^4, its cross-sectional area is 114 cm^2 and it has a depth of 203 mm.

The moment about the XX axis is 120 × 10^3(0.100 + 0.203/2) = 24.18 kN m. The total load is 300 + 120 = 420 kN. Thus the stress at the outermost edges of the column is given by equation [25] as:

$$\sigma = \frac{F}{A} \pm \frac{M_e y}{I} = \frac{420 \times 10^3}{114 \times 10^{-4}} \pm \frac{24.18 \times 10^3 \times 0.203/}{4560 \times 10^{-8}}$$

$$= 36.9 \times 10^6 \pm 53.8 \times 10^6$$

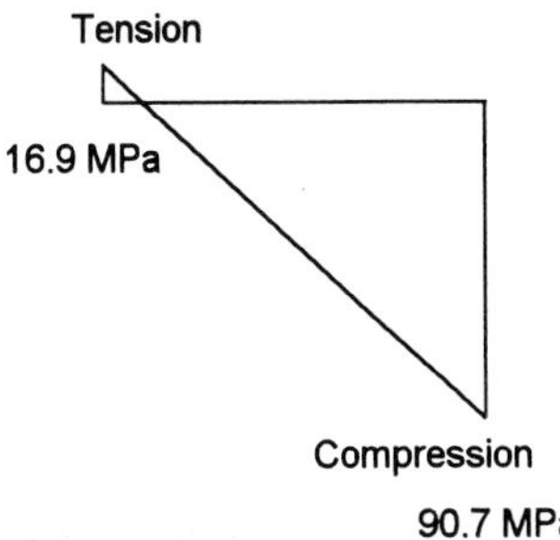

Figure 2.52 *Example*

The stress thus varies linearly from one edge at 90.7 MPa compressive to 16.9 MPa tensile at the other (Figure 2.52).

Example

At what distance from the centroid axis of a rectangular cross-section short column of 120 mm × 120 mm should a load be applied to give no tensile stresses?

Applying the middle-third rule, no tensile stresses will occur if the load is applied within 120/6 = 20 mm of the centroid axis.

Revision

20 A short 150 mm × 150 mm rectangular cross-section column supports a load of 90 kN. Determine the stress distributions across the section if this load is applied (a) centrally, (b) with an eccentricity of 20 mm from the Y–Y axis, (c) with an eccentricity of 25 mm from the Y–Y axis. For a rectangular cross-section I_{YY} = $bd^3/12$.

21 At what distance from the centroid axis of a rectangular cross-section short column of 600 mm × 600 mm should a load be applied to give no tensile stresses?

Problems

1 A beam of length 4.0 m is supported at its ends and carries a concentrated load of 20 kN at its midpoint. Determine the shear force and bending moment at distances of (a) 0.5 m and (b) 1.0 m from the right-hand end. Neglect the weight of the beam.

2 A uniform beam of length 4.0 m is supported at its ends and has a weight of 10 kN/m. It carries no other loads. Determine the shear force and bending moment at distances of (a) 0.5 m and (b) 1.0 m from the right-hand end.

3 A beam of length 6 m is supported at both ends and carries a point load of 40 kN at its midpoint. Draw the shear force and bending moment diagrams and determine the position and value of the maximum bending moment.

4 A beam of length 6 m is supported at both ends and carries a point load of 60 kN at a distance of 2 m from one end. Draw the shear force and bending moment diagrams and determine the position and value of the maximum bending moment.

5 A beam of length 4 m is supported at points 1 m and 3 m from one end. It carries point loads of 20 kN at each end. Determine the maximum shear force and the maximum bending moment.

6 A cantilever of length 4 m carries point loads of 30 kN at 1 m from the fixed end, 20 kN at 3 m from the fixed and 10 kN at the free end. Draw the shear force and bending moment diagrams and determine the positions and values of the maximum shear stress and bending moment.

7 A beam of length 10 m is supported at one end and at a point 8 m from the supported end. It carries a uniformly distributed load of 16 kN/m over its entire length and a point load of 40 kN at the unsupported end. Draw the shear force and bending moment diagrams and determine the positions and values of the maximum shear stress and bending moment.

8 A steel strip of thickness 0.8 mm is bent round a pulley of radius 200 mm. What is the maximum stress produced in the wire as a result of the bending? The steel has a modulus of elasticity of 210 GPa.

9 Determine the position of the neutral axis of a T-section if the top of the T is a rectangle 100 mm by 10 mm and the stem of the T is a rectangle 120 mm by 10 mm.

10 Determine the second moment of area of a rectangular section of breadth 50 mm and depth 100 mm.

11 Determine the position of the neutral axis of a T-section if the top of the T is a rectangle 150 mm by 10 mm and the stem of the T is a rectangle 90 mm by 10 mm.

12 A beam has a rectangular cross-section of width 60 mm and depth 100 mm. Determine the maximum bending moment that can be applied if the bending stresses are not to exceed ±150 MPa.

13 A rectangular tube section has an overall width of 80 mm and a depth of 150 mm. The inner walls have a width of 60 mm and a depth of 130 mm, the walls of the tube being 10 mm thick. Determine the maximum tensile and compressive stresses such a section will experience when subjected to a bending moment of 20 kN m.

14 A beam has to support loading which results in a maximum bending moment of 25 kN m. If the maximum permissible bending stress is 7 MPa, what will be the required section modulus?

15 For the beam sections shown in Figure 2.53, determine the position of the neutral axis, the second moment of area about the neutral axis and the section modulus for the top edge.

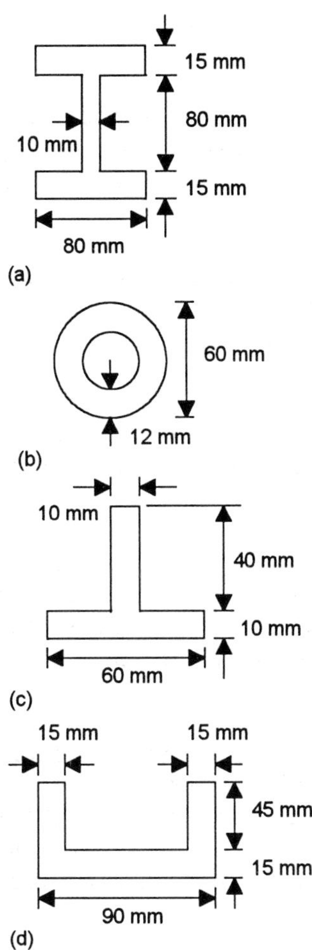

Figure 2.53 *Problem 15*

16 A uniform beam of length 6 m and section modulus 2.3×10^{-3} m^3 is supported at its ends. Point loads of 40 kN are carried at 1.5 m from each end. What will be the maximum stress experienced by the beam?

17 Determine the slenderness ratio for a 5.0 m long column having a radius of gyration of 90 mm if it is (a) pinned at both ends, (b) fixed at both ends.

18 Determine the maximum permissible length for a pin-ended universal 254 × 254 × 73 kg/m column for dead plus imposed loads.

19 What will be (a) the slenderness ratio and (b) the compressive strength of a 305 × 305 × 97 kg/m universal column if it is fixed at both ends, has a length of 4.0 m and buckles in the Y–Y direction? The steel has a yield stress of about 265 MPa.

20 Select a suitable universal column to withstand a load of 2 MN, have a length of 3.0 m and be pinned at both ends.

21 What will be the stress distribution across the section for a short universal column if it carries (a) an axial load of 120 kN, (b) an eccentric load of 120 kN at 230 mm from the X–X centroidal axis? The section has a second moment of area about the X–X axis of 143 × 10^6 mm^4 and a cross-sectional area of 11 400 mm^2.

22 What is the maximum distance from the centroidal axis of a 240 mm × 240 mm short rectangular section that a load can be applied without giving rise to any tensile stresses?

23 A short rectangular cross-section column 100 mm × 100 mm carries an eccentric load of 40 kN. What will be the stress distribution across the column if the eccentricity is 12 mm along the X–X axis from the centroidal axis? What is the maximum eccentricity for there to be no tensile stresses?

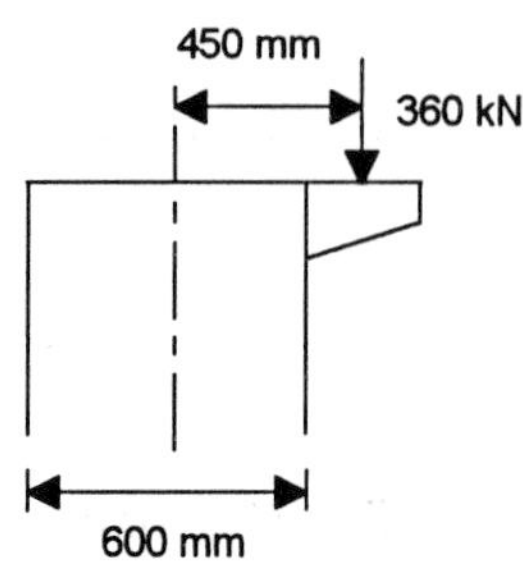

Figure 2.54 *Problem 24*

24 A short rectangular column 400 mm × 600 mm carries an eccentric load of 360 kN with an eccentricity of 450 mm along the long axis from the centroidal axis (Figure 2.54). Determine the stress distribution across the section.

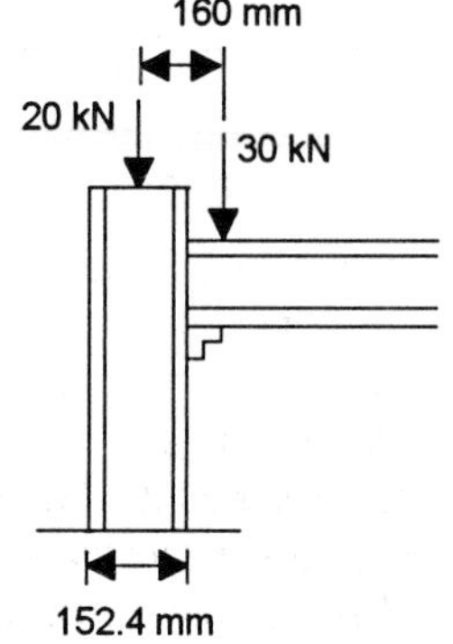

Figure 2.55 *Problem 25*

25 A short universal column 152 × 152 × 23 kg/m has an axial load of 20 kN and supports, via a bracket, a universal beam such that a load of 30 kN is applied eccentrically to the column with an eccentricity of 160 mm (Figure 2.55). Determine the stress on the opposite faces of the column. A 152 × 152 universal column has a depth of 152.4 mm, a cross-sectional area of 29.8 cm^2 and a second moment of area about the X–X axis of 1260 cm^4.

3 Torsion

3.1 Introduction

As discussed in Chapter 1 in relation to the loading of structures, one basic form of loading is torsion in which members are twisted. In discussing torsion we use the term torque. *Torque T* is defined as the turning moment of an applied force about an axis; for a force F with a radius of rotation r about the axis (Figure 3.1):

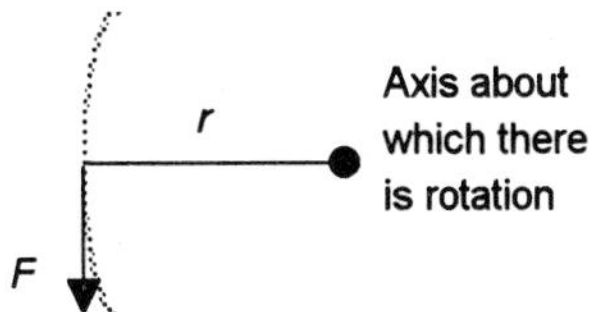

Figure 3.1 *Torque = Fr*

$$\text{torque } T = Fr \qquad [1]$$

An example of a force causing a torque is in the use of a spanner (Figure 3.2). The applied force produces a torque at the head of the bolt.

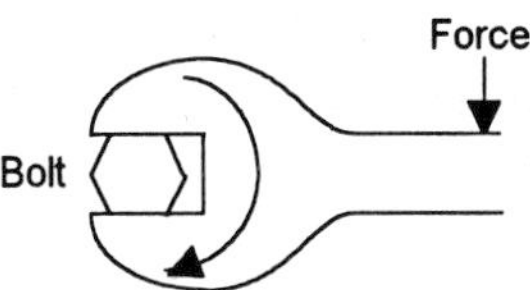

Figure 3.2 *Spanner applying a twisting moment to a bolt*

When we use a screwdriver (Figure 3.3) we apply a twisting force to the screwdriver handle which is transferred along the length of the screwdriver shaft to the screw and results in a torque being applied to it. This action occurs because the twisting force applied to the screwdriver head twists the screwdriver shaft. The twisted shaft is said to be in *torsion.* This chapter is a discussion of the twisting of shafts, the stresses produced and the power that can be transmitted by shafts. Both solid and hollow circular section shafts are considered.

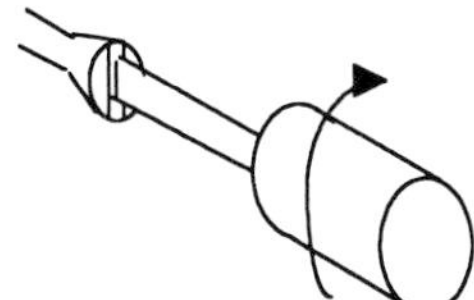

Figure 3.3 *Screwdriver turning a screw*

Application

For a motor being used to rotate a load (Figure 3.4), the shaft becomes twisted as a result of the motor rotating its end of the shaft. It twists until the resisting torque offered by the shaft material balances the torque to be transmitted. The shaft is in torsion.

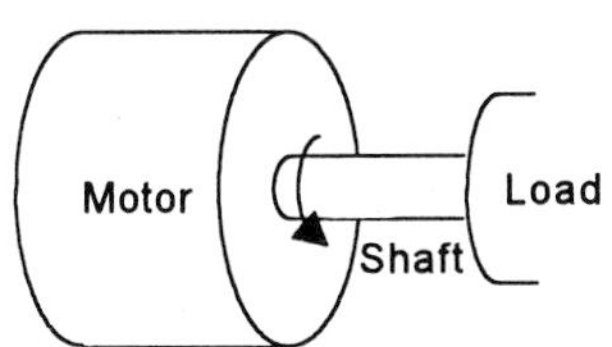

Figure 3.4 *Motor rotating a load*

3.1.1 Shear stress and strain

In discussions of torsion we will be concerned with shear stress and shear strain. The following is a review of such terms. *Shear* is said to occur if the forces applied to a block of material result in a tendency for one face of the material to slide relative to the opposite face (Figure 3.5). With shear the area over which forces act is in the same plane as the line of action of the forces, unlike direct stress where the area is at right angles to the forces. The force per unit area is called the *shear stress* τ:

$$\text{shear stress } \tau = \frac{F}{A} \qquad [2]$$

The unit of shear stress is the pascal (Pa) when the force is in newtons and the area in square metres. The deformation produced by the shear is, for Figure 3.5, XY relative to a parallel face a distance L away. The *shear strain* is this distance XY divided by L. But tan ϕ = XY/L and for the small angles generally involved, tan ϕ is approximately ϕ. Thus shear strain is expressed as the angular deformation ϕ:

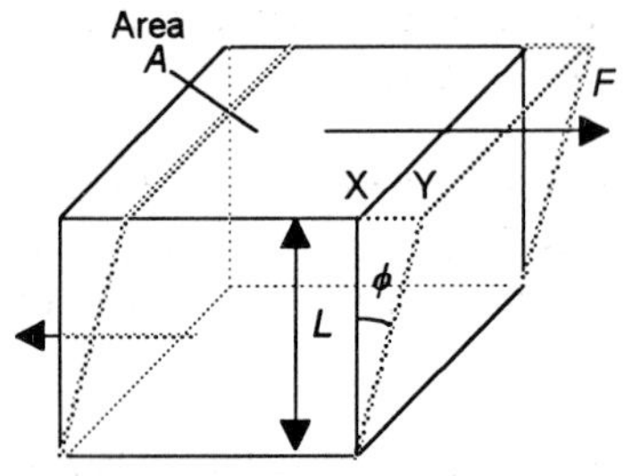

Figure 3.5 *Shear*

Application

As illustrations of shear situations, Figure 3.6(a) shows shear occurring in a riveted joint, the form of joint shown being termed a lap joint. The rivet is in shear as a result of the forces applied to the plates joined by the rivet. Figure 3.6(b) shows shear stresses being applied by a punch to deform a sheet of material and give a cupped indentation or to punch a circular hole in the material.

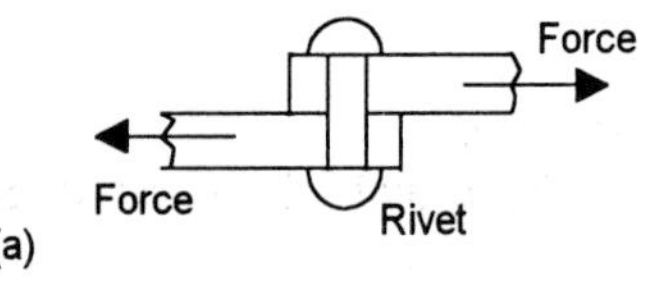

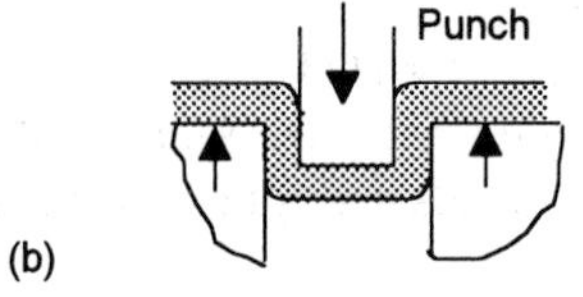

Figure 3.6 *Examples of shear*

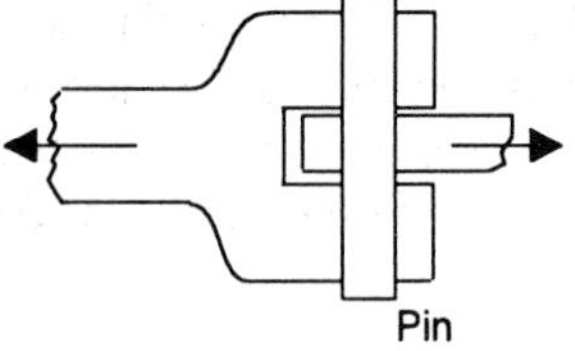

Figure 3.7 *Example*

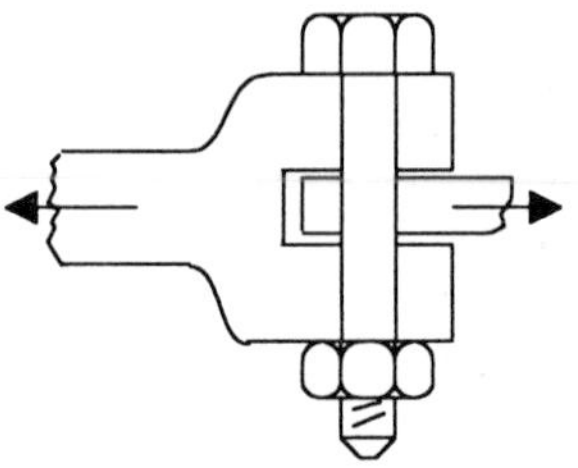

Figure 3.8 *Revision problem 1*

$$\text{shear strain} = \phi \qquad [3]$$

The unit used for ϕ is the radian and, since the radian is a ratio, shear strain can be either expressed in units of radians or as a ratio and so without units.

Shear stress is proportional to shear strain for many materials and the stress/strain ratio is called the *shear modulus G* or *modulus of rigidity*:

$$\text{shear modulus} = \frac{\text{shear stress}}{\text{shear strain}} \qquad [4]$$

The unit of the shear modulus is the same as that of shear stress, since shear strain has no units, and is thus Pa.

Example

What forces are required to shear a lap joint made using a single 25 mm diameter rivet if the maximum shear stress the rivet can withstand is 250 MPa?

The joint is as shown in Figure 3.6(a). Using equation [2]:

$$\text{force} = \text{shear stress} \times \text{area}$$

$$= 250 \times 10^6 \times \tfrac{1}{4}\pi \times 0.025^2 = 123 \text{ kN}$$

Example

What is the maximum load that can be applied to the pin coupling shown in Figure 3.7 if the pin has a diameter of 10 mm and the shear stress in the pin is not to exceed 50 MPa.

There are two surfaces in the pin being sheared by the action of the forces, thus the area to be sheared is double the cross-sectional area of the pin. Using equation [2]:

$$\text{force} = \text{shear stress} \times \text{area}$$

$$= 50 \times 10^6 \times 2 \times \tfrac{1}{4}\pi \times 0.010^2 = 7.85 \text{ kN}$$

Revision

1 What is the minimum diameter required for the bolt in Figure 3.8 if the shear stress in it is not to exceed 90 MPa when the forces applied to the components joined by the bolt are 30 kN?

2 What is the maximum diameter hole that can be punched in an aluminium plate of thickness 14 mm if the punching force is limited

to 50 kN? The shear strength, i.e. the maximum shear stress the material can withstand before failure, is 90 MPa.

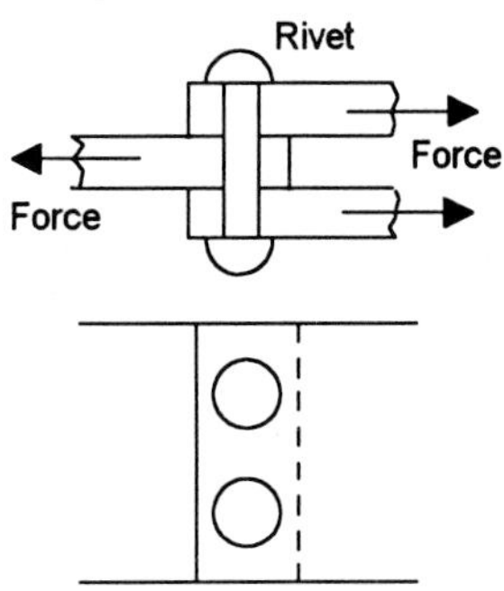

Figure 3.9 *Revision problem 3*

3 Three steel plates are joined by two rivets (Figure 3.9), with each rivet having a diameter of 15 mm. What is the maximum force that can be applied if the shear stress in the rivets is not to exceed 200 MPa?

4 A metal cube has a side of 20 mm. Opposite faces have forces applied to them to give a shear strain of 0.0005. What is the relative displacement of the opposite faces?

5 The shear modulus for a material is 80 GPa. What will be the shear strain when the shear stress is 20 MPa?

3.2 Torsion of circular shafts

Torsion is the term used for the twisting of a structural member when it is acted on by torques (Figure 3.10) so that rotation is produced about the longitudinal axis of one end of the member with respect to the other. For simplification in deriving equations for torsion we will make the following assumptions:

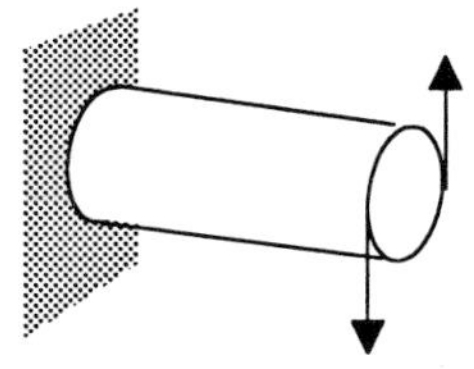

Figure 3.10 *Torsion*

1 The shaft has a uniform circular cross-section.
2 The shaft material is uniform throughout and shear stress is proportional to shear strain.
3 The shaft is straight and initially unstressed.
4 The axis of the twisting moment is the axis of the shaft.
5 Plane transverse sections remain plane after twisting. As a result, each circular section is rotated different amounts and results in shear forces (Figure 3.11).

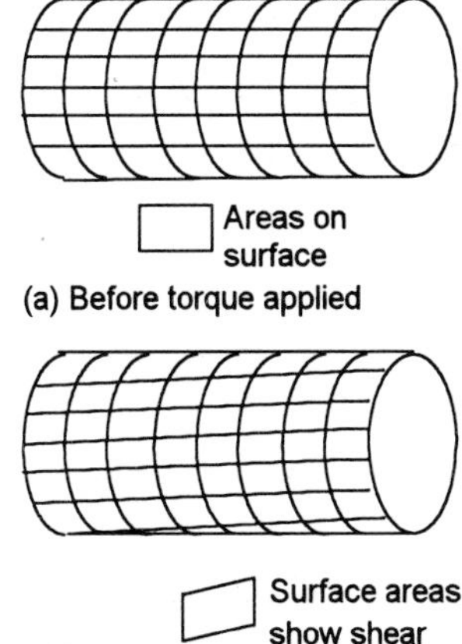

Figure 3.11 *Circular shaft: (a) before, (b) after torque applied*

Consider such a shaft of radius r and length L. If the angle of twist is θ then the situation is as shown in Figure 3.12; AC is the initial position of a line along the surface of the bar and BC is its new position as a result of the end of the bar being rotated through the arc AB. The arc AB subtends an angle θ and so arc AB = $r\theta$. But AB also equals $L\phi$, where ϕ is the resulting shear strain; if you think of a strip on the surface of the shaft it becomes sheared as shown in Figure 3.13.

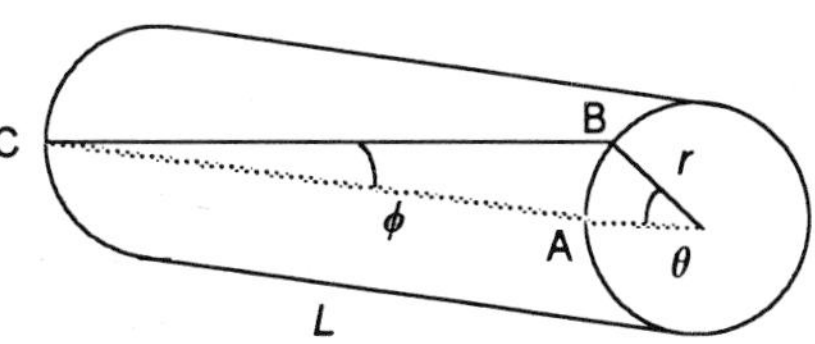

Figure 3.12 *Twisting of a cylindrical shaft*

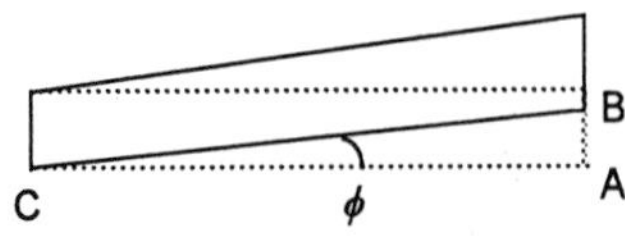

Figure 3.13 *Shear of strip on the surface*

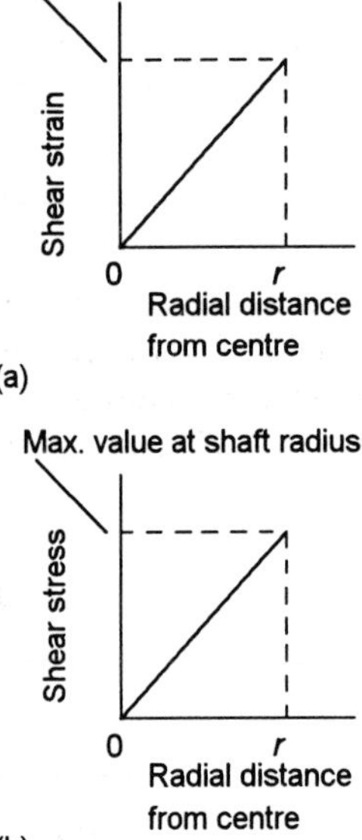

Figure 3.14 *Shear strain and shear stress*

Thus:

$$\phi L = r\theta$$

and so the shear strain is:

$$\phi = \frac{r\theta}{L} \qquad [5]$$

This states that for a given angle of twist per unit length, i.e. θ/L, the shear strain is proportional to the distance r from the central axis (Figure 3.14(a)). The shear stress τ at this radius is thus, using equation [4]:

$$\tau = \frac{Gr\theta}{L} \qquad [6]$$

where G is the shear modulus. The shear stress is, for a particular angle of twist per unit length and material, proportional to the distance r from the central axis (Figure 3.14(b)).

As a simplification, consider the torsion of a thin-walled tube element of the shaft with a radius r and thickness Δr (Figure 3.15). The tube is considered to be thin enough for the assumption to be made that the shear stress is uniform throughout the thickness of the tube wall. Thus for the element of area ΔA the shear force is:

$$\text{shear force} = \text{area} \times \text{shear stress} = \Delta A \times \tau$$

where τ is the shear stress acting on the element. The shear stress acts circumferentially to the tube across the area of the cross-section. Thus the shear force on the total area of the tube wall is:

$$\text{shear force} = 2\pi r\Delta r \times \tau$$

This shear force is acting at a radius r in a tangential direction. It gives a torque T about the axis of the tube of:

$$T = \text{shear force} \times \text{radius} = 2\pi r\Delta r \times \tau \times r = 2\pi r^2\Delta r \times \tau \qquad [7]$$

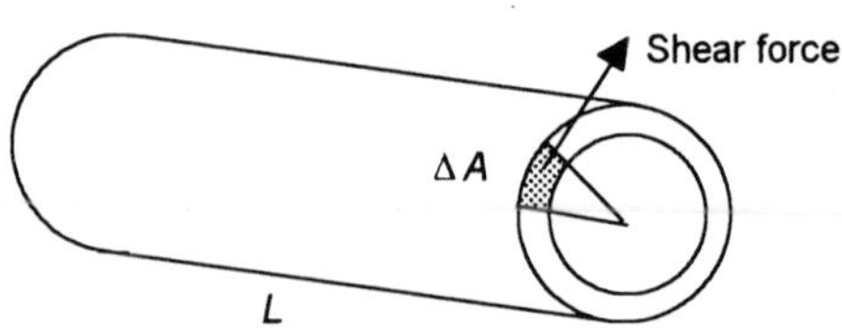

Figure 3.15 *Shear force acting on an element of a thin-walled tube of thickness Δt and radius r*

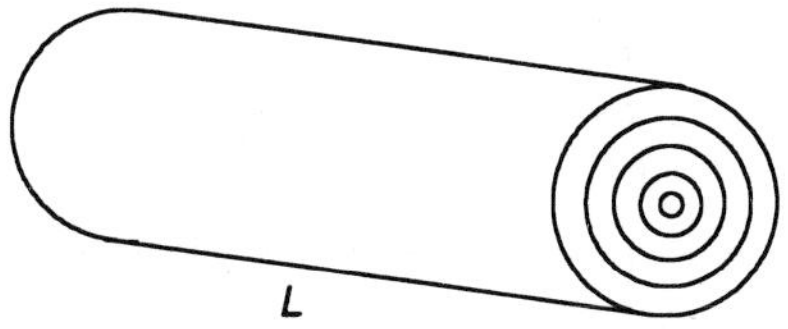

Figure 3.16 *Solid tube as composed of a large number of thin-walled tubes of different radii*

Using equation [6] we have $\tau = Gr\theta/L$ and so we can write equation [7] as:

$$T = 2\pi r^2 \Delta r \times \frac{Gr\theta}{L} = \frac{G\theta}{L} \times 2\pi r^3 \Delta r \qquad [8]$$

Equation [8] gives the shear force for a thin-walled tube. We can think of a solid rod as being composed of a large number of such tubes (Figure 3.16) and thus the shear force is the sum of all the contributions from each of the tubes which go to make up the solid tube. Thus, when we make the tubes infinitesimally thin:

$$T = \int_0^r \frac{G\theta}{L} \times 2\pi r^3 \,dr = \frac{G\theta}{L} \int_0^r 2\pi r^3 \,dr \qquad [9]$$

This is written as:

$$T = \frac{G\theta J}{L} \qquad [10]$$

where J is the *polar second moment of area* about the shaft axis and defined by:

$$\text{polar second moment of area } J = \int_0^r 2\pi r^3 \,dr \qquad [11]$$

Application
Human bone has a low torsional strength. This generally does not matter because we do not often subject our bones to significant torsional torques. However, when we attach skis to our feet we are effectively attaching long lever arms so that comparatively small forces at the end of a ski can result in quite a large torque. Hence the common cause of broken bones.

Equations [6] and [10] are often written in the following form, it being referred to as the *general equation for the torsion of circular cross-section shafts*:

$$\frac{T}{J} = \frac{\tau}{r} = \frac{G\theta}{L} \qquad [12]$$

3.2.1 Polar second moment of area

The polar second moment of area J of a shaft about its axis is defined by equation [11]. Thus, with a solid shaft of diameter D:

Application
In this chapter we have only considered the torsion of circular sections. However, torsion does occur with other sections. For example, universal section beams and columns can be subject to torques and so suffer torsion. Tables of such sections list values of the torsional constant *J*, i.e. the polar second moment of area, so that torsional stresses can be determined.

$$J = 2\pi \int_0^{D/2} r^3 \, dr = \frac{\pi D^4}{32} \qquad [13]$$

For a hollow shaft with external diameter *D* and internal diameter *d*:

$$J = 2\pi \int_{d/2}^{D/2} r^3 \, dr = \frac{\pi}{32}(D^4 - d^4) \qquad [14]$$

The units of *J* are m^4 when the diameters are in metres.

Example

What is the minimum diameter of a solid shaft if it is to transmit a torque of 30 kN and the shear stress in the shaft is not to exceed 80 MPa?

For a solid shaft the polar second moment of area about its axis is given by equation [13] as $J = \pi D^4/32$ and thus equation [12] gives:

$$T = \frac{J\tau}{r} = \frac{2J\tau}{D} = \frac{\pi D^3 \tau}{16}$$

and so:

$$D^3 = \frac{16T}{\pi\tau} = \frac{16 \times 30 \times 10^3}{\pi \times 80 \times 10^6}$$

Thus the minimum diameter shaft is 0.124 m.

Example

Sketch a graph of the shear stress across a section of a 50 mm diameter solid shaft when subject to a torque of 200 N m.

For a solid shaft the polar second moment of area about its axis is given by equation [13] as $J = \pi D^4/32 = \pi \times 0.050^4/32 = 6.14 \times 10^{-7}$ m^4. Thus equation [12] gives:

$$\tau = \frac{Tr}{J} = \frac{200r}{6.14 \times 10^{-7}} = 3.26 \times 10^8 r \text{ Pa}$$

When $r = 0$ then the shear stress is 0; when $r = 25$ mm then the shear stress is 8.15 MPa. Since the shear stress is proportional to the distance *r* from the axis, the graph is as shown in Figure 3.17.

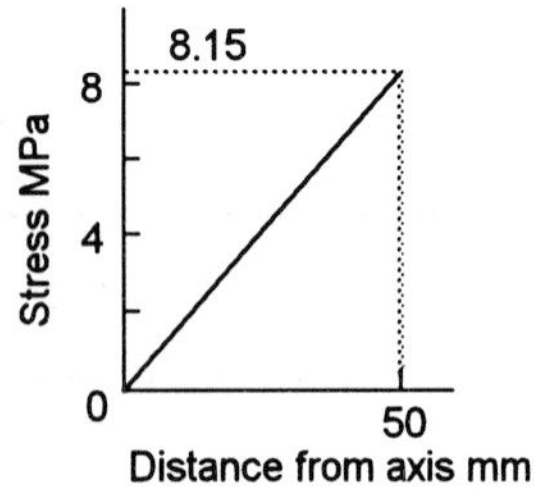

Figure 3.17 *Example*

Example

A steel tube of length 4 m has an external diameter of 10 mm and an internal diameter of 6 mm. Determine the maximum and minimum shear stresses in the tube when, with one end fixed, the other is rotated through 30°. The steel has a shear modulus of 80 GPa.

Using equation [12], with $\theta = 30° = \pi/6$ radians:

$$\tau = \frac{G\theta r}{L} = \frac{80 \times 10^9 \times (\pi/6)r}{4} = 1.05 \times 10^{10} r \text{ Pa}$$

The maximum shear stress occurs at the maximum radius of 5 mm and thus is 52.5 MPa and the minimum shear stress occurs at the minimum radius of 3 mm and thus is 31.5 MPa.

Revision

6 A solid steel shaft has a diameter of 60 mm. Determine the maximum torque that can be applied to the shaft if the maximum permissible shear stress is 40 MPa.

7 A solid steel shaft has a diameter of 60 mm. Determine the maximum torque that can be applied to the shaft if the maximum permissible twist per unit length is 1° per metre. The steel has a shear modulus of 80 GPa.

8 Determine the external diameter of a tube needed to transmit torque of 30 kN m if it has an external diameter which is twice its internal diameter and the shear stress is not to exceed 80 MPa.

9 A tubular shaft has an internal diameter which is half its external diameter and is subject to a torque of 50 kN m. What external diameter will be required if the shear stress in the shaft is not to exceed 80 MPa?

3.2.2 Polar section modulus

For a shaft of radius R, equation [13] gives for the relationship between the torque T and the maximum shear stress τ_{max}:

$$\frac{T}{J} = \frac{\tau_{max}}{R}$$

and so:

$$T = \frac{J}{R}\tau_{max} = Z_p \tau_{max} \qquad [15]$$

The *polar section modulus of section* Z_p is defined as:

$$Z_p = \frac{J}{R} \qquad [16]$$

For a solid shaft where $J = \pi D^4/32$, with $D = 2R$, then $Z_p = \pi D^3/16$.

Example

Determine the polar section modulus for a uniform solid shaft of diameter 40 mm.

For a solid shaft where $J = \pi D^4/32$, with $D = 2R$, then $Z_p = J/R = \pi D^3/16 = \pi \times 0.040^3/16 = 1.26 \times 10^{-5}$ m³.

Revision

10 A shaft has a polar section modulus of 1.5×10^{-4} m³ and is subject to a torque of 5 kN m. What will be the maximum shear stress in the shaft?

3.3 Transmission of power

For a shaft of radius r, the distance travelled by a point on its surface when it rotates through one revolution is $2\pi r$. When it rotates through n revolutions the distance travelled is $2\pi rn$. If it is rotating at n revolutions per second then the distance travelled per second is $2\pi rn$. If the shaft is being rotated by a torque T, the force acting at a radius r is $F = T/r$. The work done per second will be the product of the force and the distance moved per second by the point to which the force is applied. Thus:

$$\text{work done per second} = F \times 2\pi rn = 2\pi nT$$

But the work done per second is the power. Thus the power transmitted per second is:

$$\text{power} = 2\pi nT \qquad [17]$$

In one revolution the angle rotated by the point on the shaft surface is 2π radians. If n revolutions are completed per second then the angle rotated in one second is $2\pi n$. The angular velocity ω is thus $2\pi n$. Thus:

$$\text{power transmitted} = \omega T \qquad [18]$$

With the angular velocity in rad/s and the torque in N m, the power is in watts.

We can thus use equation [18] to determine the power that can be transmitted by an electric motor rotating its shaft. However, in the case of an internal combustion engine, the torque transmitted through the output shaft is not constant but varies during the cycle (Figure 3.18). The power that can be transmitted in this case is thus calculated from the mean torque during the cycle. But when we need to calculate the diameter required for the shaft in order that the maximum shear stress should not be exceeded, the peak value of the torque has to be used.

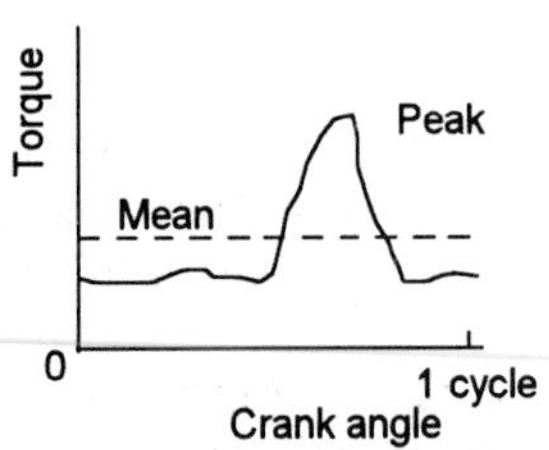

Figure 3.18 *Torque delivered by an internal combustion engine*

Example

Determine the power that can be transmitted by a solid steel shaft of diameter 100 mm which is rotating at 5 rev/s if the shear stress in the shaft is not to exceed 70 MPa.

For a solid shaft, $J = \pi D^4/32 = \pi \times 0.100^4/32 = 9.82 \times 10^{-6}$ m^4. The maximum torque that can act is given by equation [13] as $J\tau/R = 2J\tau/D = 2 \times 9.82 \times 10^{-6} \times 70 \times 10^6/0.100 = 13.7$ kN m. Thus, using equation [17]:

$$\text{max. power} = 2\pi nT = 2\pi \times 5 \times 13.7 \times 10^3 = 430 \text{ kW}$$

Example

The drive shaft of a car is a tube with an external diameter of 50 mm and an internal diameter of 47 mm. Determine the maximum shear stress in the shaft when it is transmitting a power of 70 kW and rotating at 80 rev/s.

The maximum torque is given by equation [17] as:

$$\text{torque } T = \frac{P}{2\pi n} = \frac{70 \times 10^3}{2\pi \times 80} = 139 \text{ N m}$$

For the tube, $J = \pi(D^4 - d^4)/32 = \pi(0.050^4 - 0.047^4)/32 = 1.35 \times 10^{-7}$ m^4. The maximum shear stress is given by equation [13] as

$$\text{max. shear stress} = \frac{Tr}{J} = \frac{139 \times 0.025}{1.35 \times 10^{-7}} = 25.7 \text{ MPa}$$

Revision

11 A solid shaft of diameter 80 mm rotates at 5 rev/s and transmits a power of 50 kW. Determine the maximum shear stress in the shaft.

12 A tubular shaft has an external diameter of 150 mm and an internal diameter of 100 mm. Determine the maximum power that can be transmitted by the shaft when rotating at 3 rev/s if the shear stress in the shaft is not to exceed 50 MPa.

Problems

1 Two plates lap each other and are riveted together by ten rivets, each of diameter 8 mm. If the shear stress in a rivet is not to exceed 5 MPa, what is the maximum shear force which the joint can withstand?

2 A metal cube has a side of 30 mm. Opposite faces have forces applied to them to give a shear strain of 0.0002. What is the relative displacement of the opposite faces?

3 A punch of diameter 35 mm is used to punch a hole through a metal sheet of thickness 4 mm. What force will be required if the shear stress needed to shear the material is 50 MPa?

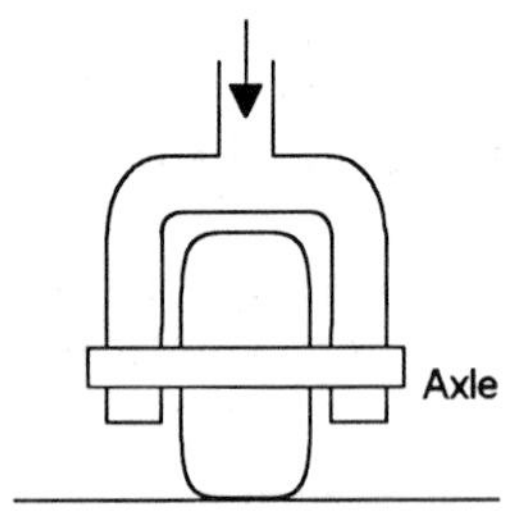

Figure 3.19 *Problem 4*

4 What is the maximum shear stress in the wheel axle shown in Figure 3.19 if the axle has a diameter of 40 mm and the load acting on the wheel is 50 kN?

5 Determine the minimum diameter a solid shaft can have if it is to transmit a torque of 1.2 kN m and the shear stress in the shaft is not to exceed 50 MPa.

6 Determine the maximum torque that can be transmitted by a solid shaft of diameter 40 mm if the shear stress is not to exceed 100 MPa.

7 A tubular shaft has an external diameter of 50 mm and an internal diameter of 25 mm. When subject to a torque of 1.5 kN m it is twisted through 1° per metre. Determine (a) the maximum shear stress in the shaft, (b) the shear modulus of the material, (c) the maximum shear strain.

8 A tubular shaft and a solid shaft are made from the same material, have the same length and the same external diameter and are subject to the same torque. The internal diameter of the tubular shaft is 0.6 times its external diameter. How do (a) the maximum shear stresses, (b) the angles of rotation, (c) the masses compare?

9 A tubular drive shaft has an external diameter of 160 mm and an internal diameter of 80 mm. What will be the maximum shear stress in the shaft when it is subject to a torque of 40 kN m?

10 Compare the torque that can be transmitted by a tubular shaft with that of a solid shaft if they both have the same material, mass, length and allowable shear stress.

11 A tubular drive shaft has an external diameter of 90 mm and an internal diameter of 64 mm. Determine the maximum and minimum shear stresses in the shaft when it is subject to a torque of 5 kN m.

12 Determine the maximum shear stress produced in a 6 mm diameter bolt when it is tightened by a spanner which applies a torque of 7 N m.

13 A shaft has a polar section modulus of 2.0×10^{-4} m^3. What will be the maximum torque that can be used with the shaft if the shear stress in the shaft is not to exceed 50 MPa?

14 A tubular shaft has an inner diameter of 30 mm and an external diameter of 42 mm and has to transmit a power of 60 kW. What will

be the limiting frequency of rotation of the shaft if the shear stress in the shaft is not to exceed 50 MPa?

15 The motor of an electric fan delivers 150 W to rotate the fan blades at 18 rev/s. Determine the smallest diameter of solid shaft that can be used if the shear stress in the shaft is not to exceed 80 MPa.

16 A solid shaft has a diameter of 30 mm and is used to transmit a power of 25 kW at 1000 rev/min. Determine (a) the maximum shear stress in the shaft and (b) the angle of twist over a length of 2 m. The material has a shear modulus of 80 GPa.

17 A tubular steel shaft has an external diameter of 50 mm and a wall thickness of 6.5 mm. Determine the torque that can be transmitted by the shaft if the shear stress in the shaft is not to exceed 15 MPa and the power that can be transmitted when the shaft rotates at 50 rev/s.

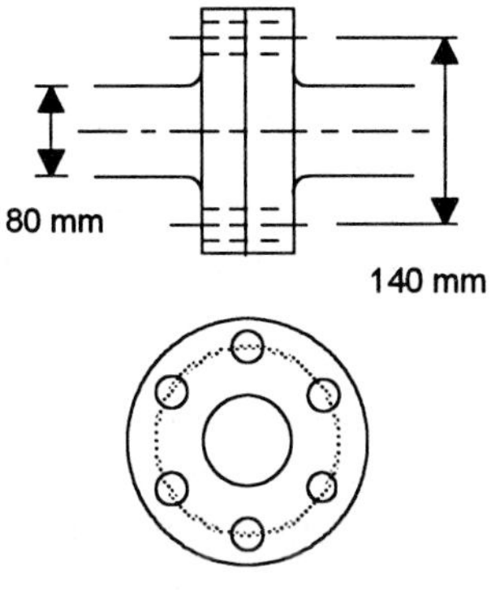

Figure 3.20 *Problem 18*

18 Two solid circular steel shafts of the same diameter of 80 mm are connected by a flanged coupling having six bolts each of 20 mm diameter and on a pitch circle of diameter 140 mm (Figure 3.20). The shear stress in a bolt must not exceed 70 MPa and that in the shafts must not exceed 120 MPa. What is the maximum power that can be transmitted at 5 rev/s? Note that there are two limiting conditions, that due to shear stress in the bolts and that due to shear stress in the shaft and it is necessary to determine which one determines the maximum torque and hence maximum power.

4 Linear and angular motion

4.1 Introduction

This chapter is concerned with the behaviour of dynamic mechanical systems when there is uniform acceleration. The terms and basic equations associated with linear motion with uniform acceleration and angular motion with uniform angular acceleration, Newton's laws of motion, moment of inertia and the effects of friction are revised and applied to the solution of mechanical system problems.

The terms scalar quantity and vector quantity are used in this chapter, so as a point of revision:

> *Scalar quantities* are those that only need to have their size to be given in order for their effects to be determined, e.g. mass. *Vector quantities* are those that need to have both their size and direction to be given in order for their effects to be determined, e.g. force where we need to know the direction as well as the size to determine its effect.

4.2 Linear motion

The following are basic terms used in the description of linear motion, i.e. motion that occurs in a straight line path rather than rotation which we will consider later in this chapter:

1 *Distance and displacement*
The term distance tends to be used for distances measured along the path of an object, whatever form the path takes; the term displacement, however, tends to be used for the distance travelled in a particular straight line direction (Figure 4.1). For example, if an object moves in a circular path the distance travelled is the circumference of the path whereas the displacement might be zero if it ends up at the same point it started from. Distance is thus a scalar quantity possessing only magnitude whereas displacement is a vector quantity with both magnitude and direction.
For motion in a straight line, the displacement in the direction of the line is numerically the same as the distance along that line.

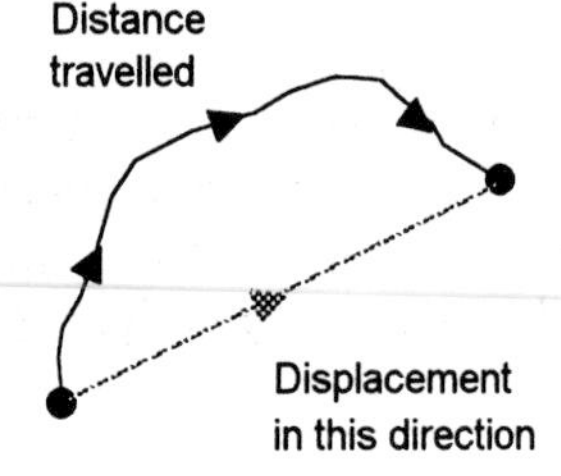

Figure 4.1 *Distance and displacement*

2 *Speed and velocity*
Speed is the rate a distance s varies with time t and is a scalar quantity since the speed of a body is independent of the direction of

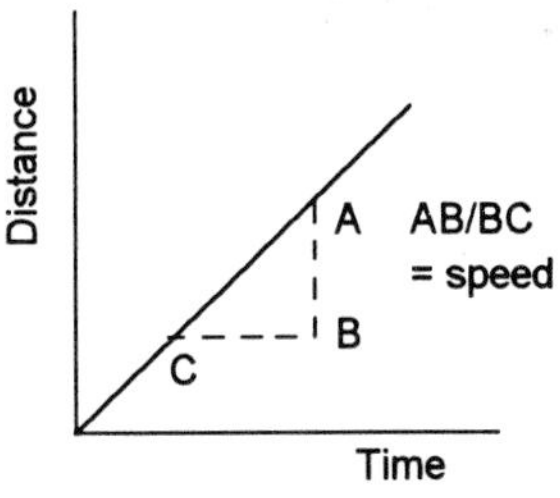

Figure 4.2 *Speed*

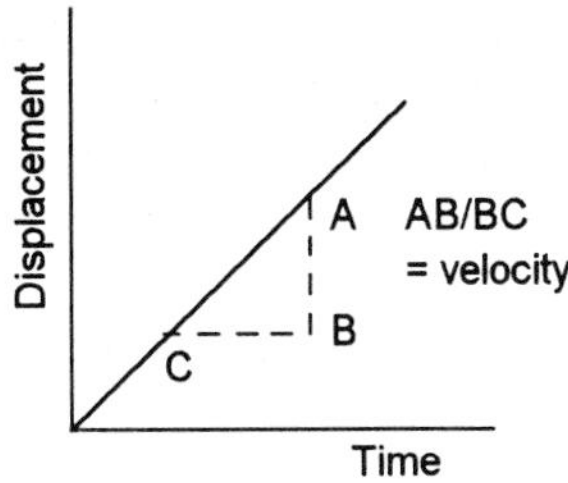

Figure 4.3 *Velocity*

motion. On a graph of distance against time, speed is the slope of the graph (Figure 4.2).

Velocity in a particular direction is the rate at which displacement s in that direction changes with time t and is a vector quantity since the velocity of a body is dependent on the direction of motion. On a graph of displacement, in a particular direction, against time, velocity is the slope of the graph (Figure 4.3).

For motion in a straight line, the velocity in the direction of the line is numerically the same as the speed along that line.

3 *Average speed and average velocity*

The average speed over some time interval is the distance covered in that time interval divided by the duration of the time interval (Figure 4.4):

$$\text{average speed} = \frac{\text{distance travelled}}{\text{time taken}} = \frac{\Delta s}{\Delta t} \qquad [1]$$

where Δs is the change in distance s when the time t changes by Δt.

The average velocity in a particular direction over some time interval is the displacement in that direction divided by the duration of the time interval (Figure 4.5):

$$\text{average velocity} = \frac{\text{displacement occurring}}{\text{time taken}} = \frac{\Delta s}{\Delta t} \qquad [2]$$

where Δs is the change in displacement s in a particular direction when the time t changes by Δt.

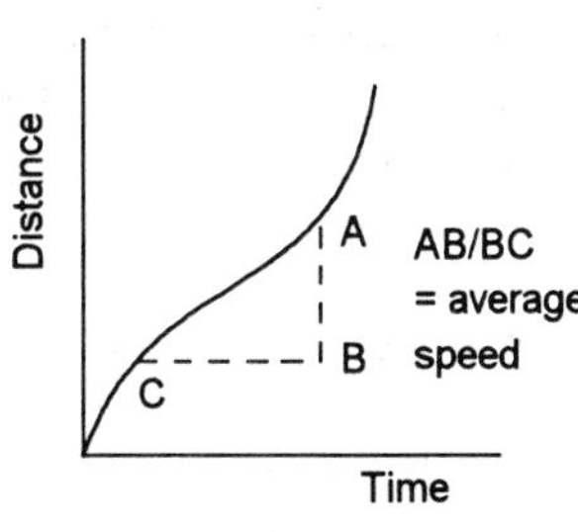

Figure 4.4 *Average speed*

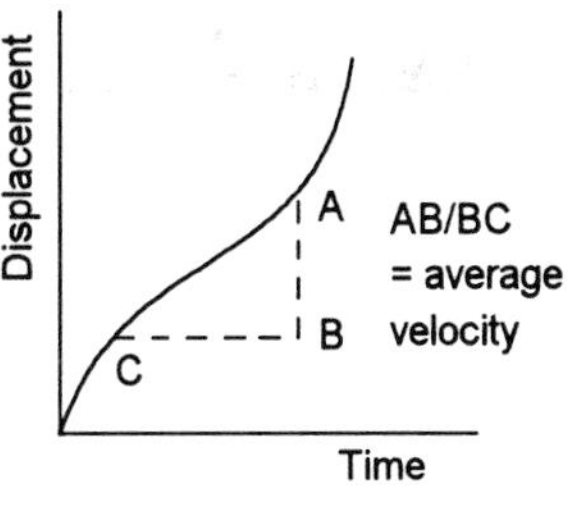

Figure 4.5 *Average velocity*

4 *Instantaneous speed and instantaneous velocity*

The speed at an instant of time can be considered to be the average speed over a time interval when we make the time interval vanishingly small. It is thus the slope of a tangent to a graph of distance against time at a particular point on the graph (Figure 4.6). It can be expressed as:

$$\text{speed} = \lim_{\Delta t \to 0} \frac{\Delta s}{\Delta t} = \frac{ds}{dt} \qquad [3]$$

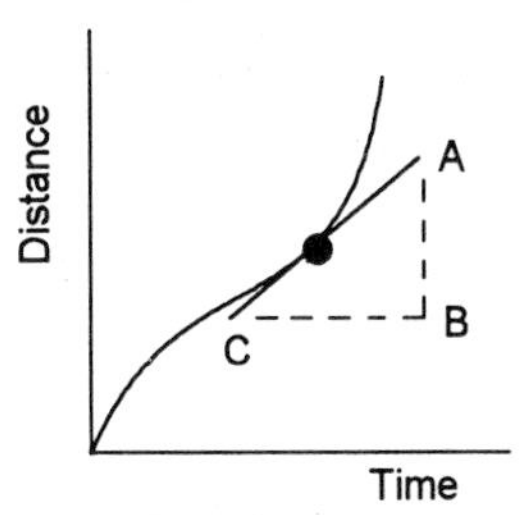

Figure 4.6 *Instantaneous speed = AB/BC*

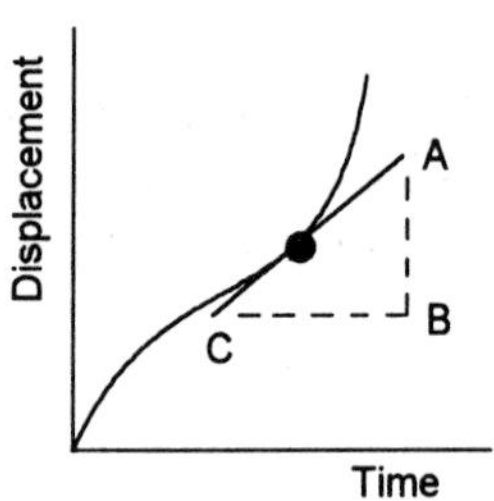

Figure 4.7 *Instantaneous velocity = AB/BC*

The velocity at an instant of time can be considered to be the average velocity over a time interval when we make the time interval vanishingly small. It is thus the slope of a tangent to a graph of displacement, in a particular direction, against time at a particular point on the graph (Figure 4.7). It can be expressed as:

$$\text{velocity } v = \lim_{\Delta t \to 0} \frac{\Delta s}{\Delta t} = \frac{ds}{dt} \qquad [4]$$

5 *Uniform speed and uniform velocity*
Uniform speed occurs when equal distances are covered in equal intervals of time, however small we consider the time intervals. Uniform velocity occurs when equal distances are being covered in the same straight line direction, however small we consider the time intervals.

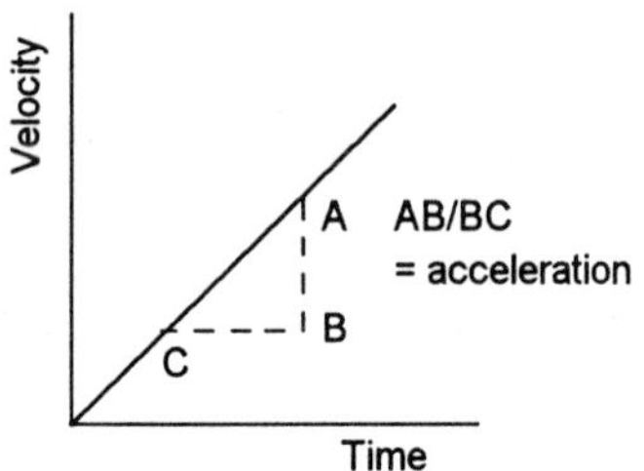

Figure 4.8 *Acceleration*

6 *Acceleration*
Acceleration is the rate of change of velocity v with time t. On a graph of velocity against time, acceleration is the slope of the graph (Figure 4.8). Acceleration is a vector quantity. The term *retardation* is often used to describe a negative acceleration, i.e. when the object has a decreasing velocity. Note that acceleration can occur without any alteration in speed, e.g. an object moving in a circular path with uniform speed. There is an acceleration because the velocity is changing and this can result from a change in its magnitude, i.e. the speed, and/or a change in direction.

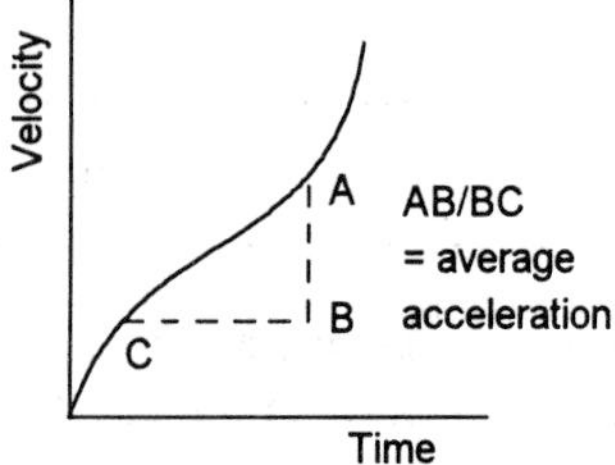

Figure 4.9 *Average acceleration*

7 *Average acceleration*
Average acceleration is the change in velocity occurring over some interval divided by the duration of that time interval (Figure 4.9):

$$\text{average acceleration} = \frac{\text{change of velocity}}{\text{time taken for change}} = \frac{\Delta v}{\Delta t} \qquad [5]$$

where Δv is the change in velocity in a particular direction occurring in a time interval Δt.

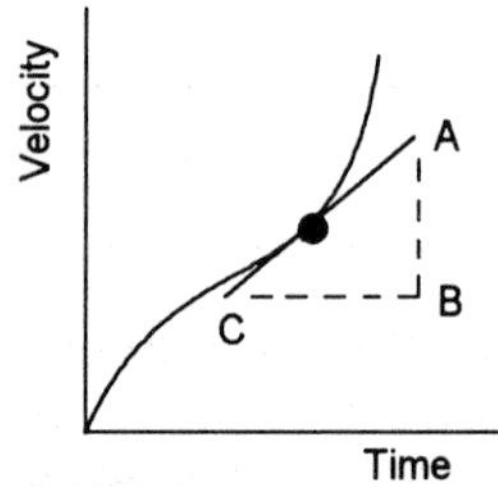

Figure 4.10 *Instantaneous acceleration = AB/BC*

8 *Instantaneous acceleration*
The acceleration at an instant of time can be considered to be the average acceleration over a time interval when we make the time interval vanishingly small (Figure 4.10):

$$\text{acceleration} = \lim_{\Delta t \to 0} \frac{\Delta v}{\Delta t} = \frac{dv}{dt} \qquad [6]$$

9 *Uniform acceleration*
Uniform acceleration occurs when the velocity changes by equal amounts in equal intervals of time, however small the time interval.

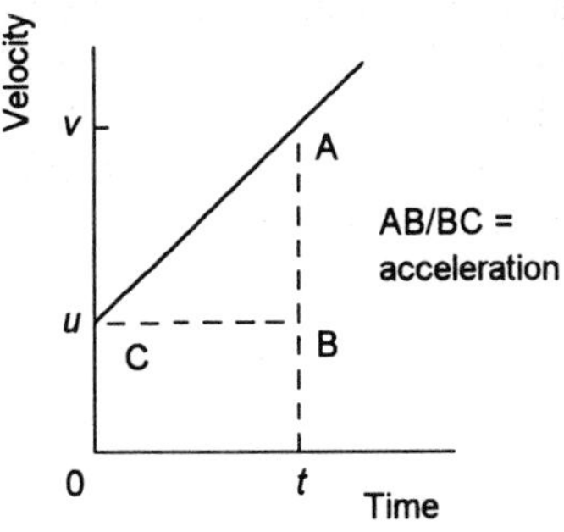

Figure 4.11 *Acceleration*

4.2.1 Motion with constant acceleration

Consider an object which is travelling in a straight line and accelerated uniformly from an initial velocity u to a final velocity v in a time t (Figure 4.11). The acceleration a is the change in velocity divided by the time interval concerned and so is:

$$a = \frac{v - u}{t}$$

Hence we can write:

$$v = u + at \qquad [7]$$

The average velocity over this time is $\frac{1}{2}(u + v)$ and thus the distance s travelled along the straight line path is:

$$s = \text{average velocity} \times \text{time} = \tfrac{1}{2}(u + v)t \qquad [8]$$

Substituting for v by means of equation [7] gives:

$$s = \tfrac{1}{2}(u + u + at)t$$

and so:

$$s = ut + \tfrac{1}{2}at^2 \qquad [9]$$

We can arrive at equation [9] in a more general manner by using equation [4]. Since $v = \mathrm{d}s/\mathrm{d}t$ then, integrating both sides of the equation between the values at zero time and time t gives:

$$\int_0^s \frac{\mathrm{d}s}{\mathrm{d}t}\,\mathrm{d}t = \int_0^t v\,\mathrm{d}t$$

and so:

$$s = \int_0^t v\,\mathrm{d}t \qquad [10]$$

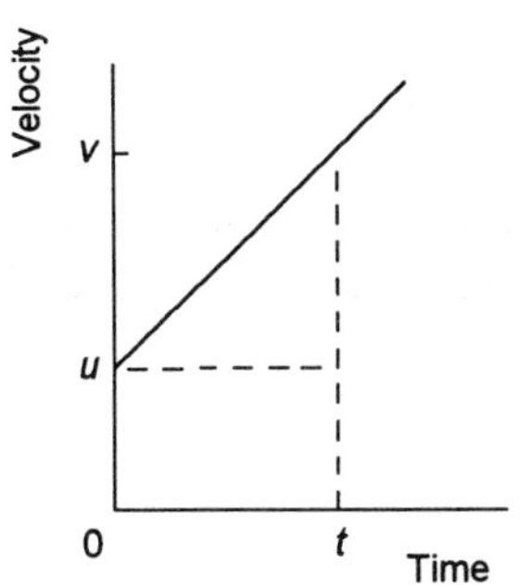

Figure 4.12 *Uniform acceleration*

This integral describes the area under the graph of velocity against time between zero time and time t. For motion with uniform acceleration, the graph will be of the form shown in Figure 4.12. The area is composed of two elements, that of a rectangle with area ut and that of a triangle with area $\frac{1}{2}(v - u)t$. Since $(v - u)/t$ is the acceleration a, the total area is:

$$s = ut + \tfrac{1}{2}at^2$$

We can derive a further equation for motion with uniform acceleration. If we square equation [7] we obtain:

$$v^2 = (u + at)^2 = u^2 + 2uat + a^2t^2 = u^2 + 2a(ut + \frac{1}{2}at^2)$$

Hence, using equation [9]:

$$v^2 = u^2 + 2as \quad [11]$$

Example

An object moves along a straight line from point A to point B with a uniform acceleration of 2 m/s^2. If the time taken is 10 s and the velocity at B is 25 m/s, determine the initial velocity at point A and the distance AB.

Using equation [7], i.e. $v = u + at$, then:

$$25 = u + 2 \times 10$$

Hence u = 5 m/s. Using equation [9], i.e. $s = ut + \frac{1}{2}at^2$, then:

$$s = 5 \times 10 + \frac{1}{2} \times 2 \times 10^2 = 150 \text{ m}$$

Alternatively we could have used equation [11], i.e. $v^2 = u^2 + 2as$, then:

$$25^2 = 5^2 + 2 \times 2 \times s$$

and so s = 150 m.

Example

A car is being driven at a speed of 20 m/s along a straight road when the driver is aware of a broken down car 50 m in front and blocking the road. If the driver immediately applies the brakes and they give a uniform retardation of 1.5 m/s, can the car stop before hitting the broken down car?

Using equation [11], i.e. $v^2 = u^2 + 2as$, then since the final velocity is 0 the stopping distance required is given by:

$$0 = 20^2 + 2 \times (-1.5) \times s$$

as 133.3 m and so the car does not come to a halt in time.

Example

The velocity v of an object moving along a straight line path is related to the time elapsed t by $v = 10 - 6t$ m/s. Determine (a) the acceleration and (b) the displacement after a time of 4 s.

(a) Acceleration is the rate of change of velocity with time, i.e. dv/dt, thus, differentiating the equation $v = 20 - 6t$ gives an acceleration of $a = dv/dt = -6$ m/s^2.

(b) Since $v = ds/dt$, then integrating gives:

$$\int_0^s ds = \int_0^t v\,dt$$

Thus:

$$s = \int_0^4 (20 - 6t)\,dt = [20t - 3t^2]_0^4 = 32 \text{ m}$$

Revision

1 An object initially at rest is accelerated at a constant 4 m/s^2 for 8 s. What will be the distance covered in that time?

2 An object has an initial velocity of 3 m/s and 10 s later, following uniform acceleration, has a velocity of 5 m/s. What is the distance covered in that time?

3 A car is accelerated at a uniform 2 m/s^2 from a velocity of 7.5 m/s until it reaches 22.5 m/s. Calculate the time taken and the distance travelled in that time.

4 The velocity v of an object moving along a straight line path is related to the time elapsed t by $v = 3 - 2t$ m/s. Determine (a) the acceleration and (b) the displacement after a time of 1 s.

4.2.2 Vertical motion under gravity

When an object freely falls under gravity it moves, when the effects of the medium in which it is falling are neglected, with a uniform acceleration called the *acceleration due to gravity*. This varies from one locality to another and at the surface of the earth at sea level is approximately 9.81 m/s^2. It is customary in considering falling objects to consider upward directions from the point of projection as positive and downward directions as negative. As a consequence, upward directed velocities and accelerations are positive and downward directed velocities and accelerations are negative. Thus, on this convention, the acceleration due to gravity is negative.

Example

An object is projected vertically upwards with an initial velocity of 40 m/s. Calculate (a) the maximum height reached, (b) the time taken to reach the maximum height, (c) the time taken for the object to fall back to its initial point of projection.

(a) Using equation [11], i.e. $v^2 = u^2 + 2as$, with $a = -9.81$ m/s^2 and the velocity at the greatest height as 0:

$$0 = 40^2 - 2 \times 9.81 \times s$$

Thus the greatest height s is 81.5 m.

(b) Using equation [7], i.e. $v = u + at$, with $a = -9.81$ m/s^2 and the velocity at the greatest height as 0:

$$0 = 40 - 9.81t$$

Hence the time t is 4.1 s.

(c) The time taken to fall from the greatest height to the initial point of projection is given by equation [9], i.e. $s = ut + \frac{1}{2}at^2$ as:

$$-81.5 = 0 - \tfrac{1}{2} \times 9.81t^2$$

Hence the time taken to fall is 4.1 s. It is the same time as it took to rise to the maximum height. Thus the total time taken is 8.2 s.

Revision

5 A brick falls off the top of a building. If the height of the building is 20 m, determine the time the brick takes to reach the ground.

6 A ball is thrown vertically upwards with a velocity of 15 m/s. Determine the greatest height reached above the point of projection.

7 A ball is thrown vertically downwards with an initial velocity of 4 m/s from the edge of a cliff. If the ball hits the ground at the base of the cliff after 2 s, determine the height of the cliff.

Activity

The following is a simple method of estimating your reaction time. Hold a ruler vertically by the lower end. Then let go and 'immediately' grab it again. You will find that the rule has fallen through some distance in the time it takes for you to stop the ruler falling. From the ruler read off the distance s fallen in the time it has taken you to react. The rule is falling freely under gravity and so the time taken for it to fall though distance s is given by $s = \frac{1}{2}gt^2$. Hence obtain an estimate for your reaction time t.

4.3 Two dimensional linear motion

Displacement, velocity and acceleration are vector quantities and thus can be represented by arrow-headed lines, the length of the line being proportional to the magnitude of the vector and the direction indicated by the arrow being the direction of the vector. As such, they can be added, or subtracted, by the methods used for vectors, e.g. the triangle or parallelogram rules. The following example illustrates this.

Example

An aeroplane sets a course due north with a speed of 200 km/h. However, there is a wind of 50 km/h blowing from the south-west. Determine the actual velocity of the plane.

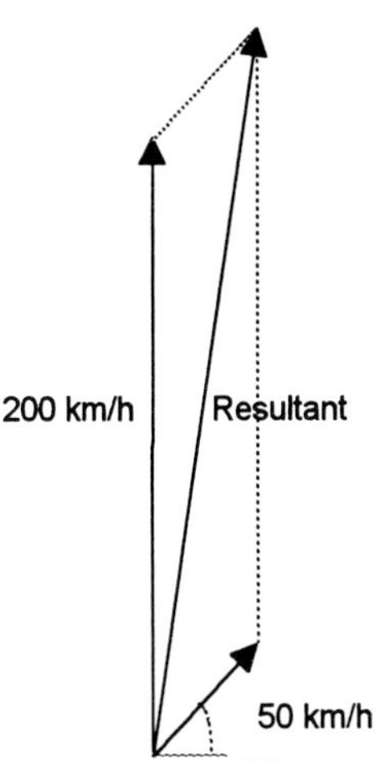

Figure 4.13 *Example*

This example involves the addition of two velocities, 200 km/h in a northerly direction and 50 km/h in a direction from the south-west. Figure 4.13 shows how these velocities can be added by the use of the parallelogram rule. Drawing the parallelogram involves drawing an arrow-headed line to represent one of the velocities, then drawing the arrow-headed line representing the other velocity so that it starts from the start point of the first velocity drawn. The opposite sides of the parallelogram can then be drawn with the resultant velocity being the line drawn as the diagonal from the start points of the two velocities already drawn. The resultant can be determined from a scale drawing or by using the sine and cosine rules. The result is 238 km/h in a direction 8.5° east of north.

Revision

8 Determine the resultant velocity when velocities of 16 km/h in an easterly direction and 10 km/h in a direction 38° east of north act on an object.

4.3.1 Resolution

A vector quantity can be resolved, using the parallelogram rule, into two components at right angles to each other. Thus the velocity v in Figure 4.14 can be replaced by the two velocities $v \cos \theta$ and $v \sin \theta$.

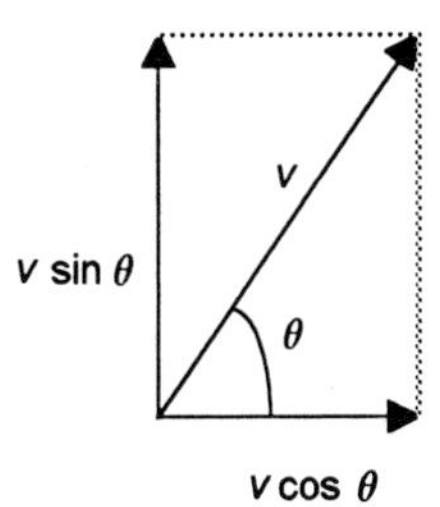

Figure 4.14 *Resolution*

Example

Determine the components in an easterly and a northerly direction of a velocity of 10 m/s in a direction 30° east of north.

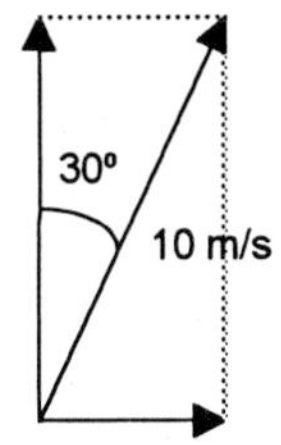

Figure 4.15 *Example*

Figure 4.15 shows the velocity and the components. The component of the velocity in the easterly direction is 10 cos 60° = 5.0 m/s and that in the northerly direction is 10 sin 60° = 8.7 m/s.

Revision

9 Determine the components in an easterly and a northerly direction of a velocity of 4 m/s in a direction 20° east of north.

4.3.2 Projectiles

Consider an object which is projected with a velocity u at an angle θ above the horizontal (Figure 4.16). The horizontal component of the velocity is $u \cos \theta$ and the vertical component is $u \sin \theta$. There is no gravitational force acting in the horizontal direction and so there is no acceleration imposed on the motion in that direction. The horizontal velocity thus remains, when we neglect air resistance, constant. The vertical component is, however, subject to gravity and so there is a constant acceleration.

> For a projectile: the horizontal motion is with constant velocity but the vertical motion is one of uniform acceleration.

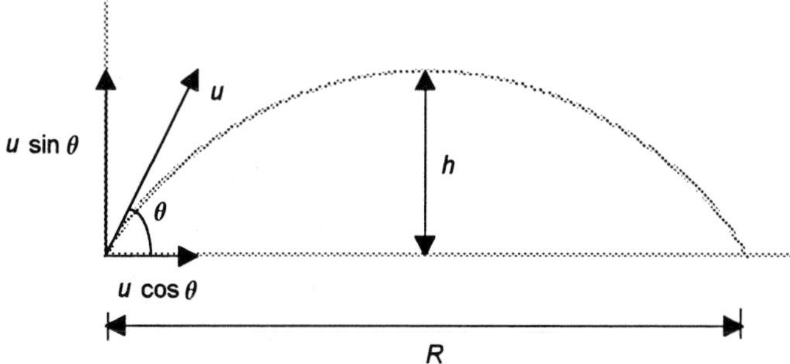

Figure 4.16 *Projectile*

Thus, for a projectile after a time t from its projection, the horizontal distance x from the start is:

$$x = u \cos \theta \times t \qquad [12]$$

There is a vertical force of gravity and so the motion in the vertical direction suffers an acceleration. Thus after a time t, equation [9], i.e. $s = ut + \frac{1}{2}at^2$, gives for the vertical distance y travelled:

$$y = u \sin \theta \times t - \frac{1}{2}gt^2 \qquad [13]$$

where g is the acceleration due to gravity.

The maximum height h reached by the projectile is when the vertical component of the velocity is zero. Thus, using equation [11], i.e. $v^2 = u^2 + 2as$, gives:

$$0 = (u \sin \theta)^2 - 2gh$$

Hence:

$$\text{greatest height reached } h = \frac{u^2 \sin^2\theta}{2g} \qquad [14]$$

At the end of the flight the vertical displacement is zero. Thus, if T is the time of flight, equation [9], i.e. $s = ut + \frac{1}{2}at^2$, gives:

$$0 = u \sin\theta \times T - \tfrac{1}{2}gT^2$$

Hence:

$$\text{time of flight } T = \frac{2u\sin\theta}{g} \qquad [15]$$

The horizontal distance, i.e. the range, R covered in time T is the horizontal velocity multiplied by the time and thus:

$$R = u\cos\theta \times \frac{2u\sin\theta}{g} = \frac{2u^2\sin\theta\cos\theta}{g}$$

Since $2\cos\theta\sin\theta = \sin 2\theta$, then:

$$\text{range } R = \frac{u^2\sin 2\theta}{g} \qquad [16]$$

Application
There are many situations where projectiles might be required to have their maximum range. For example, a shot putter at an athletics meeting wants to project the shot as far as possible. He/she has control over the speed of projection and the angle to the horizontal at which to project the shot. If we consider the problem as a projectile being projected from ground level then the maximum horizontal distance will be achieved with a launch angle of 45°. However, launching from a height of, say, 2 m results in the maximum range being achieved with a launch angle of about 42°.

For maximum range, $\sin 2\theta = 1$, i.e. $2\theta = 90°$ and $\theta = 45°$, and thus the maximum range is u^2/g. For a given speed of projection there are, in general, two possible angles of projection to give a particular range. Equation [16] can be written as:

$$\sin 2\theta = \frac{Rg}{u^2} \qquad [17]$$

For a given value of $\sin 2\theta$ there are two values of the angle less than 180°. If 2θ is one value then the other value is $180° - 2\theta$.

Example

A gun fires a shell with a velocity of 400 m/s at an elevation above the horizontal of 35°. Determine the maximum height reached by the shell and its range.

Using equation [14]:

$$h = \frac{u^2\sin^2\theta}{2g} = \frac{400^2\sin^2 35°}{2 \times 9.81} = 2683 \text{ m}$$

Using equation [16]:

$$R = \frac{u^2 \sin 2\theta}{g} = \frac{400^2 \sin 70°}{9.81} = 15\,326 \text{ m}$$

Example

A stone is thrown from the edge of a cliff with a velocity of 50 m/s at an elevation above the horizontal of 15° (Figure 4.17). If the stone strikes the sea at a point 240 m from the foot of the cliff, determine the height of the cliff.

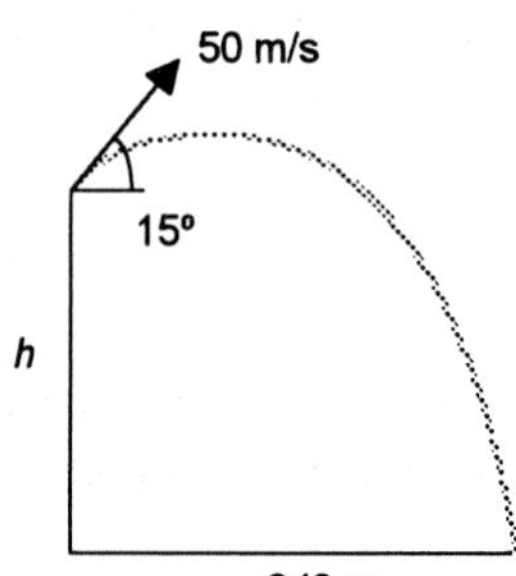

Figure 4.17 *Example*

The horizontal velocity is 50 cos 15° m/s and thus the time taken to travel 240 m is:

$$\text{time} = \frac{240}{50 \cos 15°} = 4.97 \text{ s}$$

The vertical motion has an initial velocity of 50 sin 15° m/s and thus, using equation [9], i.e. $s = ut + \frac{1}{2}at^2$:

$$h = 50 \sin 15° \times 4.97 - \tfrac{1}{2} \times 9.81 \times 4.97^2 = -56.8 \text{ m}$$

Revision

10 A projectile is given an initial velocity of 10 m/s at an elevation from the horizontal of 30°. Determine the greatest height reached and the range.

11 What are the angles of projection at which a projectile with an initial velocity magnitude of 10 m/s can have a range of 9 m?

12 A ball is thrown with an initial velocity of 25 m/s at an elevation to the horizontal of 40°. Determine the greatest height reached, the time of flight and the range.

4.4 Angular motion

The following are basic terms used to describe angular motion.

1 *Angular displacement*
The angular displacement θ is the angle swept out by the rotation (Figure 4.18) and, in all calculation, is specified in radians (rad).

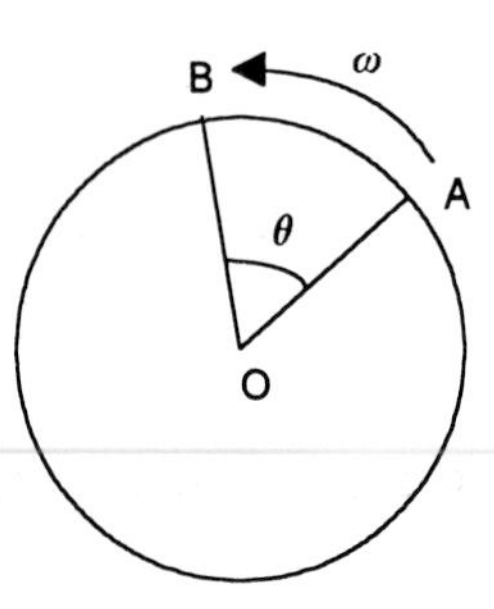

Figure 4.18 *Angular motion*

> One complete rotation through 360° is an angular displacement of 2π radians. Thus 1 rad = $360°/2\pi$.

2 *Angular velocity*
The average angular velocity over some time interval is the change in angular displacement during that time divided by the time:

$$\text{average angular velocity} = \frac{\text{change in angular displacement}}{\text{time for change}} \qquad [18]$$

The unit is rad/s. Thus if a body is rotating at f revolutions per second then it completes $2\pi f$ rad in 1 s and so has an average angular velocity given by:

$$\omega = 2\pi f \qquad [19]$$

The instantaneous angular velocity ω is the change in angular displacement with time when the time interval tends to zero. It can be expressed as:

$$\omega = \frac{d\theta}{dt} \qquad [20]$$

3 *Angular acceleration*
The average angular acceleration over some time interval is the change in angular velocity during that time divided by the time:

$$\text{average angular acceleration} = \frac{\text{change in angular velocity}}{\text{time for change}} \qquad [21]$$

The unit is rad/s^2. The instantaneous angular acceleration a is the change in angular velocity with time when the time interval tends to zero. It can be expressed as:

$$a = \frac{d\omega}{dt} \qquad [22]$$

4.4.1 Motion with constant angular acceleration

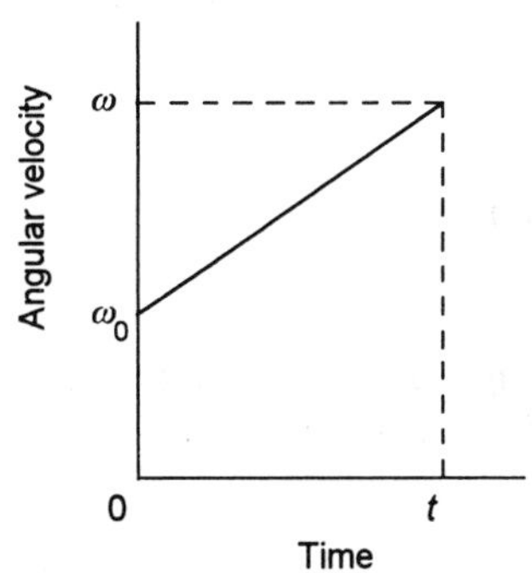

Figure 4.19 *Uniformly accelerated motion*

For a body rotating with a constant angular acceleration a, when the angular velocity changes uniformly from ω_0 to ω in time t, as in Figure 4.19, equation [21] gives:

$$a = \frac{\omega - \omega_0}{t}$$

and hence:

$$\omega = \omega_0 + at \qquad [23]$$

The average angular velocity during this time is ½($\omega + \omega_0$) and thus if the angular displacement during the time is θ:

$$\frac{\theta}{t} = \frac{\omega + \omega_0}{2}$$

Substituting for ω using equation [23]:

$$\frac{\theta}{t} = \frac{\omega_0 + at + \omega_0}{2}$$

Hence:

$$\theta = \omega_0 t + ½at^2 \qquad [24]$$

Squaring equation [23] gives:

$$\omega^2 = (\omega_0 + at)^2 = \omega_0^2 + 2a\omega_0 + a^2t^2 = \omega_0^2 + 2a(\omega_0 + \tfrac{1}{2}at^2)$$

Hence, using equation [24]:

$$\omega^2 = \omega_0^2 + 2a\theta \qquad [25]$$

Example

An object which was rotating with an angular velocity of 4 rad/s is uniformly accelerated at 2 rad/s. What will be the angular velocity after 3 s?

Using equation [23]:

$$\omega = \omega_0 + at = 4 + 2 \times 3 = 10 \text{ rad/s}$$

Example

The blades of a fan are uniformly accelerated and increase in frequency of rotation from 500 to 700 rev/s in 3.0 s. What is the angular acceleration?

Since $\omega = 2\pi f$, equation [23] gives:

$$2\pi \times 700 = 2\pi \times 500 + a \times 3.0$$

Hence $a = 419$ rad/s^2.

Example

A flywheel, starting from rest, is uniformly accelerated from rest and rotates through 5 revolutions in 8 s. What is the angular acceleration?

The angular displacement in 8 s is $2\pi \times 5$ rad. Hence, using equation [24], i.e. $\theta = \omega_0 t + ½at^2$:

$$2\pi \times 5 = 0 + ½a \times 8^2$$

Hence the angular acceleration is 0.98 rad/s^2.

Revision

13 A flywheel rotating at 3.5 rev/s is accelerated uniformly for 4 s until it is rotating at 9 rev/s. Determine the angular acceleration and the number of revolutions made by the flywheel in the 4 s.

14 A flywheel rotating at 20 rev/min is accelerated uniformly for 10 s until it is rotating at 40 rev/min. Determine the angular acceleration and the number of revolutions made by the flywheel in the 10 s.

4.4.2 Relationship between linear and angular motion

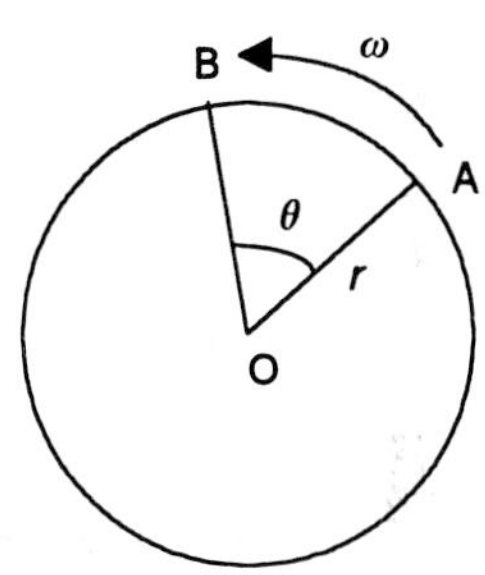

Figure 4.20 *Angular motion*

Consider the rotating line OA in Figure 4.20. When OA rotates through angle θ to OB, point A moves in a circular path and the distance moved by the point A round the circumference is:

$$s = r\theta \qquad [26]$$

If the point is moving with constant angular velocity ω then in time t the angle rotated will be ωt. Thus:

$$s = r\omega t$$

But s/t is the linear speed v of point A round the circumference. Hence:

$$v = r\omega \qquad [27]$$

We could have derived equation [27] without considering constant angular velocity and dealing with the instantaneous angular velocity. Thus if we differentiate equation [26] with respect to time:

$$\frac{\mathrm{d}s}{\mathrm{d}t} = r\frac{\mathrm{d}\theta}{\mathrm{d}t}$$

$\mathrm{d}s/\mathrm{d}t$ is the linear velocity and $\mathrm{d}\theta/\mathrm{d}t$ the angular velocity. Thus equation [27] is obtained.

If we differentiate equation [27] with respect to time:

$$\frac{\mathrm{d}v}{\mathrm{d}t} = r\frac{\mathrm{d}\omega}{\mathrm{d}t}$$

But $\mathrm{d}v/\mathrm{d}t$ is the linear acceleration a and $\mathrm{d}\omega/\mathrm{d}t$ is the angular acceleration α. Thus:

$$a = r\alpha \qquad [28]$$

Example

What is the peripheral velocity of a point on the rim of a wheel when it is rotating at 3 rev/s and has a radius of 200 mm?

Using equation [27]:

$$v = r\omega = 0.200 \times 2\pi \times 3 = 3.8 \text{ m/s}$$

Example

The wheels of a car have a diameter of 700 mm. If they increase their rate of rotation from 50 rev/min to 1100 rev/min in 40 s, what is the angular acceleration of the wheels and the linear acceleration of a point on the tyre tread?

Using equation [23], i.e. $\omega = \omega_0 + \alpha t$, then:

$$2\pi \times \frac{1100}{60} = 2\pi \times \frac{50}{60} + \alpha \times 40$$

Hence the angular acceleration is 2.75 rad/s². Using equation [28], i.e. $a = r\alpha$, then:

$$a = 0.350 \times 2.75 = 0.96 \text{ m/s}$$

Revision

15 The linear speed of a belt passing round a pulley wheel of radius 150 mm is 20 m/s. If there is no slippage of the belt on the wheel, how many revolutions per second are made by the wheel?

16 A car has wheels of diameter 550 mm and is travelling along a straight road with a constant speed of 20 m/s. What is the angular velocity of the wheel?

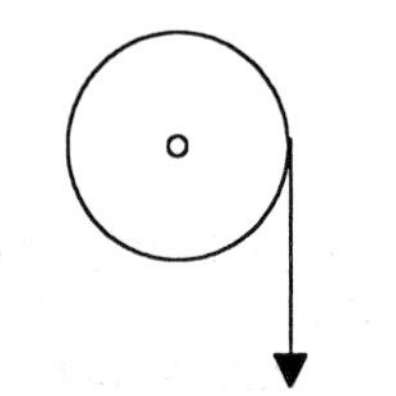

Figure 4.21 *Problem 17*

17 A cord is wrapped around a wheel of diameter 400 mm which is initially at rest (Figure 4.21). When the cord is pulled, a tangential acceleration of 4 m/s² is applied to the wheel. What is the angular acceleration of the wheel?

4.4.3 Gears and belts

Two intermeshed gears can be considered as basically two disks in contact (Figure 4.22) and for which no slippage occurs when one of them is rotated and causes the other to rotate, the gear teeth being to prevent slippage. At the point of contact between the two gears we must have the same tangential velocity. Thus, using equation [27], i.e. $v = r\omega$, we must have:

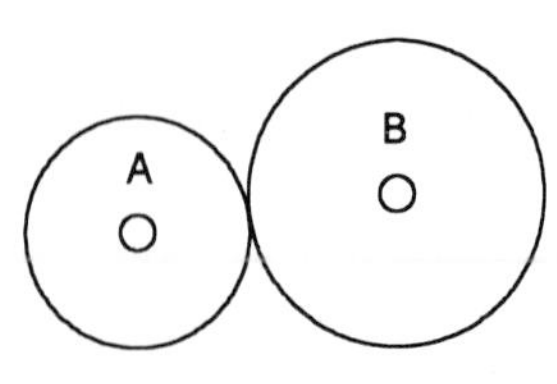

Figure 4.22 *Two gears*

$$r_A\omega_A = r_B\omega_B \qquad [29]$$

Likewise, at the point of contact the tangential acceleration will be the same. Thus, using equation [28], i.e. $a = r\alpha$, we must have:

$$r_A\alpha_A = r_B\alpha_B \qquad [30]$$

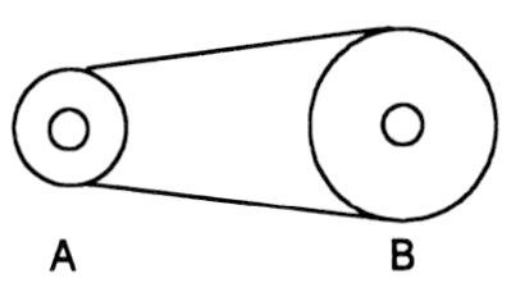

Figure 4.23 *Belt drive*

Belt drives (Figure 4.23) are similar to gears except that the motion is transferred between the two pulleys by a belt rather than the direct contact that occurs with gears. Equations [29] and [30] thus apply.

Example

Gear A is in mesh with gear B, gear A having a radius of 20 mm and gear B a radius of 50 mm. If A starts from rest and has a constant angular acceleration of 2 rad/s^2, determine the time taken for B to reach an angular velocity of 50 rad/s.

Using equation [30], i.e. $r_A\alpha_A = r_B\alpha_B$:

$$0.020 \times 2 = 0.050 \times \alpha_B$$

Thus $\alpha_B = 0.8$ rad/s^2. Using equation [23], i.e. $\omega = \omega_0 + \alpha t$:

$$50 = 0 + 0.8t$$

Hence $t = 62.5$ s.

Example

Pulley A drives pulley B by means of a belt drive, A having a radius of 100 mm and B a radius of 300 mm. Pulley A starts from rest and accelerates uniformly at 15 rad/s^2 for 12 s. Determine the number of revolutions of pulley B in that 12 s time interval.

Using equation [30], i.e. $r_A\alpha_A = r_B\alpha_B$:

$$0.100 \times 12 = 0.300 \times \alpha_B$$

Thus $\alpha_B = 4$ rad/s^2. Using equation [24], i.e. $\theta = \omega_0 t + \frac{1}{2}\alpha t^2$, then:

$$\theta = 0 + \frac{1}{2} \times 4 \times 12^2 = 72 \text{ rad}$$

The number of revolutions is thus $72/2\pi = 11.5$.

Revision

18 Gear A and gear B are intermeshed, A having a radius of 200 mm and B a radius of 150 mm. Gear A starts from rest and rotates with a constant angular acceleration of 2 rad/s^2. Determine the angular velocity and angular acceleration of B after A has completed 10 rev.

19 A 200 mm diameter drive pulley rotates at 5 rev/s. It drives, via a belt drive, another pulley. What diameter will this need to be if it is required to rotate a shaft at 2.5 rev/s?

4.4.4 Combined linear and angular motion

Consider objects which have both a linear motion and angular motion, e.g. a rolling wheel. For a wheel of radius r which is rolling, without slip, along a straight path (Figure 4.24), when the wheel rotates and rolls its centre moves from C to C′ then a point on its rim moves from O to O′. The distance CC′ equals OO′. But OO′ = $r\theta$. Thus:

horizontal distance moved by the wheel $x = r\theta$ [31]

If this movement occurs in a time t, then differentiating equation [31]:

$$\frac{dx}{dt} = r\frac{d\theta}{dt}$$

and thus:

horizontal velocity $v_x = r\omega$ [32]

where ω is the angular velocity of the wheel. Differentiating equation [32] gives:

$$\frac{dv_x}{dt} = r\frac{d\omega}{dt}$$

and thus:

horizontal acceleration = $r\alpha$ [33]

where α is the angular acceleration.

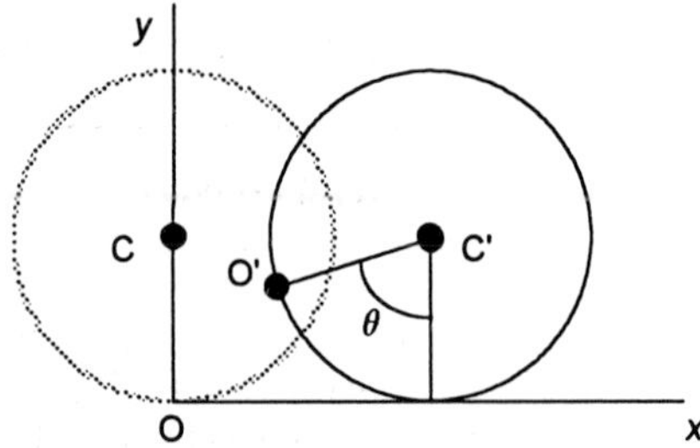

Figure 4.24 *Rolling wheel*

4.5 Force and linear motion

In considering the effects of force on the motion of a body we use Newton's laws.

> *Newton's laws* can be expressed as:
>
> *First law*
> A body continues in its state of rest or uniform motion in a straight line unless acted on by a force.
>
> *Second law*
> The rate of change of momentum of a body is proportional to the applied force and takes place in the direction of the force.
>
> *Third law*
> When a body A exerts a force on a body B, B exerts an equal and opposite force on A (this is often expressed as: to every action there is an opposite and equal reaction).

Thus the first law indicates that if we have an object moving with a constant velocity there can be no resultant force acting on it. If there is a resultant force, then the second law indicates that there will be a change in momentum with:

$$\text{force } F \propto \text{rate of change of momentum, i.e. } \frac{\mathrm{d}(mv)}{\mathrm{d}t} \qquad [34]$$

For a mass m which does not change with time:

$$\text{force } F \propto m\frac{\mathrm{d}v}{\mathrm{d}t}$$

$\mathrm{d}v/\mathrm{d}t$ is the acceleration. The units are chosen so that the constant of proportionality is 1 and thus the second law can be expressed as:

$$F = ma \qquad [35]$$

If a body of mass m is allowed to freely fall then it would fall with the acceleration due to gravity g. The force acting on the freely falling object is thus mg and this is termed its *weight*.

$$\text{weight} = mg \qquad [36]$$

An object resting in equilibrium on a table has zero acceleration and thus, according to Newton's second law, the net force on the object must be zero. Hence the weight of the body which acts downwards on the table must be exactly balanced by an upward acting force exerted by the table

Application
Reaction forces occur whenever there is an interaction between two objects. For example, in the case of a loaded cantilever, a reaction force occurs at the fixed end. In the case of an object hung on the end of a string from some fixed support, there is a reaction force at the support.

on the body, this force being termed the *reaction*. If an object is on a slope then there will be a reaction force resisting the component of the weight which is at right angles to the surface of the slope.

Application
Consider the problem of determining the minimum stopping distances required for a car when an emergency stop is required. The value of the braking force F depends on conditions such as the design of the car and the state of wear of the brakes, but a reasonable assumption that we can make is that it is proportional to the weight of the car, i.e. braking force = kmg where k is a constant and m the mass of the car. Thus, applying Newton's second law gives $ma = -kmg$ and so the retardation $a = -kg$. Using $v^2 = u^2 + 2as$, then for a final velocity of 0 we have $u^2 = 2kgs$, where s is the distance travelled in coming to rest and u the velocity of the car before braking. There is also the distance travelled during the reaction time t_r of the driver. This will be ut_r. Thus the total stopping distance is $ut_r + u^2/2kg$. Typically the reaction time is about 1 s and the braking force is about 2/3 of the weight of the car, i.e. $k = 2/3$. Hence the stopping distance is approximately $u + 3u^2/4g$.

In a mechanical system involving a number of connected bodies, we can draw a diagram for each body with all the forces acting on it, such a diagram being called a *free-body diagram* because we have isolated the body from its surroundings. All the interactions between the body and other adjacent components of the system have been taken care of by showing all the forces acting on a body in the free-body diagram and we can treat the body as though it was a free body on its own. Thus, if the system is in equilibrium we must have equilibrium for each body in the system and so can apply the conditions for equilibrium to each body and the forces described by its free-body diagram.

Force is a vector quantity and thus, when adding or subtracting forces, the parallelogram or triangle laws have to be used. A common method in problems involving more than one force is to resolve the forces acting on an object into the horizontal and vertical directions. Then the forces acting in the same directions can be added.

Example

A car of mass 1200 kg accelerates from a velocity of 30 km/h to 70 km/h in a time of 25 s. Determine the force required to achieve this acceleration.

We need to convert the velocities into m/s. Thus the acceleration is:

$$a = \frac{v_2 - v_1}{t} = \frac{(70-30)\times 10^3}{3600\times 25} = 0.44 \text{ m/s}^2$$

Using Newton's second law:

$$F = ma = 1200 \times 0.44 = 528 \text{ N}$$

Example

A body of mass 3 kg slides down an smooth plane inclined at 30° to the horizontal. Determine the acceleration of the body and the normal reaction exerted by the plane on the body.

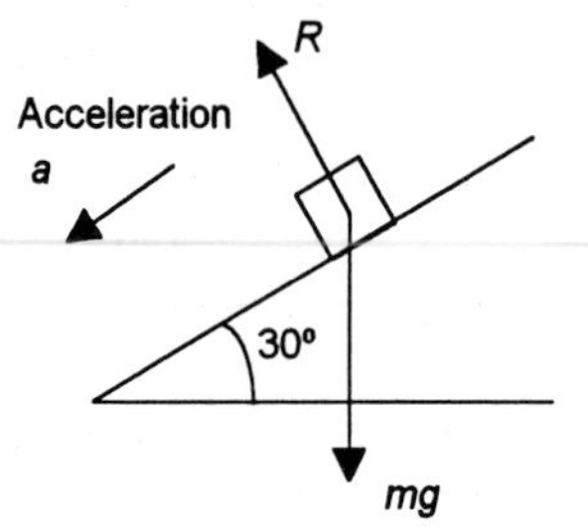

Figure 4.25 *Example*

Figure 4.25 shows the forces acting on the body; because the plane is termed as being smooth we assume there are no frictional forces. The weight of the body is mg, where g is the acceleration due to gravity. R is the reaction of the plane arising from the body pressing down on it. It is at right angles to the plane. If we resolve the weight

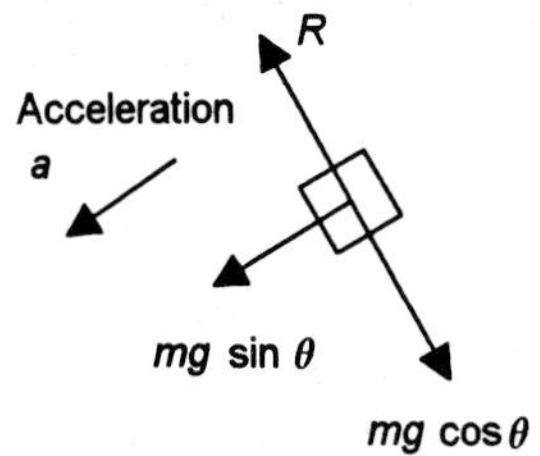

Figure 4.26 *Example*

of the body into a force component parallel to the plane and one at right angles to it, then the free body diagram for the forces acting on the body is as shown in Figure 4.26. Thus the force accelerating the body down the plane is $mg \sin \theta$, where θ is the angle of the plane from the horizontal. Thus applying equation [35], i.e. $F = ma$:

$$ma = mg \sin \theta$$

and so the acceleration $a = g \sin \theta = 9.81 \times \sin 30° = 4.9 \text{ m/s}^2$.

The reaction R is equal to $mg \cos \theta$ since there is no motion in the direction at right angles to the plane. Thus $R = 3 \times 9.81 \times \cos 30° = 25.5$ N.

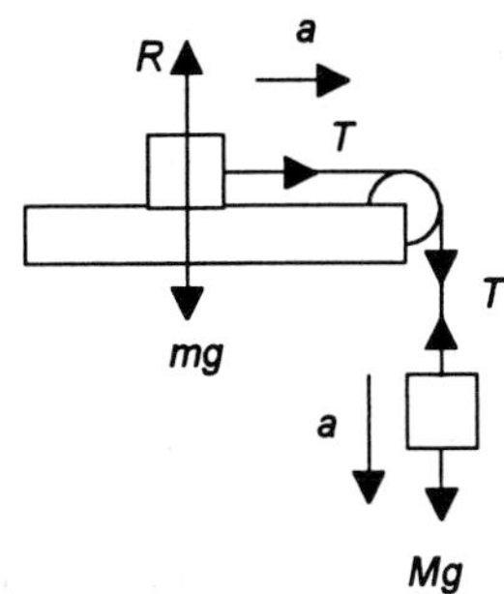

Figure 4.27 *Example*

Example

A block of mass 6 kg rests on a horizontal table top and is connected by a light inextensible string that passes over a pulley at the edge of the table to another block of mass 5kg which is hanging freely (Figure 4.27). Determine the acceleration of the system and the tension in the string.

Figure 4.28 shows the free body diagrams for the two masses. For the block on the table, the force causing acceleration is T and thus:

$$T = ma = 6a \qquad [37]$$

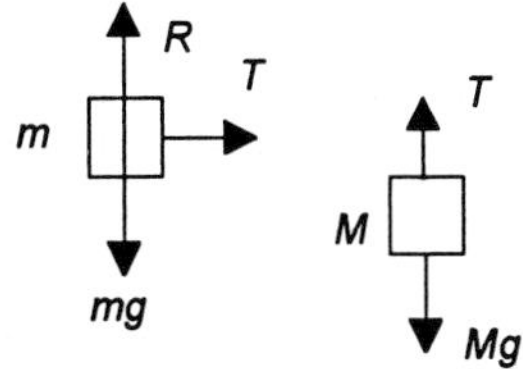

Figure 4.28 *Example*

For the hanging block, the net force causing acceleration is $Mg - T$ and thus:

$$5g - T = 5a \qquad [38]$$

Using equation [37] to eliminate T in equation [38] gives:

$$5g - 6a = 5a$$

and so $a = 5g/11 = 4.46 \text{ m/s}^2$. Equation [36] thus gives the tension as $T = 6a = 6 \times 4.46 = 26.76$ N.

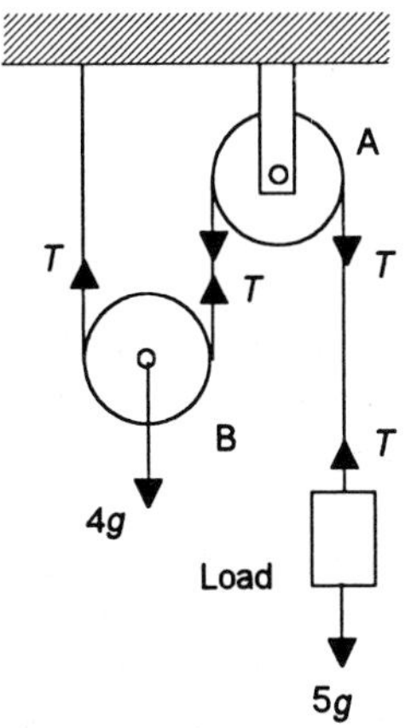

Figure 4.29 *Example*

Example

Figure 4.29 shows a pulley system. A is a fixed smooth pulley and pulley B, also smooth, has a mass of 4 kg. A load of mass 5 kg is attached to the free end of the rope. Determine the acceleration of the load and the tension in the rope when the load is released.

When the load is released, it will move in a particular time through twice the distance the pulley B will move through. Thus if a is the acceleration of pulley B, then the acceleration of the load will be $2a$. For the load, the net downward force is $5g - T$ and thus:

$$5g - T = 5 \times 2a$$

For pulley B, the net upward force is $2T - 4g$ and thus:

$$2T - 4g = 4a$$

Hence, eliminating T from these equations gives a = 2.45 m/s². Pulley B has thus an acceleration of 2.45 m/s² upwards and the load an acceleration of 4.90 m/s² downwards. The tension on the string is 24.6 N.

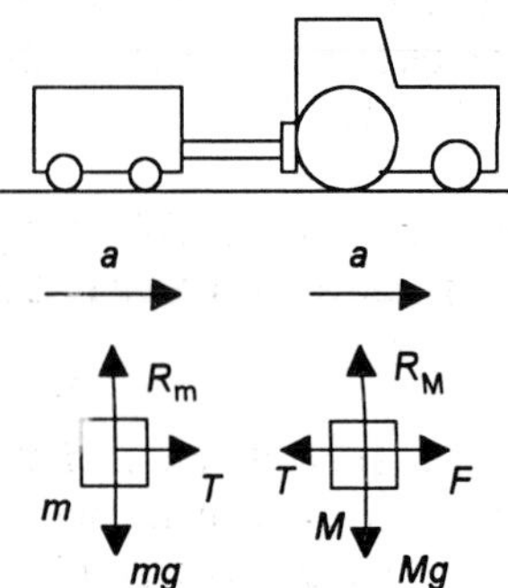

Figure 4.30 *Example*

Example

A tractor of mass M pulls, by means of a tow bar, a trailer of mass m. If the tractor exerts a steady driving force of F, determine the tension T in the tow bar in terms of M, m and F.

Figure 4.30 shows the arrangement and the free-body diagrams for the tractor and the trailer. For the trailer we have:

$$T = ma$$

For the tractor we have:

$$F - T = Ma$$

Hence:

$$T = ma = m\left(\frac{F-T}{M}\right)$$

$$MT = mF - mT$$

$$T = \frac{mF}{M+m}$$

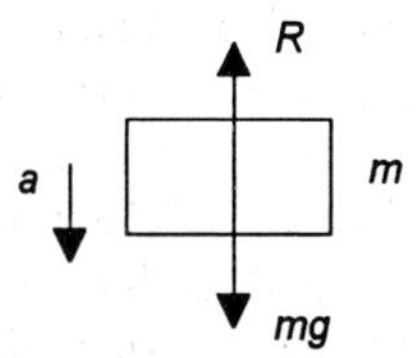

Figure 4.31 *Example*

Example

A brick with a mass of 2 kg falls through still water with an acceleration of 2 m/s². What will be the force resisting its motion?

Figure 4.31 shows the free-body diagram. Hence, Newton's second law gives:

$$R - mg = ma$$

$$R - 2 \times 9.8 = 2 \times 2$$

and so the resistive force R = 23.6 N

Revision

20 A car of mass 1200 kg tows by means of a towbar a trailer of mass 300 kg along a level road. Determine the tension in the tow bar if the car engine produces a thrust of 900 N and there is a resistance to motion of 150 N.

21 A 100 kg person is in a lift which accelerates upwards at 0.6 m/s². What force will be exerted on the person by the floor?

22 A train engine exerts a force of 35 kN on a train of mass 240 Mg and pulls it up a slope of 1 in 120 against a resistive force of 14 kN. Determine the acceleration up the slope of the train.

23 Masses of 3 kg and 2 kg are connected by a light inextensible string which passes over a smooth, fixed pulley. Determine the acceleration of the masses and the tension in the string.

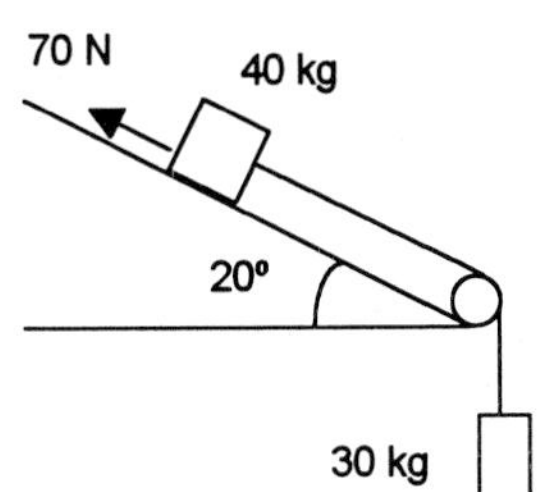

Figure 4.32 *Revision problem 24*

24 An object of mass 40 kg is on a smooth incline at an elevation of 20° to the horizontal (Figure 4.32). It is connected by a cord over a pulley at the bottom of the slope to a vertically hanging load of 30 kg and slides down the slope against a resistance of 70 N. Determine its acceleration. The pulley can be assumed to be smooth and the cord as having negligible mass and being inextensible.

4.5.1 Friction

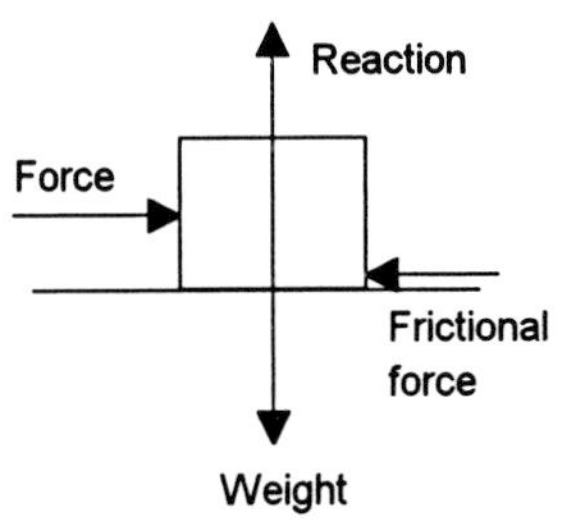

Figure 4.33 *Frictional force*

When there is a tendency for one surface to slide over another, a frictional force occurs which acts along the common tangent at the point of contact and so at right angles to the contact reaction force. The frictional force is always acting in such a direction as to oppose the motion of sliding. Thus when an object, resting on a horizontal surface is acted on by a force F, there will be an opposing frictional force (Figure 4.33). If the force is gradually increased from zero, the object does not move until the applied force reaches some particular value. When the object is not moving, though a force is being applied, we must have no net force and so the frictional force must be opposite and equal to the applied force and thus increase as the applied force increases. When the object starts to slide under the action of the applied force then there is a net force and so the applied force is greater than the frictional force, the frictional force must have stopped increasing and reached a maximum or limiting value.

The are a number of basic laws governing friction:

Application
If you attempted to drive a car along a road when there was no frictional force between the car wheels and the road surface then the wheels would just rotate freely and the car would not progress along the road. Frictional forces between the car wheels and the road are necessary if there is to be forward motion.

1 The frictional force is independent of the areas in contact.

2 The frictional force depends on the nature of the materials in contact.

3 The limiting frictional force F is proportional to the normal reaction force N between the two surfaces in contact, the constant of proportionality being termed the *coefficient of friction* μ.

$$F = \mu R \qquad [39]$$

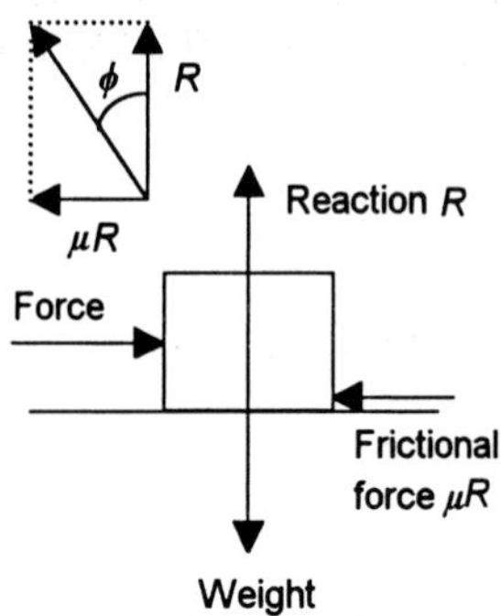

Figure 4.34 *Motion along a horizontal plane*

Application
In order to increase his/her overall speed down a snow slope, a skier will apply wax to the skis. Typically the wax used will decrease the coefficient of friction between the skis and the snow to about 0.05. The forces acting on a skier down an incline are the weight component down the slope minus the frictional force up the slope minus a drag force on the skier's body due to air resistance.

The term *coefficient of static friction* is used when we are concerned with the frictional forces involved when trying to start an object sliding and the term *coefficient of dynamic* or *kinetic friction* when we are concerned with the frictional forces involved when an object is already moving. For unlubricated metals sliding on metals, the coefficient of friction is typically between about 0.15 and 0.3, for wood sliding on wood between about 0.25 and 0.5. In general, the static coefficient of friction is greater than the dynamic coefficient of friction.

Consider motion along a horizontal plane (Figure 4.34). The frictional force is μR and thus the resultant of the frictional force and the normal reaction force is a force at an angle ϕ to the reaction force, this angle being termed the *angle of friction* since $\tan \phi = \mu R/R$ and so :

$$\tan \phi = \mu \qquad [40]$$

Example

An object of mass m rests on a plane at an elevation of θ to the horizontal (Figure 4.35). Determine the force P parallel to the plane that is needed to push the object up the plane with an acceleration a if the coefficient of friction is μ.

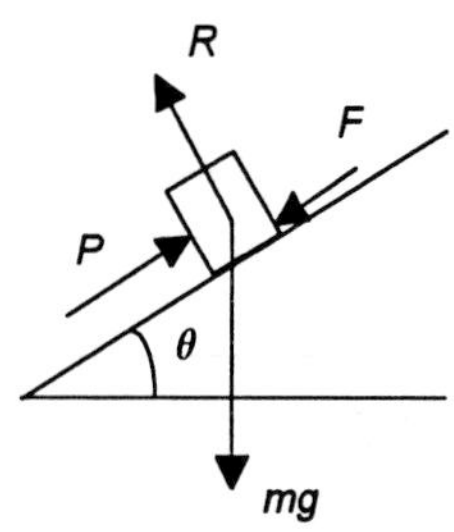

Figure 4.35 *Example*

The forces acting parallel to the plane are the applied force P, the frictional force F and the component of the weight acting down the plane, i.e. $mg \sin \theta$. Thus:

$$\text{net force down the plane} = P - F - mg \sin \theta$$

Since $F = \mu R$ and $R = mg \cos \theta$, then:

$$P - \mu mg \cos \theta - mg \sin \theta = ma$$

and so:

$$a = (P/m) - \mu g \cos \theta - g \sin \theta$$

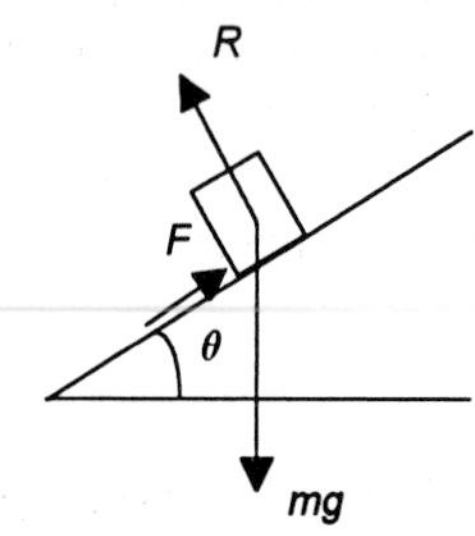

Figure 4.36 *Example*

Example

An object of mass m rests on a plane at an elevation of θ to the horizontal (Figure 4.36). Determine the angle to which the plane can be tilted until the block just begins to slide down it.

The force parallel to the plane is the component of the weight in that direction, i.e. $mg \sin \theta$, and the frictional force μR. The block will just begin to slide down the plane when:

$$mg \sin \theta = \mu R$$

But $R = mg \cos \theta$, thus the condition for motion to start is:

$$mg \sin \theta = \mu mg \cos \theta$$

and so it is:

$$\tan \theta = \mu$$

Example

An object of mass 0.5 kg rests on a rough plane inclined at 40° to the horizontal. If the coefficient of friction is 0.5, determine the smallest force that can be applied parallel to the plane to prevent it from sliding down the plane.

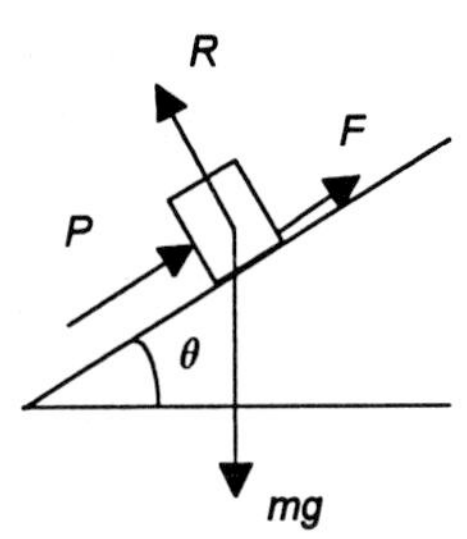

Figure 4.37 *Example*

Figure 4.37 shows the arrangement. The net force parallel to the plane must be zero. The frictional force acts up the plane since the object is on the point of sliding down the plane. We have $F = \mu R = \mu mg \cos \theta$. Thus:

$$P + F = mg \sin \theta$$

$$P + \mu mg \cos \theta = mg \sin \theta$$

and so:

$$P = mg(\sin \theta - \mu \cos \theta)$$

$$= 0.5 \times 9.81(\sin 40° - 0.5 \cos 40°) = 1.27 \text{ N}$$

Example

An object of mass 2.5 kg is on a horizontal plane and being pulled by a string inclined at 60° to the horizontal (Figure 4.38). If the object is just on the point of sliding along the plane when the tension in the string is 16 N, what is the coefficient of friction?

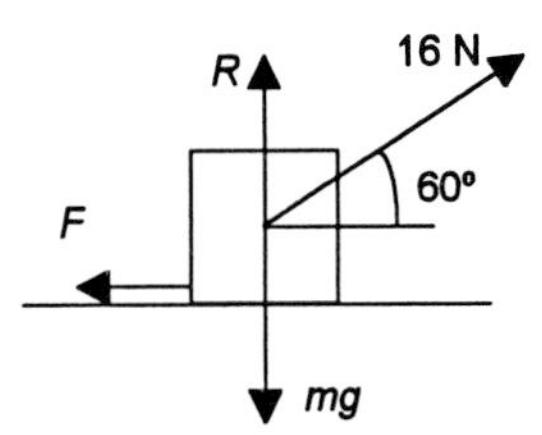

Figure 4.38 *Example*

For the vertical forces to be in equilibrium we must have:

$$R + 16 \sin 60° = mg$$

Thus $R = 2.5 \times 9.81 - 16 \sin 60° = 10.7$ N. When the object is just on the point of moving the net horizontal force must be zero. Thus:

$$16 \cos 60° = F = \mu R = \mu \times 10.7$$

Hence the coefficient of friction is 0.75.

Revision

25 An object of mass 6 kg rests in equilibrium on a rough plane inclined at 30° to the horizontal. Determine the minimum value of the coefficient friction between the object and the plane.

26 An object of mass 0.5 kg is on a rough plane at an elevation of 20° to the horizontal and is acted on by a force of 6 N up and parallel to the plane. If the coefficient of friction is 0.7, what is the acceleration of the object up the plane?

27 An object of weight 24 N rests on a rough plane inclined at 30° to the horizontal and is held by the tension in a string parallel to the plane. Determine the tension in the string if the coefficient of friction is 0.2.

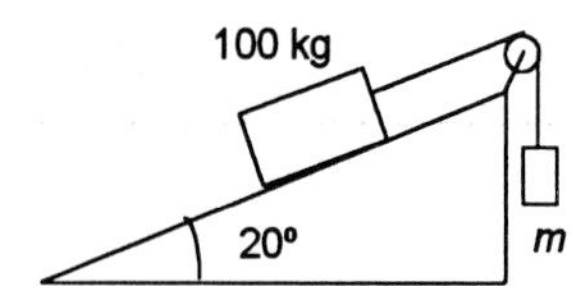

Figure 4.39 *Revision problem 30*

28 An object of mass 60 kg is at rest on a rough plane which is inclined at 45° to the horizontal. Determine the force needed to maintain the object at rest if the force is applied (a) parallel to the plane, (b) horizontally. The coefficient of friction is 0.25.

29 An ice-hockey puck has, immediately after being hit, a velocity of 15 m/s and then travels a distance of 75 m before coming to rest. What is the coefficient of friction between the puck and the ice?

30 For the arrangement shown in Figure 4.39, what value should m have if the 100 kg block does not start to move (a) up the plane, (b) down the plane?

Activity

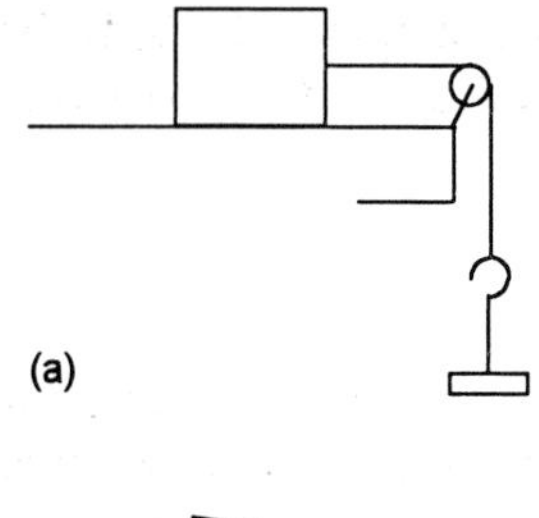

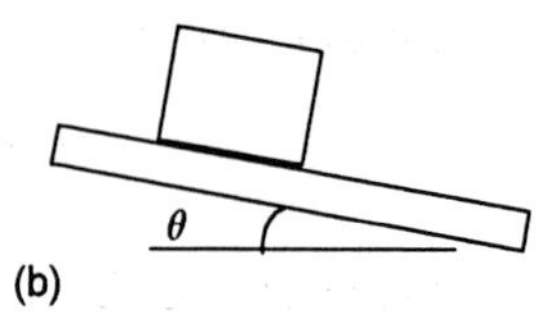

Figure 4.40 *Activity*

Determine the static coefficient of friction between two surfaces, e.g. a wooden block sliding on a wooden surface, by the following two methods.

Using the arrangement shown in Figure 4.40(a), determine the weights that have to be used on the hanger in order that the block just starts to move. Hence, calculate the coefficient.

Using the arrangement shown in Figure 4.40 (b), determine the angle to which the plane has to be raised for the block to just begin sliding down the plane. Hence, calculate the coefficient.

4.6 Torque and angular motion

Consider a force F acting on a small element of mass δm of a rigid body, the element being a distance r from the axis of rotation at O (Figure 4.41). The torque T acting on the element is Fr and since $F = \delta m \times a$ we can express the torque as:

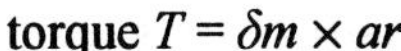

torque $T = \delta m \times ar$

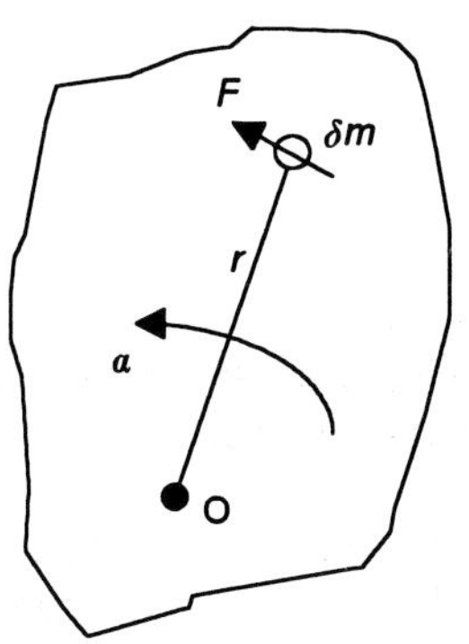

Figure 4.41 *Rotation of a rigid body*

But $a = r\alpha$, where α is the angular acceleration. Thus $T = \delta m \times r^2\alpha$. We can express this as:

$$\text{torque } T = I\alpha \qquad [41]$$

where I is termed the *moment of inertia* and for the small element of mass m at radius r is given by:

$$\text{moment of inertia } I = \delta m \times r^2 \qquad [42]$$

Equation [41] gives the torque required to give an angular acceleration to just an element of the mass of the rigid body. The torque required to give a rotational acceleration for the entire rigid body will thus be the sum of the torques required to accelerate each element of mass in the body and given by equation [41] when we consider the sum of the moments of inertia of the entire body. Thus:

$$\text{torque to give body angular acceleration } T = I\alpha \qquad [43]$$

where:

$$\text{moment of inertia of body } I = \int r^2 \, dm \qquad [44]$$

The moment of inertia of a body is the inertial property of a rotating body in much the same way as mass is the inertial property of a body for linear motion. An inertial property describes the inertia of the body that has to be overcome to get it moving.

4.6.1 Moment of inertia

For a small mass at the end of a light pivoted arm with radius of rotation r, we can consider the entire mass of the body to be located at the same distance r and so the moment of inertia is given by equation [44] as:

$$I = mr^2 \qquad [45]$$

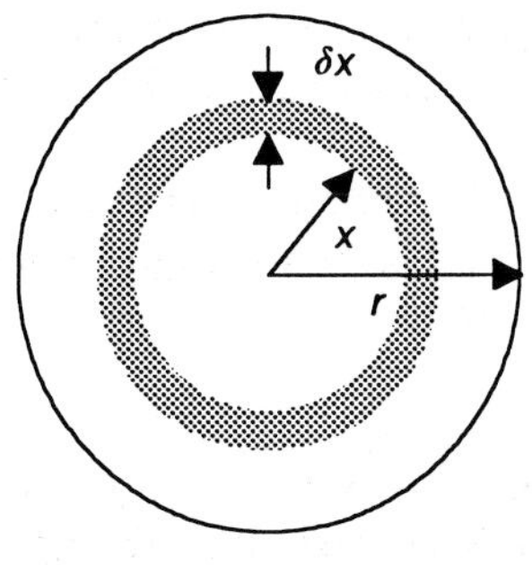

Figure 4.42 *Uniform disc*

For a uniform disc of radius r and total mass m, the moment of inertia about an axis through the disc centre can be found by considering the disc to be composed of a number of rings (Figure 4.42). For a ring of thickness δx and radius x, its area is effectively its circumference multiplied by the thickness and so is $2\pi x\,\delta x$. Since the mass per unit area of the disc is $m/\pi r^2$ then the mass δm of this ring element is:

$$\delta m = \frac{m}{\pi r^2} 2\pi x \delta x = \frac{2m}{r^2} x \delta x$$

Thus equation [44] gives for the moment of inertia of the disc:

$$I = \int_0^r x^2 \, dm = \int_0^r \frac{2m}{r^2} x^3 \, dx$$

$$I = \tfrac{1}{2} m r^2 \quad [46]$$

In a similar manner we can derive the moments of inertia of other bodies. Figure 4.43 gives the moments of inertia for some commonly encountered bodies.

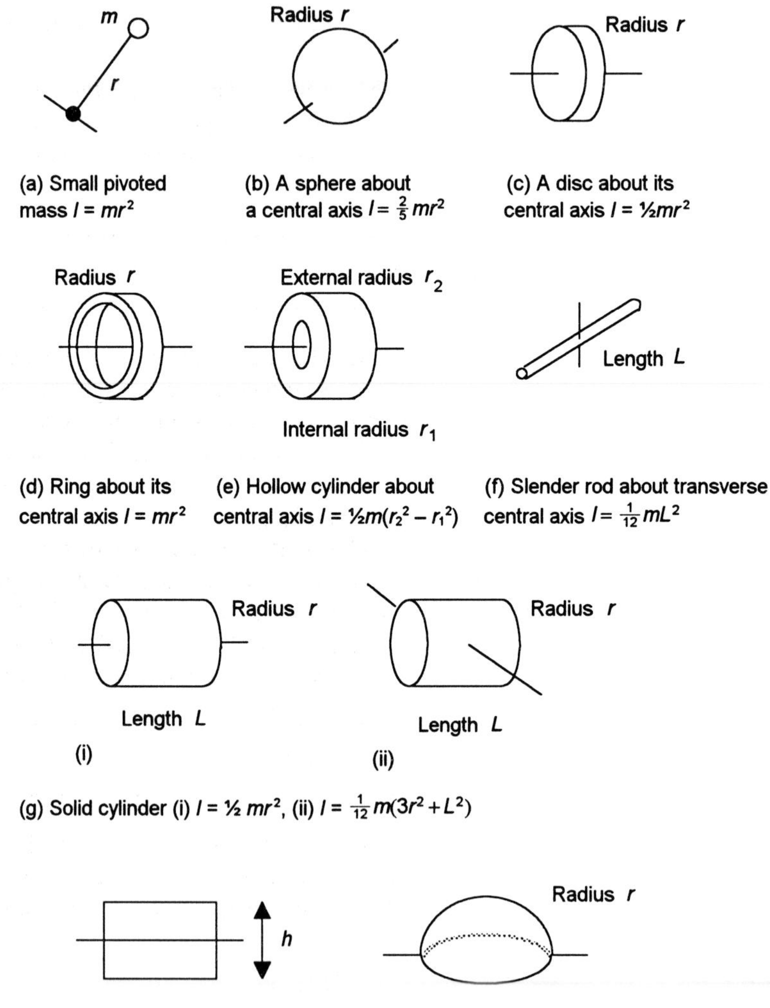

Figure 4.43 *Moments of inertia about the indicated axes*

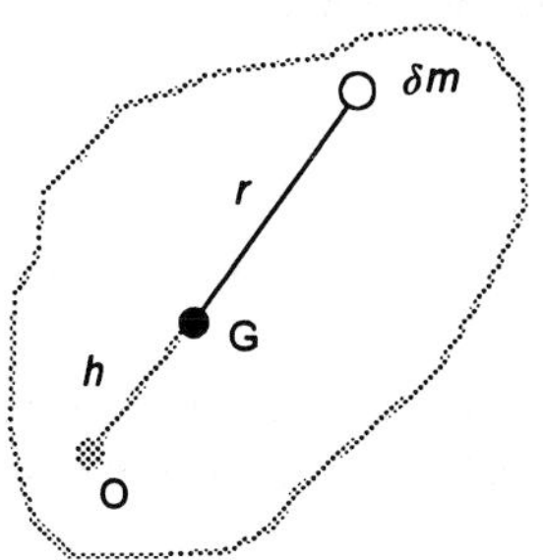

Figure 4.44 *Moment of inertia about a parallel axis*

The value of the moment of inertia for a particular body depends on the axis about which it is calculated. If I_G is the moment of inertia about an axis through the centre of mass G, consider what the moment of inertia would be about a parallel axis through O, a distance h away (Figure 4.44). The moment of inertia of an element dm of the mass about the axis through O is:

$$\text{moment of inertia of element} = (r + h)^2\,\delta m$$

$$= (r^2 + h^2 + 2rh)\,\delta m$$

The total moment of inertia about the axis through O is thus:

$$I_O = \int r^2\,\mathrm{d}m + \int h^2\,\mathrm{d}m + \int 2rh\,\mathrm{d}m$$

The moment of inertia about an axis through the centre of mass G is $I_G = \int r^2\,\mathrm{d}m$. The integral $\int h^2\,\mathrm{d}m = mh^2$. The integral $2h\int r\,\mathrm{d}m$ is the total moment of the mass about the axis through the centre of mass and is thus zero. Hence:

$$I_O = I_G + mh^2 \qquad [47]$$

This is known as the *theorem of parallel axes.*

It is sometimes necessary to obtain the moment of inertia of a composite body about some axis. This can be done by summing the moments of inertia of each of the elemental parts about the axis, using the parallel axis theorem where necessary. This is illustrated in one of the following examples.

Consider replacing a body by one of the same mass but it all concentrated at just one point, the point being so located that the same moment of inertia is obtained. If the total mass of a body is m and we consider it all to be concentrated at a small point a distance k from the pivot axis then its moment of inertia would be

$$I = mk^2 \qquad [48]$$

k is called the *radius of gyration*. Thus, for a disc of radius r and mass m the moment of inertia about its central axis is $\frac{1}{2}mr^2$ and thus the radius of gyration for the disc is $k^2 = \frac{1}{2}r^2$ and so $k = r/\sqrt{2}$.

Example

Determine the moment of inertia about an axis through its centre of a disc of radius 100 mm and mass 200 g.

The moment of inertia of the disc is given by:

$$I = \tfrac{1}{2}mr^2 = \tfrac{1}{2} \times 0.200 \times 0.100^2 = 1.0 \times 10^{-3}\text{ kg m}^2$$

Example

Determine the moment of inertia about an axis of an object having a mass of 2.0 kg and a radius of gyration from that axis of 300 mm.

The moment of inertia is given by equation [48] as:

$$I = mk^2 = 2.0 \times 0.300^2 = 0.18\text{ kg m}^2$$

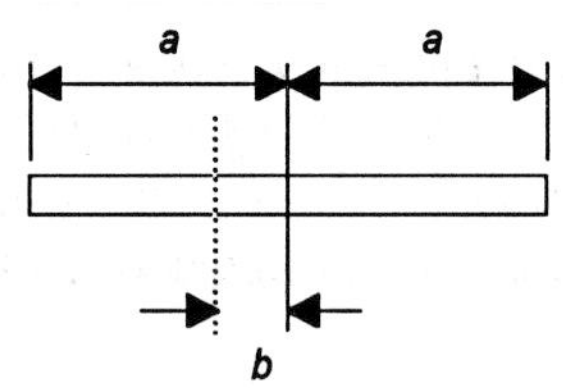

Figure 4.45 *Example*

Example

Determine the moment of inertia of a slender rod of mass m about an axis at right angles to its length of $2a$ and a distance b from its centre (Figure 4.45).

The moment of inertia of a slender rod of length L about an axis through its centre is $mL^2/12$. Thus, for $L = 2a$, we have

$$I = \frac{ma^2}{3}$$

Using the parallel axis theorem, the moment of inertia about a parallel axis a distance b from the centre is thus:

$$I = \frac{ma^2}{3} + mb^2$$

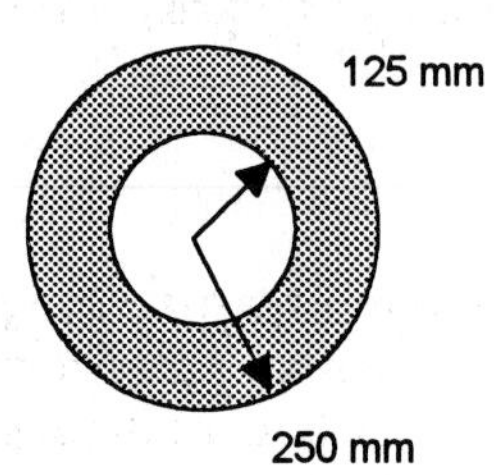

Figure 4.46 *Example*

Example

Determine the moment of inertia of the plate shown in Figure 4.46 about an axis at right angles to it and through its centre. The plate is 10 mm thick and has a radius of 250 mm and contains a central hole with a radius of 125 mm. The material has a density of 8000 kg/m³.

This can be tackled by considering the plate as a composite body, i.e. a solid disc of radius 250 mm minus a disc of radius 125 mm. For the 250 mm diameter disc, the moment of inertia is:

$$I = \tfrac{1}{2}MR^2 = \tfrac{1}{2} \times \pi \times 0.250^2 \times 0.010 \times 8000 \times 0.250^2$$

$$= 0.491\text{ kg m}^2$$

For the 125 mm diameter disc:

$$I = \tfrac{1}{2}MR^2 = \tfrac{1}{2} \times \pi \times 0.125^2 \times 0.010 \times 8000 \times 0.125^2$$

$$= 0.031\text{ kg m}^2$$

Thus the moment of inertia of the disc with the hole is 0.491 – 0.031 = 0.460 kg m^2.

Revision

31 Determine the moment of inertia of an object having a mass of 4 kg and a radius of gyration of 200 mm.

32 Determine the moment of inertia of a sphere about an axis through its centre if it has a radius of 200 mm and is made of material with a density of 8000 kg/m^3.

33 A disc with an outside radius of 250 mm and a mass of 11.78 kg has a moment of inertia of 0.460 kg m^2 about an axis at right angles to its plane and through its centre. Determine its moment of inertia about an axis at right angles to its plane and passing through a point on its circumference.

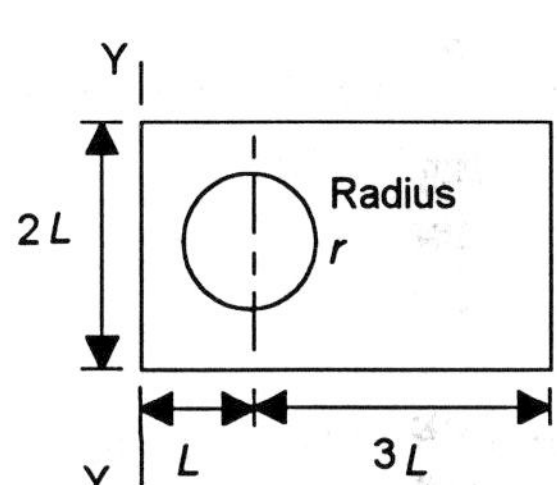

Figure 4.47 *Revision Problem 35*

34 Determine the moment of inertia of a disc, about an axis perpendicular to the face of the disc and through its centre, which has a constant thickness of 50 mm, a diameter of 750 mm and a central hole of diameter 150 mm. The disc material has a density of 7600 kg/m^3.

35 Determine the moment of inertia of the rectangular plate shown in Figure 4.47 about the Y–Y axis shown if the plate has a mass m per unit area of plate.

4.6.2 Rotational problems

The following examples illustrate how rotational problems involving moments of inertia can be tackled.

Example

A drum with a mass of 60 kg, radius of 400 mm and radius of gyration of 250 mm has a rope of negligible mass wrapped round it and attached to a load of 20 kg (Figure 4.48). Determine the angular acceleration of the drum when the load is released.

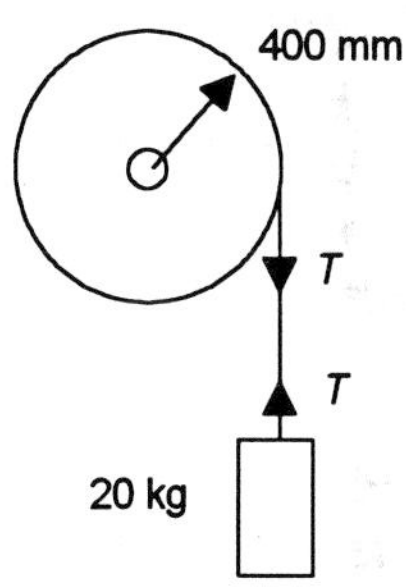

Figure 4.48 *Example*

For the forces on the load we have:

$$20g - T = 20a \qquad [49]$$

For the drum, the torque is $T \times 0.400$ and thus, using torque = $I\alpha$:

$$T \times 0.400 = I\alpha$$

Since $I = mk^2 = 60 \times 0.250^2$ then:

$$T \times 0.400 = 60 \times 0.250^2 \times \alpha \qquad [50]$$

The point of contact between the rope and the drum will have the same tangential acceleration and thus, using $a = r\alpha$:

$$a = 0.400\alpha$$

Thus equation [49] can be written as:

$$20g - T = 20 \times 0.400\alpha$$

and so, eliminating T between this equation and equation [50]:

$$(20g - 20 \times 0.400\alpha) \times 0.400 = 60 \times 0.250^2 \times \alpha$$

The angular acceleration is thus 11.3 rad/s^2.

Example

A record turntable is a uniform flat plate of radius 120 mm and mass 0.25 kg. What torque is required to uniformly accelerate the turntable to 33.3 rev/min in 2 s?

The angular acceleration required is given by $\omega = \omega_0 + \alpha t$ as:

$$2\pi \times \frac{33.3}{60} = 0 + \alpha \times 2$$

Hence $\alpha = 1.74$ rad/s^2. The moment of inertia of the turntable is:

$$I = \tfrac{1}{2}mr^2 = \tfrac{1}{2} \times 0.25 \times 0.120^2 = 1.8 \times 10^{-3} \text{ kg m}^2$$

The torque required to accelerate the turntable is thus:

$$T = I\alpha = 1.8 \times 10^{-3} \times 1.74 = 3.13 \times 10^{-3} \text{ N m}$$

Example

A solid sphere rolls, without slipping, down a rough inclined plane which is at an elevation of θ to the horizontal (Figure 4.49). Determine its acceleration.

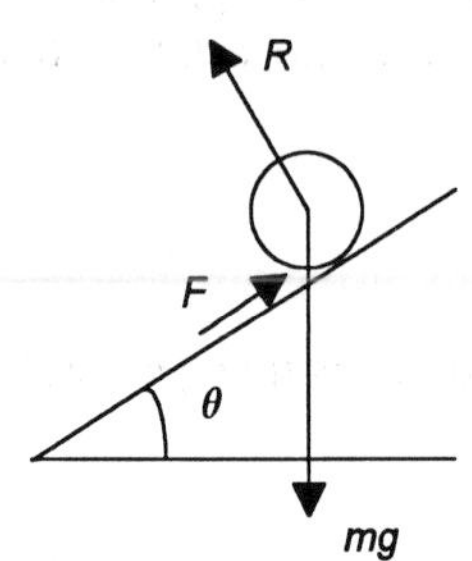

Figure 4.49 *Example*

For the linear motion of the centre of mass of the sphere, the forces acting parallel to and down the plane are:

$$mg \sin\theta - F = ma \qquad [51]$$

where a is the linear acceleration down the plane and F is the frictional force. As the sphere descends it rotates about its centre. The torque giving this rotation is that due to the frictional force and is thus Fr, where r is the radius of the sphere. Thus:

$$Fr = Ia$$

where a is the angular acceleration of the sphere. The moment of inertia I of the sphere is $2mr^2/5$. Thus:

$$Fr = \tfrac{2}{5}mr^2 \times a$$

But $a = ra$ and so:

$$a = ra = r\frac{5F}{2mr} = \frac{5F}{2m}$$

Using equation [51] and substituting for F gives:

$$mg\sin\theta - \frac{2ma}{5} = ma$$

Hence:

$$a = \tfrac{5}{7}g\sin\theta$$

Example

A cable drum rests on a rough horizontal surface (Figure 4.50). It has a moment of inertia I about its axis of rotation and a mass m. The outer part of the drum has a radius r_2 and the radius round which the cable is wound is r_1. Determine the linear acceleration of the drum when the cable is pulled with a horizontal force of P and the drum rolls without slipping.

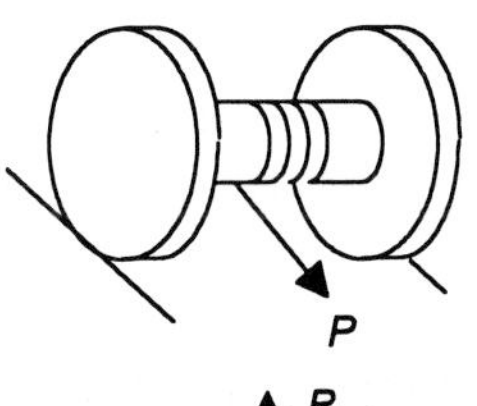

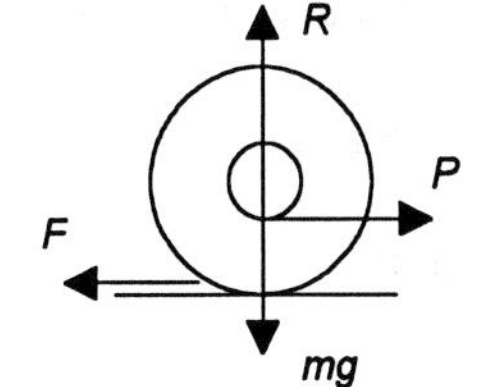

Figure 4.50 *Example*

For the horizontal linear motion of the centre of mass of the cable drum:

$$P - F = ma \qquad [52]$$

where F is the frictional force and a the linear acceleration. For vertical equilibrium we have:

$$R = mg$$

The drum rotates under the net action of the torque applied by the cable and that resulting from friction between the drum and the horizontal surface. Thus:

$$Pr_1 - Fr_2 = Ia$$

where a is the angular acceleration of the drum. Using equation [52] to eliminate F gives:

$$Pr_1 - (P - ma)r_2 = Ia$$

$$P(r_2 - r_1) = mar_2 - I\alpha$$

Since with no slip we have $a = -r_2\alpha$, the minus sign being because the direction of the acceleration for the drum is in the opposite direction to that for the cable, then:

$$P(r_2 - r_1) = mar_2 + Ia/r_2$$

$$a = \frac{P(r_2 - r_1)}{mr_2 + I/r_2}$$

Revision

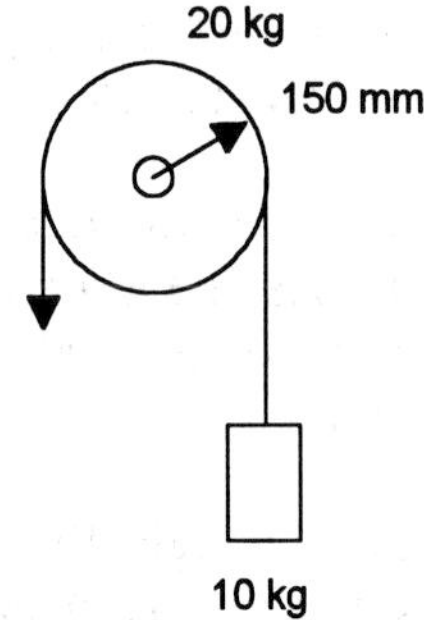

Figure 4.51 *Revision problem 37*

36 A disc has a mass of 2.5 kg and a radius of gyration of 100 mm. Determine the torque required to give the disc an angular acceleration of 15 rad/s^2.

37 A uniform pulley has a mass of 20 kg and a radius of 150 mm. A rope passes over it and is attached to a load of 10 kg (Figure 4.51). What force must be applied to the other end of the rope to give the load an upward acceleration of 0.2 m/s?

38 A uniform cylinder rolls, without slipping, from rest down a rough inclined plane which is at an elevation of 30° to the horizontal. Determine its linear acceleration down the plane.

39 A uniform cylinder of mass 6 kg and radius 100 mm is made to roll without slipping along a rough horizontal surface as a consequence of a horizontal force of 80 N being applied along its central axis. Determine the angular acceleration of the cylinder.

Problems

1 An object has an initial velocity of 3 m/s and is given a uniform acceleration of 2 m/s^2 for 6 s. What will be the resulting velocity?

2 An object initially at rest is given a uniform acceleration of 2 m/s^2 for 4 s. What will be the distance covered in that time?

3 An object with an initial velocity of 30 m/s is slowed down with a uniform acceleration of –4 m/s^2 until the velocity is 10 m/s. What will be the time taken?

4 An object, starting from rest, has a uniform acceleration of 2 m/s^2. What will be the distance covered in (a) the first four seconds, (b) the fourth second of the motion?

5 An object moves along a straight line from point A to point B. It has a velocity of 3 m/s at A and accelerates between A and B with a uniform acceleration of 0.5 m/s^2 to attain a velocity of 5 m/s^2 at B.

What is the distance between A and B and the time taken for the object to move from A to B?

6 What initial velocity must an object have if it is to come to rest in 100 m as a result of a uniform retardation of 8 m/s^2?

7 A cam is designed so that its cam follower increases its velocity at a uniform rate from 3 m/s to 5 m/s when it moves vertically through a distance of 16 mm. Determine the time taken and the acceleration.

8 An object accelerates from rest with a uniform acceleration of 4 m/s^2 and then decelerates at 8 m/s^2 until at rest again, covering a total distance from rest-to-rest of 5000 m. What is the time taken?

9 A car accelerates along a straight road with uniform acceleration. When it passes point A it has a speed of 12 m/s and when it passes point B it has a speed of 32 m/s. Points A and B are 1100 m apart. Determine the time taken to move from A to B.

10 The velocity v of an object moving along a straight line path is related to the time t by $v = 4 + 2t$ m/s. Determine the acceleration.

11 The distance s travelled by an object moving along a straight line path is related to the time t by $s = 10t + 4t^2$. Determine, after 2 s, (a) the velocity, (b) the acceleration.

12 A stone is thrown vertically upwards from a point some distance above the ground level with an initial velocity of 20 m/s. If the stone hits the ground after 5 s, how far below the point of projection is the ground?

13 An object is thrown vertically upwards with an initial velocity of 14 m/s. Determine the height of the object above the point of projection after (a) 1 s, (b) 2 s.

14 Determine the resultant velocity of an object if it has velocities of 24 m/s in a northerly direction and 7 m/s in an easterly direction.

15 A raindrop falls in still air with a velocity of 3 m/s. What will be its velocity when there is a horizontal air current of 4 m/s?

16 Determine the components in northerly and easterly directions of a velocity of 2 m/s in a north-easterly direction.

17 What will be the greatest range on a horizontal plane for a projectile when the magnitude of the velocity of projection is 20 m/s?

18 Neglecting air resistance, what is the maximum range of a bullet from a rifle if it leaves the rifle with a velocity of 400 m/s.

19 A projectile has an initial velocity of 40 m/s at an elevation of 60° from the horizontal. Determine the speed of the projectile after 3 s.

20 A stone is thrown horizontally from the top of a tower with a velocity of 12 m/s. If the tower has a height of 10 m, determine how far from the foot of the tower the stone will land.

21 A bullet has an initial velocity of 600 m/s at an elevation of 25° to the horizontal. What will be its range?

22 A flywheel rotating at 210 rev/min is uniformly accelerated to 250 rev/min in 5 s. Determine the angular acceleration and the number of revolutions made by the flywheel in that time.

23 A flywheel rotating at 0.5 rev/s is uniformly accelerated to 1.0 rev/s in 10 s. Determine the angular acceleration and the number of revolutions made by the flywheel in that time.

24 A grinding wheel is rotating at 50 rev/s when the power is switched off. It takes 250 s to come to rest. What is the average angular retardation?

25 A wheel of diameter 350 mm rotates with an angular velocity of 6 rad/s. What is the speed of a point on its circumference?

26 A flywheel of diameter 360 mm increases its angular speed uniformly from 10.5 to 11.5 rev/s in 11 s. Determine (a) the angular acceleration of the wheel, (b) the linear acceleration of a point on the wheel rim.

27 A 160 mm diameter drive pulley is rotating at 400 rad/s and drives, via a belt drive, a second pulley of diameter 360 mm. Determine the uniform angular acceleration required for the drive pulley if the driven pulley is to be accelerated to an angular rotation of 200 rad/s after 6 s of acceleration.

28 A 200 mm radius pulley is rotating at 1200 rev/min and is used to drive, via a belt drive, a second pulley. Determine the radius required for the second pulley if it is to rotate a shaft at 800 rev/min.

29 A gear A with radius 50 mm is rotating at 30 rad/s and drives a second, intermeshed, gear B of radius 200 mm. If the gear B must rotate through 30 rad after 2 s, what is the angular acceleration needed for gear A?

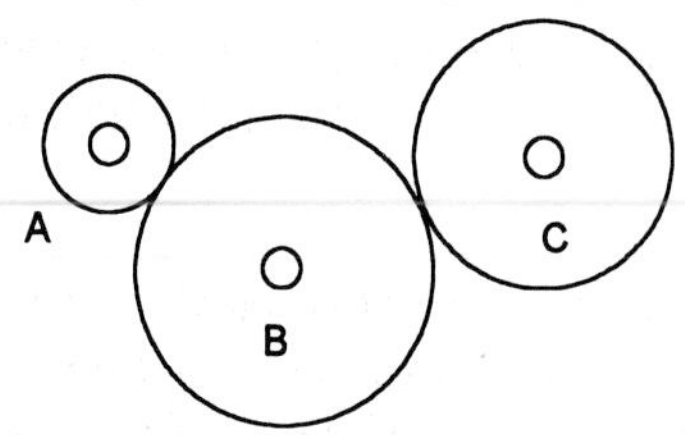

Figure 4.52 *Problem 30*

30 For the three gears shown in Figure 4.52, gear A has a radius of 60 mm, gear B a radius of 140 mm and gear C a radius of 120 mm. When gear A is rotating at 200 rev/min, what will be the angular velocity of gear C?

31 An object of mass 4 kg rests on a smooth plane which is inclined at 60° to the horizontal. The object is connected to one end of a light inextensible string which passes over a smooth fixed pulley at the top of the plane and from the vertical end of which another object of mass 2 kg is attached. Determine the acceleration of the objects and the tension in the string.

32 An object of mass 5 kg rests on a smooth horizontal table and is connected by a light inextensible string passing over a smooth fixed pulley at the edge of the table to a vertically suspended mass of 3 kg. Determine the acceleration of the objects and the tension in the string.

33 An object rests on a smooth horizontal table. On its left it is attached by a light inextensible string passing over a smooth fixed pulley on the edge of the table to a vertically suspended mass of 10 kg and on the left by another light inextensible string passing over a pulley at the edge of the table to a vertically suspended mass of 2 kg. Both the strings are attached to the object along the same horizontal line. Determine the mass of the object on the table if, when the system is released from rest, it accelerates at 2 m/s^2.

34 A car of mass M pulls a trailer of mass m along a level road. If the car engine exerts a forward force F and the tension in the tow bar is T, show that the acceleration of the car and trailer is $F/(M + m)$ and the tension in the tow bar is $F - Ma$.

35 A car of mass 900 kg pulls a trailer of mass 600 kg along a level road. The car experiences a resistance of 200 N and the trailer a resistance of 300 N. If the car exerts a forward force of 3000 N, determine the acceleration of the car and trailer and the tension in the tow bar.

36 An object of mass m rests on a smooth horizontal table and is connected to a freely hanging object of mass M by a light inextensible string passing over a smooth fixed pulley at the edge of the table. If the object on the table starts from rest a distance d from the pulley at the edge of the table, determine the time taken for it to reach the pulley.

37 A child of mass 25 kg sits on a toboggan of mass 15 kg on a snow slope inclined at 10° to the horizontal. If the toboggan slides down the slope with an acceleration of 1.2 m/s^2, what is the coefficient of friction?

38 An object of mass 9 kg is held, just on the point of moving, on a rough plane which is inclined at 30° to the horizontal by a force of 12 N acting parallel to the plane. What is the coefficient of friction?

39 An object of mass 5 kg is on a rough plane which has an angle of inclination to the horizontal which is gradually increased. When the angle reaches 13.5° the object begins to slide down the plane. What is the coefficient of friction? What would be the minimum force parallel to the plane which would be necessary to pull the block up the plane?

40 An object of mass m rests on a rough plane inclined at 30° to the horizontal and is attached, by means of a light inextensible string passing over a pulley at the top of the slope to a mass M which hangs vertically. The coefficient of friction is $1/\sqrt{3}$. Show that when the system is released that there will be an acceleration of $(M - m)g/(M + m)$.

41 A flywheel has a mass of 2 kg and a radius of gyration about an axis at right angles to its face and through its centre of 100 mm. Determine its moment of inertia about this axis.

42 A flywheel is a solid disc of mass 120 kg and radius 150 mm. Determine its moment of inertia about an axis at right angles to its plane and through its centre.

43 Determine the radius of gyration for an object having a moment of inertia of 0.5 kg m^2 and a mass of 2 kg.

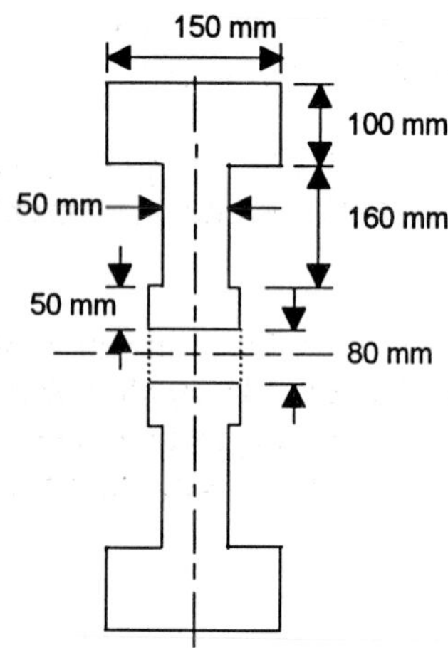

Figure 4.53 *Problem 44*

44 Determine the moment of inertia about the rotational axis of the circular flywheel with the cross-section shown in Figure 4.53.

45 A dumbbell consists of a slender rod of length 1.20 m and mass 10 kg with uniform spheres attached to the ends, each sphere having a mass of 45 kg and a radius of 100 mm. Determine the moment of inertia of the dumbbell about an axis passing through the centre of the rod and at right angles to it.

46 A solid uniform disc has a mass of 120 kg and a radius of 150 mm. With the disc initially at rest, a constant torque of 10 N m is applied for 2 s. Determine the angular velocity of the disc after 2 s.

47 A wheel is mounted on a horizontal axle of radius 10 mm which is supported by bearings which can be assumed to be frictionless. A cord is wrapped round the axle and attached to a load of 5 kg. With the cord taut, when the load is released it falls through a distance of 1 m in 10 s. Determine the moment of inertia of the wheel and axle.

48 A solid, uniform, drum of mass 40 kg and radius 500 mm has a rope wrapped round its periphery. What will be the angular acceleration of the drum when a force of 200 N is applied to the rope.

49 A flywheel has a mass of 360 kg and a radius of gyration of 600 mm. If it is rotating at 10 rev/s, determine the uniform torque that has to be applied to bring it to rest in 30 s.

50 A torque of 1.2 N m is applied for 30 s to the axis of a flywheel to accelerate it up to a speed of 48 rad/s. The flywheel then takes 1200 s to come to rest. If the flywheel has a mass of 30 kg, what will be the frictional torque at the bearings and the moment of inertia of the flywheel?

51 A wheel of mass 50 kg and radius of gyration 0.60 m is rotating at 200 rev/min. Determine the time the wheel will take to come to rest when a retarding torque of 80 N m is applied to the wheel.

52 A wheel has a moment of inertia about its axis of 0.12 kg m^2. What torque will be required to accelerate it to 10 000 rev/min in a time of 30 s?

53 A flywheel has a mass of 4000 kg and a radius of gyration of 0.80 m. What is (a) the moment of inertia about the flywheel axis, (b) the torque that will be needed to accelerate the flywheel from rest to a a speed of 5 rev/s in a time of 60 s if there is a resisting torque of 350 N m?

5 Energy transfer

5.1 Introduction

This chapter is concerned with the energy transfers that can occur with mechanical systems. Energy can be transferred from one form to another by work being done or by heat transfer. Here we restrict the discussion to transfers involving work. There are many forms that energy can take and in this chapter potential energy, linear and angular kinetic energy and strain energy are discussed and the principles applied to the solution of mechanical system problems.

5.1.1 Conservation of energy

There is a basic principle that is used in all discussions of energy and that is that energy is never lost, it is only transformed from one form to another or transferred from one object to another. This is the *principle of the conservation of energy*. In any process we never increase the total amount of energy, all we do is transform it from one form to another.

> The principle of the *conservation of energy* is that energy is never created or lost but only converted from one form to another. In all such conversions, the total amount of energy remains constant.

5.2 Work

Work is said to be done when the energy transfer takes place as a result of a force pushing something through a distance (Figure 5.1), the amount of energy transferred W being the product of the force F and the displacement s of the point of application of the force in the direction of the force.

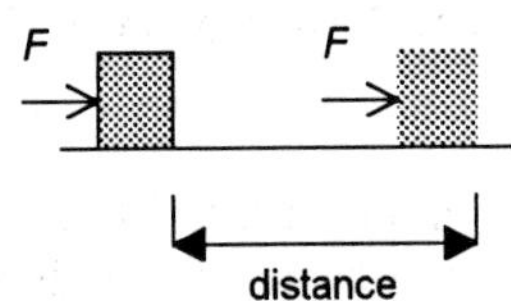

Figure 5.1 *Work*

> Work done by a constant force $W = Fs$ [1]

With force in newtons and distance in metres, the unit of work is the joule (J) with 1 J being 1 N m.

Consider the work done by a force F when the resulting displacement s is at some angle θ to the force (Figure 5.2). We can look at this in two equivalent ways. We can consider the displacement in the direction of the force F is $s \cos \theta$ and so the work done is:

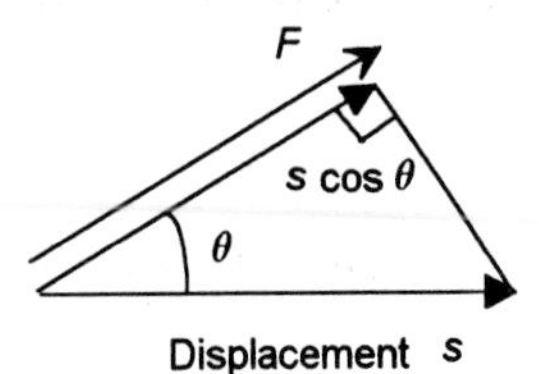

Figure 5.2 *An oblique force*

$$\text{work done} = F \times s \cos \theta \qquad [2]$$

Alternatively, we can consider the force component acting in the direction of the displacement. The force can be resolved into two components, namely $F \cos \theta$ in the direction of the displacement and $F \sin \theta$ at right angles to it. There is no displacement in the direction of the $F \sin \theta$ component and so it does no work. Hence the work done by the oblique force is solely due to the $F \cos \theta$ component and so is, as before:

$$W = (F \cos \theta) \times s.$$

Example

The locomotive of a train exerts a constant force of 120 kN on a train while pulling it at 40 km/h along a level track. What is the work done in 15 minutes?

In 15 minutes the train covers a distance of 10 km. Hence, work done = 120 × 1000 × 10 × 1000 = 1200 MJ.

Example

A barge is towed along a canal by a tow rope inclined at an angle of 20° to the direction of motion of the barge. Calculate the work done in moving the barge a distance of 100 m along the canal if the pull on the rope is 400 N.

We have an oblique force and so, since the displacement in the direction of the force is 100 cos 20°, the work done = 400 × 100 cos 20° = 37.6 kJ.

Example

What is the work done, by a force parallel to a slope, in moving a body of mass m at uniform velocity a distance s up a rough slope at an angle θ to the horizontal? The coefficient of friction between the body and the slope is μ.

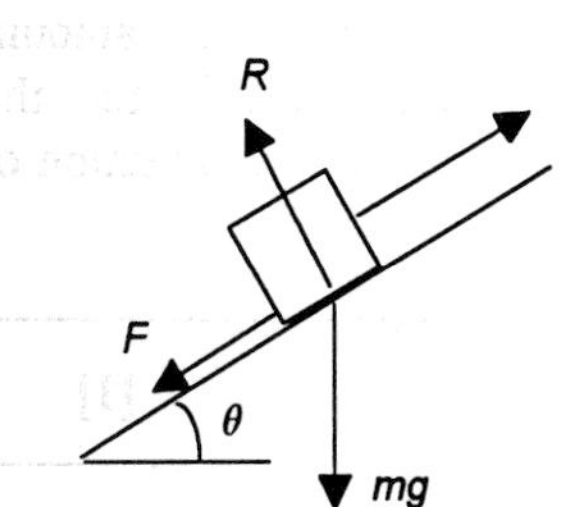

Figure 5.3 *Example*

Figure 5.3 shows the situation. The component of the weight acting down the slope is $mg \sin \theta$ and so the work that has to be done against gravity is $(mg \sin \theta) \times s$.

The frictional force acting down the slope F is μR. Since $R = mg \cos \theta$, then $F = \mu mg \cos \theta$. The work done against the frictional force is thus $(\mu mg \cos \theta) \times s$.

The total work done W is thus:

$$W = (mg \sin \theta) \times s + (\mu mg \cos \theta) \times s = mgs(\sin \theta + \mu \cos \theta)$$

Revision

1 Determine the work done when a constant force of 100 N is applied to an object and moves it through a distance of 3 m in (a) the

direction of the force, (b) at an angle of 30° to the direction of the force.

2 A horizontal force is used to pull an object of mass 2 kg a distance of 8 m along a rough horizontal floor, the coefficient of friction being 1/3. If the object moves with a constant velocity, what is the work that has to be done against friction?

3 A body of mass 10 kg is pulled at a uniform velocity a distance of 20 m up a rough slope by a force parallel to the slope. The coefficient of friction is 0.1. If the slope is at 30° to the horizontal, what will be the work done?

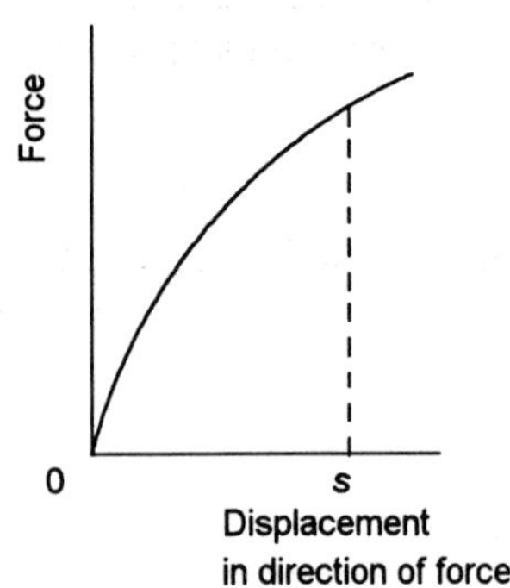

Figure 5.4 *Force–displacement graph*

5.2.1 Work as area under force–distance graph

Consider the work done when the force moving an object is not constant but varying, e.g. in the manner shown in Figure 5.4. We can tackle such a problem by considering the displacement over some distance as being made up of a small number of displacements for each of which the force can be considered constant. Figure 5.5 illustrates this. For each small displacement the work done is the product of the force and the displacement and so is equal to the area of the strip. The total work done in giving a displacement from 0 to s is thus the sum of the areas of all the strips between 0 and s and so is equal to the area under the graph.

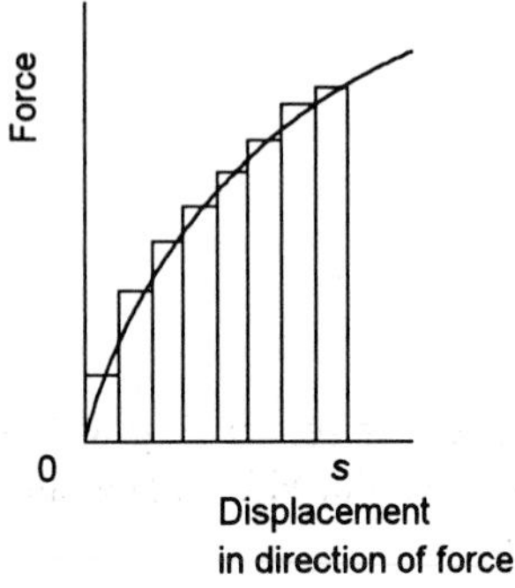

Figure 5.5 *Force–displacement graph*

Work is the area under the force–displacement graph	[3]

Thus, for any force variation with distance s, the general equation for the work done by a varying force is:

$$W = \int_0^s F\,\mathrm{d}s \qquad [4]$$

Example

A load is hauled along a track with a tractive effort F which varies with the displacement s in the direction of the force in the following manner:

F in kN	1.6	1.4	1.2	1	0.8	0.6
s in m	0	10	20	30	40	50

Determine the work done in moving the load from displacement 0 to 50 m.

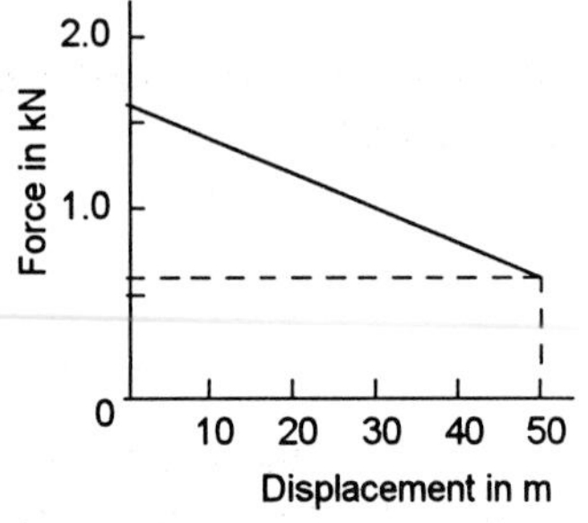

Figure 5.6 *Example*

Figure 5.6 shows the force–displacement graph. The work done is the area under the graph between displacements 0 and 50 m and hence is the area of a rectangle 600 × 50 J plus the area of the triangle ½ × (1600 – 600) × 50 J. The work done is thus 55 000 J.

Revision

4 The following shows the force–displacement data for a mechanism. Determine the work done when the mechanism moves from a displacement of zero to 0.6 m.

Force in N	Displacement in m
0	0
2	0.1
4	0.2
4	0.3
4	0.4
2	0.5
0	0.6

5.2.2 Work done by a torque

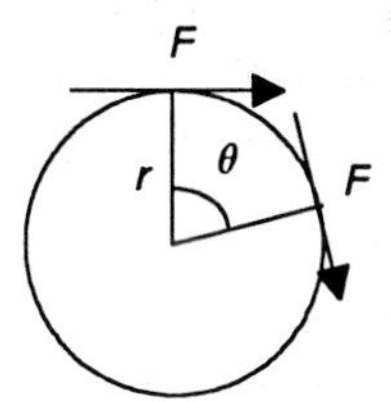

Figure 5.7 *Torque causing rotation*

Now consider an object which, starting from rest, is set rotating with an angular acceleration α as a result of a torque T produced by a tangential force F (Figure 5.7). The work done by the force is:

$$\text{work done} = \text{force} \times \text{distance} = F \times \text{arc length} = Fr\theta$$

The torque $T = Fr$ and so the work done is:

$$\text{work done} = T\theta \qquad [5]$$

Example

A starting handle for an engine has a crank arm length of 200 mm. What will be the work done when a force of 100 N is applied at right angles to the crank arm and rotates it through 120°?

The applied torque T is $Fr = 100 \times 0.200 = 20$ N m. Since $360° = 2\pi$ rad then 120° is $2\pi/3$ rad. Hence, using equation [5]:

$$\text{work done} = 20 \times 2\pi/3 = 41.9 \text{ J}$$

Example

A gear wheel with a pitch circle diameter of 130 mm is acted on by a force of 800 N acting tangentially to the pitch circle circumference. What will be the work done in rotating the gear wheel through one revolution?

The applied torque T is $Fr = 800 \times 0.130/2 = 52$ N m. One revolution is 2π rad and so the work done in one revolution is given by equation [5] as:

$$\text{work done} = T\theta = 52 \times 2\pi = 326.7 \text{ J}$$

Revision

5 What is the work done when a wheel of diameter 240 mm is rotated through 20 revolutions by a force of 50 N acting tangentially to its rim?

6 A nut is rotated through 60° by a spanner. What will be the work done if the force applied to the spanner is 80 N and the line of action of the force is 250 mm from the centre of the nut?

5.2.3 Work as area under a torque–angle graph

If the torque applied to rotate a body is not constant then, as with the earlier consideration of the work done by a force which is not constant (Section 5.2.1), we can tackle such a problem by considering the rotation through some angle as being made up of a small number of angular displacements for each of which the force can be considered constant. Figure 5.8 illustrates this. For each small angular displacement the work done is the product of the torque and the angular displacement and so is equal to the area of the strip. The total work done in giving an angular displacement from 0 to θ is thus the sum of the areas of all the strips between 0 and θ and so is equal to the area under the graph.

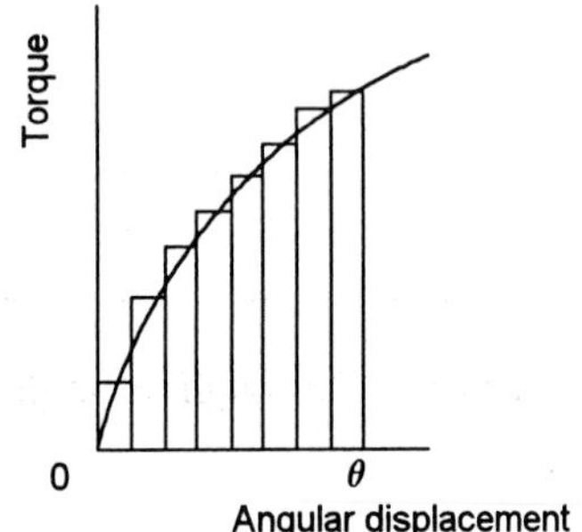

Figure 5.8 *Torque–angle graph*

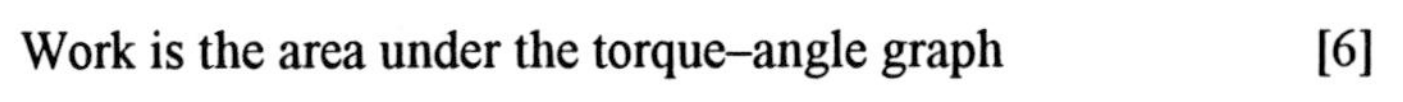

Work is the area under the torque–angle graph [6]

Thus, for any torque variation with angle θ, the general equation for the work done by a varying force is:

$$W = \int_0^\theta T \, d\theta \quad [7]$$

Example

The torque–angle diagram for half a rotation of the crank shaft of an engine is shown in Figure 5.9. Determine the work done for that half rotation.

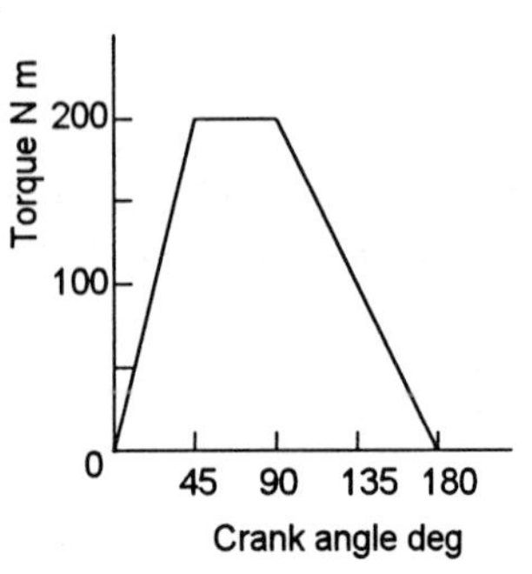

Figure 5.9 *Example*

The work done is the area under the graph with the angle having to be converted into radians. Thus:

$$\text{work done} = \tfrac{1}{2}\, 200 \times \pi/4 + 200 \times \pi/4 + \tfrac{1}{2}\, 200 \times \pi/2 = 393 \text{ J}$$

Revision

7 Determine the work done per revolution for a shaft for which the torque rises at a constant rate from 0 at 0° to 30 N m at 90°, is then constant at 30 N m from 90° to 270°, then decreases at a constant rate to 0 at 360°.

5.2.4 Power

Power is the rate at which energy is transferred. With the energy in units of joule and time in seconds then power is in J/s or watt (W).

For an object on which work is done for a time t and results in a displacement s in that time in the direction of the force, the power P developed for a constant force F is:

$$P = \frac{W}{t} = \frac{Fs}{t}$$

But the average velocity $v = s/t$ and so:

$$P = Fv \qquad [8]$$

For an object which is rotated by a constant torque T through an angle θ in time t:

$$P = \frac{W}{t} = \frac{T\theta}{t}$$

But the angular velocity $\omega = \theta/t$ and so:

$$P = T\omega \qquad [9]$$

If the rotation is at n revolutions per second then $\omega = 2\pi n$ and so equation [9] can be written as:

$$P = 2\pi nT \qquad [10]$$

Example

The locomotive of a train exerts a constant force of 120 kN on a train while pulling it at 40 km/h along a level track. What is the power?

Power = $Fv = 120 \times 10^3 \times 40 \times 10^3/3600 = 1.3$ MW

Example

What power will be required for a pump to extract water from a mine at 5 m^3/s and pump it through a vertical height of 20 m. Water has a density of 1000 kg/m^3. Take g as 9.8 m/s^2.

A volume of 5 m^3 of water has a mass of 5000 kg and a weight of 5000g N. This weight of water has to be moved through a distance of 20 m in 1 s, i.e. an average velocity of 20 m/s. Hence:

$$\text{Power} = Fv = 5000 \times 9.8 \times 20 = 980 \times 10^3 \text{ W} = 980 \text{ kW.}$$

Example

The shaft of a car rotates at 1000 rev/min and transmits power of 40 kW. What is the torque being applied to the shaft?

Since power = $T\omega = 2\pi nT$ then $40 \times 10^3 = 2\pi(1000/60)T$ and so the torque T = 382 N m.

Revision

8 A jet engine produces a thrust of 60 kN to maintain it in flight at 900 km/h. What is the power used?

9 A cutting tool operates against a constant resistive force of 2000 N. If the tool moves through a distance of 150 mm in 6 s, what is the power used?

10 What power is required to rotate a shaft of diameter 140 mm at 1 rev/s when a tangential force of 200 N is being applied to it?

11 If an engine running at 20 rev/s develops a torque of 150 N m, what will be the power delivered?

Activity

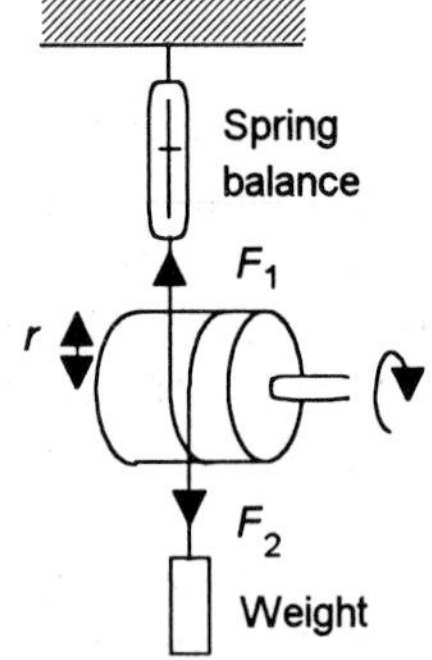

Figure 5.10 *Activity*

The output power of a motor can be determined by a form of mechanical brake; Figure 5.10 shows one such form. A belt or rope is wrapped round a pulley driven by the motor. One end of the belt is attached by means of a spring balance to a horizontal support and the other end to a weight. With the arrangement shown in the figure, when the pulley is rotated in a clockwise direction the reading of the balance is taken and the value of the weight noted. The torque due to brake friction applied to the pulley is $r(F_1 - F_2)$ where r is the the radius of the pulley, F_1 is equal to the weight and F_2 is equal to the reading on the spring balance. If the motor has a rotational frequency of f then the output power is $r(F_1 - F_2) \times 2\pi f$. The output power of the motor is converted by the brake into heat. Use this, or a similar method, to determine the output power of a motor.

5.3 Potential energy

The term *potential energy* is used for the energy stored in an object as a consequence of its position. It is the work that has to be done to move it from some reference position to a given position and so becomes the energy associated with the object at that given position. The most common form of potential energy is that due to the force of gravity,

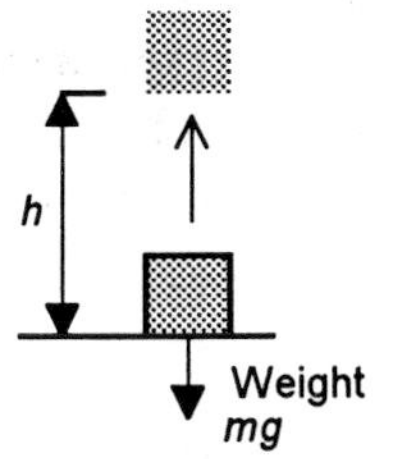

Figure 5.11 *Potential energy*

though there are other forms such as that due to a compressed or stretched spring or a compressed gas.

5.3.1 Gravitational potential energy

When work is done to raise a body of mass m against gravity through a vertical height h (Figure 5.11) the work done is the product of the force that has to be applied to overcome the weight of the body and the distance covered in the direction of the force and so is mgh. This is the energy transferred to the body as a result of its position in the gravitational field and is termed *gravitational potential energy*. Thus:

$$\text{gravitational potential energy} = mgh \quad [11]$$

Example

What is the gravitational potential energy of an object of mass 3.0 kg relative to the floor when it is lifted vertically from the floor to a height of 1.2 m above the floor?

The potential energy relative to the floor is $mgh = 3.0 \times 9.8 \times 1.2 =$ 35.28 J.

Revision

12 What is the gravitational potential energy of an object of mass 5 kg relative to the floor when it is lifted through a vertical height of 2 m above the floor.

5.3.2 Strain energy

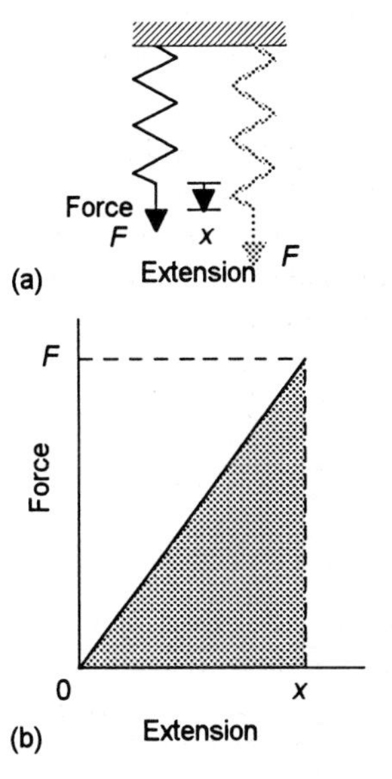

Figure 5.12 *Strain energy*

There are other forms of potential energy. Suppose we had the object attached to a spring and apply a force to the object which results in the spring being extended (Figure 5.12(a)). Work is done because the point of application of the force is moved through a distance. Thus the object gains potential energy as a result of the work that has been done. This form of potential energy is termed *elastic potential energy or strain energy*.

For a spring, or a strip of material, being stretched, the force F is generally proportional to the extension x (Figure 5.12(b)) and so the average force is $\frac{1}{2}F$ and work done = $\frac{1}{2}Fx$. This is the energy stored in the spring as a result of it being extended; it is equal to the area under the force–extension graph from an extension of 0 to x. Thus:

$$\text{strain energy} = \tfrac{1}{2}Fx \quad [12]$$

For an elastic material that obeys Hooke's law, the extension is proportional to the applied force, i.e.

$F = kx$ [13]

where k is a constant, generally known as the *stiffness*. Thus, we can write:

strain energy = $\frac{1}{2}kx^2$ [14]

Application
If you stretch a rubber band and then release it, as you are well aware, the twanging rubber band releases energy. Steel hawsers used to moor ships become stretched and store energy which if suddenly released can spring back very dangerously.

Example

What is the energy stored in a spring when a force of 200 N is needed to stretch it by 20 mm?

The strain energy is $\frac{1}{2}Fx = \frac{1}{2} \times 200 \times 0.020 = 2$ J.

Application
For a close-coiled helical spring when the wire diameter d is small compared with the mean radius R of the spring, the stiffness k is $Gd^4/64R^3n$, where G is the modulus of rigidity of the spring material, and n is the number of coils in the spring. Thus, the energy stored in such a spring when stretched to an extension x is $\frac{1}{2}(Gd^4/64R^3n)x^2$.

Example

What is the energy stored in a spring when it is stretched by 0.05 m if the force needed to stretch it is 100 N/m?

Using Hooke's law, equation [13], we have k = 100 N/m. Hence, equation [14] gives:

strain energy = $\frac{1}{2} \times 100 \times 0.05^2 = 0.125$ J

Revision

13 What is the energy stored in a spring as a result of stretching it by 0.2 m if the force needed to stretch it is 10 N/m?

14 What is the energy stored in an elastic string when it is stretched by a force of 8 N from a length of 1.0 m to 1.2 m?

5.4 Linear and angular kinetic energy

Consider an object which starting from rest is accelerated uniformly over a linear distance s. The work done is the product of the force applied and the distance covered in the direction of the force. But the force is $F = ma$ and so the work done is $Fs = mas$. After the distance s the object has a velocity v which is given by $v^2 = u^2 + 2as$ as $v^2 = 2as$. Thus the work done in accelerating the object to this velocity is:

work done = $mas = \frac{1}{2}mv^2$

This is the energy transferred to the body as a result of the work done and is called *kinetic energy*. Thus, for linear motion:

kinetic energy = $\frac{1}{2}mv^2$ [15]

Example

A body of mass 2 kg is moving in a straight line under the action of a constant force. If its velocity increases from 2 m/s to 5 m/s, what will be the work done by the force?

The work done is the increase in kinetic energy and thus is $\frac{1}{2}\, 2 \times 5^2 - \frac{1}{2}\, 2 \times 2^2 = 21$ J

Example

A car of mass 1200 kg is moving along a straight horizontal road. If the thrust of the engine is 500 N, determine the resistance to motion when the car accelerates from 10 m/s to 15 m/s in a distance of 200 m.

The work done results in an increase in kinetic energy of $\frac{1}{2} \times 1200 \times 15^2 - \frac{1}{2} \times 1200 \times 10^2 = 75$ kJ. But the work done is $Fs = (500 - F) \times 200$, where F is the resistance to motion. Hence:

$$(500 - F) \times 200 = 75 \times 1000$$

and $F = 125$ N.

Revision

15 A bullet of mass 40 g is travelling at 300 m/s when it hits a fixed wooden block and penetrates a distance of 40 mm in coming to rest. What is the average resistance force of the wood?

16 What is the work done by a constant force if its application to a body of mass 10 kg results in it moving from rest to a velocity of 3 m/s?

5.4.1 Angular kinetic energy

The work done in rotating a body through an angle θ by a tangential torque is $T\theta$. But $T = I\alpha$ and thus we can write:

$$\text{work done} = I\alpha\theta$$

As a result of this rotation from rest through angle θ, the object has an angular velocity ω. Using $\omega^2 = \omega_0^2 + 2\alpha\theta$, we have $\omega^2 = 0 + 2\alpha\theta$ and so:

$$\text{work done} = \tfrac{1}{2}I\omega^2$$

This is the energy transferred to the body as a result of the work done and is called the angular kinetic energy. Thus, for angular motion:

$$\text{angular kinetic energy} = \tfrac{1}{2}I\omega^2 \qquad [16]$$

Application
There are many applications where a motor/engine is used and has to deliver an output torque which is not constant, e.g. in a press where the torque suddenly increases during the pressing operation. A consequence of an increase in torque can be a significant decrease in the motor speed. In order to reduce the speed variations, a flywheel can be added. This stores energy, as angular kinetic energy, during periods of excess torque and releases it during the periods when extra torque is required.

Example

A wheel of moment of inertia 12 kg m^2 rotates with an angular speed of 10 rad/s. What is its angular kinetic energy?

Using equation [16], the angular kinetic energy = $\frac{1}{2}I\omega^2 = \frac{1}{2} \times 12 \times 10^2 = 600$ J.

Example

A solid cylindrical flywheel has a mass of 100 kg and a radius of 300 mm. What will be its angular kinetic energy when it is rotating at 10 rev/s?

The flywheel has a moment of inertia is $\frac{1}{2}mr^2 = \frac{1}{2} \times 50 \times 0.15^2 = 0.5625$ kg m^2. Its angular velocity is $2\pi \times 10 = 62.83$ rad/s. Thus its angular kinetic energy is $\frac{1}{2}I\omega^2 = \frac{1}{2} \times 0.5625 \times 62.83^2 = 1110$ J.

Revision

17 A solid cylindrical flywheel of mass 200 kg and radius 750 mm rotates at 20 rev/s. What is its angular kinetic energy?

18 What work is required to accelerate a wheel of moment of inertia 0.4 kg m^2 to an angular velocity of 2 rad/s?

5.5 Conservation of mechanical energy

When a mass m at rest falls a height h there is a loss in potential energy of mgh. But there will be a gain in kinetic energy of $\frac{1}{2}mv^2$, where v is the velocity at the conclusion of the fall. We must have, for energy conservation:

loss of PE = gain in KE

Application
Hydroelectric schemes can involve a large reservoir of water held back behind a dam with water then being allowed to fall through large pipes from the reservoir to impact on turbines below. The potential energy of the water behind the dams is transformed to kinetic energy at the turbines.

so that the sum of the energy of the body remains a constant, i.e.

PE + KE = a constant [17]

> If for a body there is no force other than gravity doing work then the sum of the gravitational potential energy and kinetic energy of the body remains constant.

If we have, in say an object sliding down an incline, the mechanical energy of a system reduced by work done against friction then:

initial (PE + KE)
= work done against friction + final (PE + KE) [18]

Application

Impact tests are designed to simulate the response of a material to a high rate of loading and involve a test piece being struck a sudden blow. There are two main forms of test, the *Izod* and *Charpy* tests. Both involve a pendulum (Figure 5.13) swinging down from a specified height h_0 to hit the test piece and fracture it. The height h to which the pendulum rises after striking and breaking the test piece is a measure of the energy used in the breaking. If no energy were used the pendulum would swing up to the same height h_0 it started from, i.e. the potential energy mgh_0 at the top of the pendulum swing before and after the collision would be the same. The greater the energy used in the breaking, the greater the 'loss' of energy and so the lower the height to which the pendulum rises. If the pendulum swings up to a height h after breaking the test piece then the energy used to break it is $mgh_0 - mgh$.

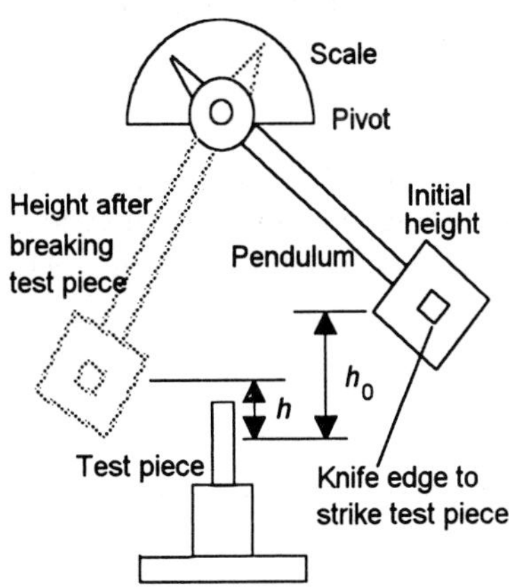

Figure 5.13 *Impact testing*

Example

What is the kinetic energy gained by an object of mass 2 kg when it has fallen through a height of 3 m?

The loss in gravitational potential energy is $mgh = 2 \times 9.8 \times 3 =$ 58.8 J. This must equal the gain in kinetic energy.

Example

An object is released from rest on a slope inclined at an angle to the horizontal such that the sine of the angle is 3/5. Determine the velocity acquired by the object after travelling a distance of 6 m down the slope.

Initially the kinetic energy is zero and the potential energy relative to its finishing position is $mgh = mg \times 6 \sin \theta = mg \times 6 \times 3/5$. At the finishing position the potential energy is zero and so to conserve mechanical energy we must have a kinetic energy of $mg \times 6 \times 3/5$. Thus:

$$\tfrac{1}{2}mv^2 = mg \times 6 \times 3/5$$

and so $v = 8.4$ m/s.

Example

A car of mass 1000 kg is driven up an incline of length 750 m and inclination 1 in 25. Determine the driving force required from the engine if the speed at the foot of the incline is 25 m/s and at the top is 20 m/s and resistive forces can be neglected.

The work done by the driving force F is $F \times 750$ J. There is a change in both potential energy and kinetic energy as a result of the work done. The gain in potential energy is $1000 \times 9.81 \times 750 \sin \theta$, where θ is the angle of elevation of the incline. Since $\sin \theta$ is given as 1/25 then the gain in potential energy is 294.3 kJ. The initial kinetic energy at the foot of the incline is $\tfrac{1}{2} \times 1000 \times 25^2 = 312.5$ kJ and at the top of the incline is $\tfrac{1}{2} \times 1000 \times 20^2 = 200$ kJ. Hence there is a loss in kinetic energy of 112.5 kJ in going up the slope. The work done must equal the total change in energy and so:

$$F \times 750 = 294.3 \times 10^3 - 112.5 \times 10^3$$

and so the driving force is 242.4 N.

Example

A solid cylinder of diameter 25 mm and length 25 mm is allowed to roll down an inclined plane, the plane being at an elevation of 20° to the horizontal. Determine the linear velocity of the cylinder when it has rolled a distance of 1.2 m. The cylinder material has a density of 7000 kg/m^3.

The loss in potential energy of the cylinder in rolling down the slope must equal the gain in linear kinetic energy plus the gain in rotational kinetic energy. The loss in potential energy is mgh with m being $\frac{1}{4}\pi \times 0.025^2 \times 0.025 \times 7000 = 0.0859$ kg and $h = 1.2 \sin 20° = 0.410$ m. The loss in potential energy is thus $0.0859 \times 9.81 \times 0.410 = 0.345$ J. The gain in linear kinetic energy is $\frac{1}{2} \times 0.0859v^2$. The gain in rotational kinetic energy is $\frac{1}{2}I\omega^2$, where $I = \frac{1}{2}mr^2 = \frac{1}{2} \times 0.0859 \times 0.0125^2 = 6.71 \times 10^{-6}$ kg m^2 and $\omega = v/r = v/0.0125$. Thus the gain in rotational energy is $\frac{1}{2} \times 6.71 \times 10^{-6} \times (v/0.0125)^2 = 0.0215v^2$. Since the loss in potential energy must equal the total gain in kinetic energy:

$$0.345 = \tfrac{1}{2} \times 0.0859v^2 + 0.0215v^2$$

and so $v = 2.31$ m/s.

Example

A square trap door, 1.8 m by 1.8 m, has a mass of 10 kg and is at rest in the horizontal position (Figure 5.14). When it is released it rotates about its hinges to strike a rubber door stop when hanging vertically. What will be its angular velocity on impact with the door stop if there is a frictional torque of 20 N m at the hinges?

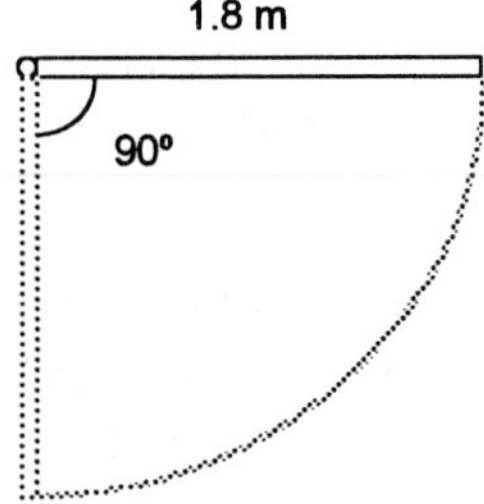

Figure 5.14 *Example*

The centre of mass of the trap door will fall through a vertical height of 0.9 m and so the loss in potential energy will be $10 \times g \times 0.9 = 88.3$ J. The gain in rotational kinetic energy is $\frac{1}{2}I\omega^2$, where the moment of inertia about the centre of mass of the door is $m \times 1.8^2/12$ and, using the theorem of parallel axes, about the edge of the door is $m \times 1.8^2/12 + m \times 0.9^2 = 1.08m$. The work done in rotating the door through 90° against the frictional torque is given by equation [5] as $T\theta = 20 \times \pi/2 = 31.4$ J. The loss in potential energy equals the gain in rotational kinetic energy plus the work done in rotating the door through 90° and so:

$$88.3 = 1.08 \times 10 \times \omega^2 + 31.4$$

and the angular velocity is 2.30 rad/s.

Example

A load of 50 kg hangs vertically from one end of a rope, the other end being wound round the rim of a drum of diameter 600 mm (Figure 5.15). The drum has a moment of inertia of 30 kg m^2. Initially the system is at rest. A torque of 200 N m is then applied to the drum and the resulting rotation lifts the load. Determine the time taken to lift the load through 30 m.

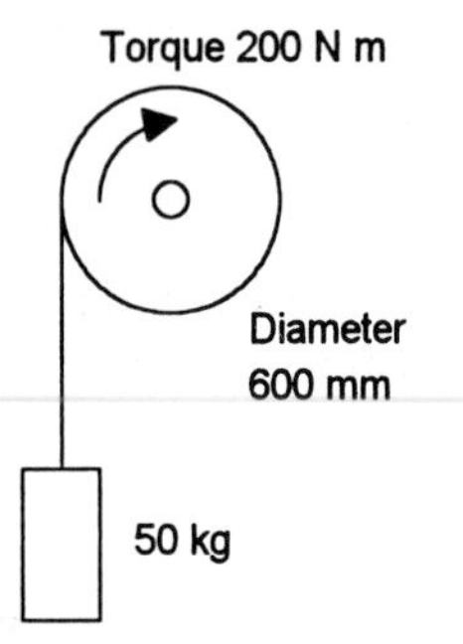

Figure 5.15 *Example*

The movement of the load by 30 m means that the rope has had to be wound up on the drum by that amount. Thus the angle through which the drum must have been rotated is given by the equation arc length = radius × angle rotated as $\theta = 30/0.300 = 100$ rad. The work done by the torque is thus $T\theta = 200 \times 100 = 20$ kJ. This work has resulted in the kinetic energy of the drum changing from its initially zero value, the kinetic energy of the load changing from its initial zero value and the potential energy of the mass changing. Thus:

$$20\,000 = \tfrac{1}{2}I\omega^2 + \tfrac{1}{2}mv^2 + mgh$$

$$= \tfrac{1}{2} \times 30 \times \omega^2 + \tfrac{1}{2} \times 50 \times v^2 + 50 \times 9.81 \times 30$$

But, if we assume no slipping, $v = r\omega$ and so:

$$20\,000 = 15 \times (v/0.300)^2 + 25v^2 + 14\,715$$

Hence $v = 5.25$ m/s. Using $v^2 = u^2 + 2as$ then $5.25^2 = 0 + 2a \times 30$ and so $a = 0.459$ m/s^2. Thus, using $v = u + at$ then $5.25 = 0 + 0.459t$ and so $t = 11.4$ s.

Revision

19 A particle of mass 8 kg is pulled 4 m up a smooth plane inclined at 30° to the horizontal. Determine the work done if the particle is pulled at a constant speed.

20 A tile slides down a roof which is at an angle to the horizontal such that its sine is 2/3. The tile reaches the edge of the roof after travelling 4.5 m and then falls vertically to the ground through a height of 7 m. With what velocity will it hit the ground?

21 A car of mass 1500 kg travels along a horizontal road against a constant resistance to motion of 500 N at a speed of 40 m/s. Determine the rate at which the engine is working.

22 A wheel with a mass of 7 kg, radius of 400 mm and radius of gyration 300 mm rolls, without slipping, from rest down an incline at an elevation to the horizontal of 30°. What will be its angular velocity when it has rolled 5 m down the slope?

23 A uniform rod of length 3 m, mass 2 kg is pivoted about that end. Initially the rod is at rest in the horizontal position. Determine its angular velocity when it is released and has swung through 90°. Neglect any frictional torque.

24 A flywheel is in the form of a solid uniform disc of radius 600 mm and mass 290 kg and is rotating at 5 rev/s. Determine the tangential force which has to be applied to the edge of the wheel to bring it to rest in 10 s.

5.6 Mechanical power transmission

In this section we take a basic look at the principles involved in the transmission of mechanical power by gears and belts.

5.6.1 Gears

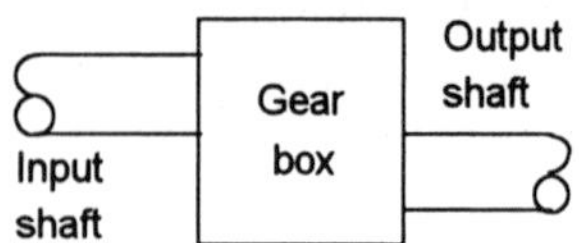

Figure 5.16 *Gear system*

A gear system can be used to change the speed of rotation of a shaft (Figure 5.16). Suppose the input shaft is rotating with an angular velocity ω_i and the output shaft with an angular velocity ω_o. The input power is ω_i, where T_i is the torque on the input shaft. The output power is $T_o\omega_o$, where T_o is the torque on the input shaft. If no power is lost, then the output power equals the input power and so:

$$T_o\omega_o = T_i\omega_i$$

and so, if we define the overall gear ratio of the system as the ratio of the angular velocity of the input shaft to the angular velocity of the output shaft:

$$\text{gear ratio} = \frac{\omega_i}{\omega_o} = \frac{T_o}{T_i} \quad [19]$$

A reduction gear box is one which reduces the angular velocities, e.g. one with a gear ratio 8 to 1, and thus converts power at a high angular speed and low torque to power at a lower angular speed and high torque.

In the above discussion, the gear system was assumed to be 100% efficient in converting the power of the input shaft to the power of the output shaft. In practice there would be some power loss. The *transmission efficiency* η is defined as:

$$\eta = \frac{\text{output power}}{\text{input power}} \quad [20]$$

Then, output power = $\eta \times$ (input power) and so:

$$T_o\omega_o = \eta T_i\omega_i$$

$$\text{gear ratio} = \frac{\omega_i}{\omega_o} = \frac{T_o}{\eta T_i} \quad [21]$$

Example

A gear box has a gear ratio of 2 to 1. If the input shaft rotates at 20 rev/s when a torque of 200 N m is applied, what will be (a) the number of revolutions per second of the output shaft and (b) the output shaft torque? Assume the system is 100% efficient.

(a) Using gear ratio = ω_i/ω_o, then, since $\omega_i = 2\pi f_i$ and $\omega_o = 2\pi f_o$ we have $f_o = f_i$/(gear ratio) = 20/2 = 10 rev/s.

(b) Using gear ratio = $\omega_i/\omega_o = T_o/T_1$, then T_o = (gear ratio) × T_i = 2 × 200 = 400 N m.

Example

If the gear box in the previous example had not been 100% efficient but 85% efficient, what would have been the output torque?

Gear ratio = $\omega_i/\omega_o = T_o/T_1$, so $T_o = 0.85 \times 2 \times 200 = 340$ N m.

Revision

25 A gear box has a gear ratio of 5 to 1. If the input shaft rotates at 40 rev/s when a torque of 300 N m is applied, what will be (a) the number of revolutions per second of the output shaft and (b) the output shaft torque if the system is 100% efficient.

(a) Using gear ratio = ω_i/ω_o, then, since $\omega_i = 2\pi f_i$ and $\omega_o = 2\pi f_o$ we have $f_o = f_i$/(gear ratio) = 20/2 = 10 rev/s.
(b) Using gear ratio = $\omega_i/\omega_o = T_o/T_1$, then T_o = (gear ratio) × T_i = 2 × 200 = 400 N m.

5.6.2 Belt drives

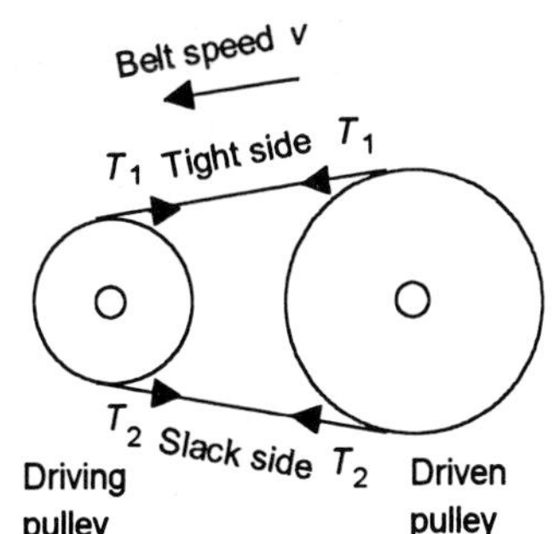

Figure 5.17 *Belt drive*

Power can be transmitted from one shaft to another by means of a continuous belt wrapped round pulleys mounted on the shafts (Figure 5.17). The belt has an initial tension when the shafts are at rest. When the driver pulley starts to rotate, frictional forces between the shafts and the belt cause the tension on one side to increase and on the other side to decrease. It is this difference in tension which is responsible for the transmission of power. If the tension on the 'tighter' side is T_1 and on the 'slacker' side is T_2 then there is a net tangential force of $(T_1 - T_2)$ acting on each pulley. If the belt has a linear speed v, the power transmitted is:

$$\text{power} = (T_1 - T_2)v \qquad [22]$$

If we assume that the belt is elastic and obeys Hooke's law, the increase in the length of the slack side must equal the reduction in the length of the tight side. The change in length of the belt will be proportional to the tension in the belt and so, if T_0 is the initial tension in the belt when it is not running, then $T_1 - T_0 = T_0 - T_2$ and so:

$$T_0 = \tfrac{1}{2}(T_1 + T_2) \qquad [23]$$

If no slipping occurs, the peripheral speed of the driving pulley must be the same as of the driven pulley. Hence, using $v = r\omega$:

driving pulley radius × driving pulley angular velocity
= driven pulley radius × driven pulley angular velocity [24]

Example

A belt is installed with an initial tension of 400 N. If the maximum permissible tight side tension is 560 N, what will be (a) the slack side tension and (b) the maximum power that can be transmitted if the smaller pulley is rotating at 10 rev/s and has a diameter of 100 mm?

(a) Using $T_0 = \frac{1}{2}(T_1 + T_2)$, then $400 = \frac{1}{2}(560 + T_2)$ and $T_2 = 240$ N.
(b) $v = r\omega = r \times 2\pi f = 0.050 \times 2\pi \times 10 = 3.14$ m/s. Hence, the maximum power that can be transmitted is:

$$\text{max. power} = (T_1 - T_2)v = (560 - 240) \times 3.14 = 1005 \text{ W.}$$

Example

The pulley on a motor shaft has a radius of 40 mm and rotates at 100 rev/s. What will be the number of revolutions per second of a shaft with a pulley of radius 160 mm which is connected to the motor shaft by a belt drive and no slipping occurs?

Since the peripheral speed will be the same for both pulleys we have driving pulley radius × driving pulley angular velocity = driven pulley radius × driven pulley angular velocity and so, as $\omega = 2\pi f$:

$$40 \times 2\pi \times 100 = 160 \times 2\pi \times f_{driven}$$

Hence the driven pulley rotates at 25 rev/s.

Revision

26 For a belt drive linking two pulleys, the tight side tension is 580 N and the slack side tension 300 N. What will be the power transmitted if the smaller pulley is rotating at 12 rev/s and has a diameter of 100 mm?

Problems

1 An object is pulled along a horizontal floor a distance of 4 m with a constant velocity. What will be the work done if the total resistance to motion is 8 N.

2 A body of mass 15 kg is pulled, by a force parallel to the slope, with a constant velocity a distance of 10 m up a slope inclined at 30° to the horizontal. What will be the work done against gravity?

3 A body of mass 50 kg is pulled, by a force parallel to the slope, with a constant velocity a distance of 5 m up a slope which rises for 3 m

for every 4 m horizontal movement. If there are frictional forces of 100 N, what will be the total work done?

4 A body of mass 50 kg is pulled, by a force parallel to the slope, with a constant velocity a distance of 15 m up a slope which rises for 3 m for every 4 m horizontal movement. If the coefficient of friction is 1/3, determine the work done.

5 The following shows the force–displacement data for a mechanism. Determine the work done when the mechanism moves from a displacement of zero to 3.0 m.

Force in N	Displacement in m
0	0
5	0.5
10	1
10	1.5
10	2
5	2.5
0	3

6 The following shows the force–displacement data for a mechanism. Determine the work done when the mechanism moves from a displacement of zero to 3.0 m.

Force in N	Displacement in m
0	0
5	0.5
10	1
15	1.5
20	2
25	2.5
30	3

7 A torque of 120 N m is applied to the shaft of a motor and rotates it through one revolution. What will be the work done?

8 What work has to be done to rotate a 300 mm diameter wheel through 10 revolutions if a force of 40 N is applied tangentially to the rim of the wheel?

9 Determine the work done per revolution for a machine that is supplied with a torque which rises at a constant rate from 100 N m at 0° to 400 N m at 90°, then decreases at a uniform rate to 100 N m at 180°, and then remains constant at 100 N m for the remaining part of the revolution.

10 A train moving along a level track has a maximum speed of 50 m/s. Determine the maximum power of the train engine if the total resistance to motion is 30 kN.

11 Determine the power required for a car to be driven along a straight road at a constant speed of 25 m/s if the resistance to motion is constant at 960 N.

12 A car is found to have a maximum speed of 140 km/h along a level road when the engine is developing a power of 50 kW. What is the resistance to motion?

13 A train draws a train of mass 250 Mg at a constant speed of 15 m/s against resistive forces of 15 kN. Determine the power needed to maintain this speed (a) along a horizontal track, (b) up an incline of 1 in 200.

14 What is the energy stored in an elastic cord of length 1.2 m when it is stretched to a length of (a) 2.2 m, (b) 2.4 m? The cord requires a force of 2 N/m to stretch it.

15 What is the energy stored in a spring when a force of 120 N is used to stretch it by 0.02 m?

16 A cylinder, of moment of inertia 12 kg m^2 and radius 200 mm, rolls along a horizontal surface with a speed of 2 m/s. What is the angular kinetic energy of the cylinder?

17 A wheel has a moment of inertia of 1.5 kg m^2. Determine its angular kinetic energy when it is rotating at 2 rad/s.

18 What is the kinetic energy acquired by a body of mass 3 kg that has fallen from a height of 10 m?

19 An object of mass 0.2 kg is thrown vertically downwards with an initial velocity of 3 m/s. What will be the velocity of the object when it has moved downwards through a distance of 4 m?

20 An object is allowed to freely fall from rest and strikes the ground with a velocity of 7 m/s. Through what height has it fallen?

21 An object is thrown vertically upwards with an initial velocity of 21 m/s. At what height will it have an upward velocity of 7 m/s?

22 An object is given an initial velocity of 6 m/s at the bottom of an incline. The incline has a slope to the horizontal such that the tangent of the angle is 4/3. If the coefficient of friction is 2/3, how far up the plane will the object slide before coming to rest?

23 Determine the power needed to raise water from a reservoir through a height of 3 m and discharge it through a nozzle of cross-sectional area 400 mm^2 at 10 m/s. Density of water = 1000 kg/m^3.

24 A pump draws water from a reservoir at the rate of 5 m^3/min and ejects it at 6 m/s through a nozzle at a height of 6 m above the level in the reservoir. Determine the power needed. Density of water = 1000 kg/m^3.

25 A flywheel has a mass of 10 Mg and a radius of gyration of 1.5 m and is running at 500 rev/min. Determine the time it will take to come to rest if a constant resisting torque of 680 N m is applied.

26 A car of mass 1200 kg stands on a slope which is at an angle to the horizontal such that the sine of the angle is 0.1. If the hand brake on the car is released and it runs down the slope, what will be its speed after travelling 200 m if the resistance to motion is 400 N?

27 In a drop-forging operation, the top die of mass 50 kg falls freely for 3 m before hitting the metal resting on the bottom die. Determine the force exerted on the metal if the top die is brought to rest in 25 mm after impact.

28 A body of mass 3 kg is connected to another body of mass 2 kg by a light inextensible string which passes over a smooth pulley. If the string is taut, what will be the speeds of the two bodies when the 3 kg object has fallen through 2 m?

29 A solid cylinder of mass 20 kg and diameter 200 mm rolls from rest down a slope which has an angle to the horizontal such that its sine is 0.5. What will be the linear speed of the cylinder when it has rolled 10 m down the slope?

30 A uniform bar of length 1 m has a mass of 10 kg and is pivoted about a horizontal axis at one end. Initially the bar is horizontal and at rest. Determine its angular velocity when it has rotated from this position through 60°. Neglect any frictional torque.

31 A gear box with an overall gear ratio of 5 to 1 has an input shaft with a power of 6 kW at 30 rev/sec. What will be the torque on the input shaft and the torque on the output shaft if the gear box is assumed to be 100% efficient?

32 A gear box with an overall gear ratio of 6 to 1 has a torque of 35 N m applied to the input shaft. What will be the torque on the output shaft if the gear box is 90% efficient?

33 The output shaft of a reduction gear rotates at one-eighth the speed of the input shaft. What will be the efficiency if the output torque is 40 N m when the input torque is 6 N m?

34 Pulleys on two parallel shafts are linked by a belt drive. If the pulley on the drive shaft has a radius of 100 mm and is rotating at 30 rev/s, what will be the belt speed if no slippage occurs?

35 The tensions on the two sides of a belt passing round a pulley are 2000 N and 500 N. If the effective radius of the pulley is 200 mm and it rotates at 10 rev/s, calculate the power transmitted by the belt.

36 A pulley of radius 400 mm is used to drive a belt. When the difference in tension between the tight and slack sides of the belt drive is 2 kN and the pulley is rotating at 5 rev/s, what is (a) the torque applied to the pulley, (b) the power transmitted?

6 Mechanical oscillations

6.1 Introduction

This chapter is about mechanical oscillations; obvious examples of such oscillations are those of a mass suspended from the end of a vertical spring (Figure 6.1(a)), the oscillations of a loaded cantilever (Figure 6.1(b)) and the angular oscillations of a simple pendulum (Figure 6.1(c)). Basically we can think of all mechanical objects as having mass and elasticity and being capable of oscillations. Such oscillations often can be detrimental and result in the failure of machine parts, excessive wear or just undesirable noise.

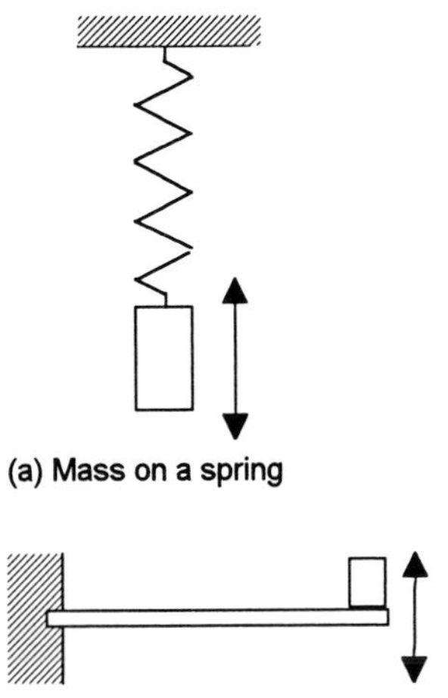

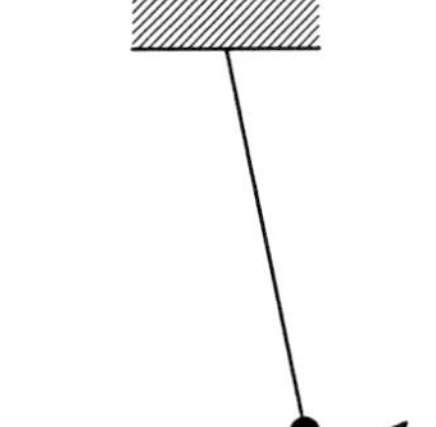

(c) Simple pendulum

Figure 6.1 *Examples of mechanical oscillation systems*

6.1.1 Basic terms

The following are basic terms used to describe oscillations:

1 *Periodic time, cycle and frequency*
Oscillatory motions are periodic, i.e. they keep on repeating themselves after equal intervals of time (Figure 6.2). The time between repetitions is called the *periodic time* and the oscillatory motion occurring within one period is called a *cycle*. The number of cycles per second is called the *frequency*. Thus, since one cycle occurs in the periodic time T, the frequency is $1/T$.

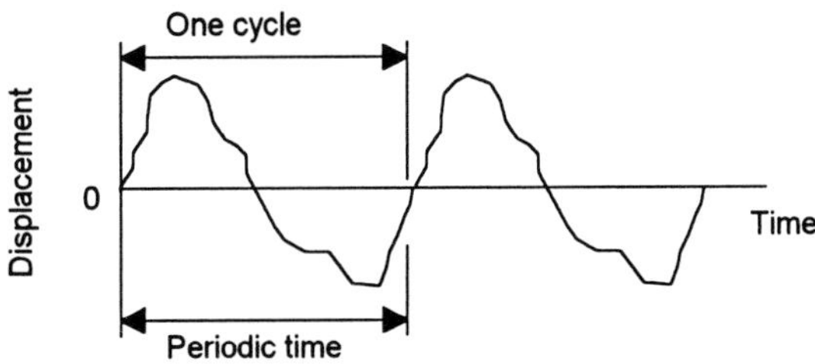

Figure 6.2 *A periodic oscillation*

2 *Free oscillations and natural frequency*
The term *free oscillation* is used when an elastic system oscillates under the action of forces inherent in the system itself with there being no externally applied forces. For example, with the spring system in Figure 6.1(a) when it is given an initial deflection to start the oscillation and then left to freely oscillate. The system will oscillate at, what is termed, a *natural frequency*, such frequencies being determined by the properties of the system.

3 *Forced oscillations*
Oscillations which take place under the effect of externally applied periodic forces are called *forced oscillations*. Thus if the support of the spring system in Figure 6.1(a) was itself oscillating it would give forced oscillations of the spring system.

4 *Damping*
Oscillating mechanical systems are all subject to damping due to energy being dissipated by friction and other resistances. With free oscillations, since no energy is externally supplied to the system when oscillating then the effect of energy being dissipated by damping is for the oscillations to die away with time. Thus if the simple pendulum in Figure 6.1(c) is set into free oscillation, the oscillation will gradually die away with time with the amplitude of the oscillation becoming progressively smaller.

6.2 Simple harmonic motion

The simplest form of periodic motion is *simple harmonic motion*. Consider a basic mechanical system of mass which when deflected from its rest position is restored to it by forces arising from elasticity in the system. Figure 5.3 shows such a system as a trolley tethered between two supports by springs. A trolley is considered for the mass in order to effectively eliminate frictional effects.

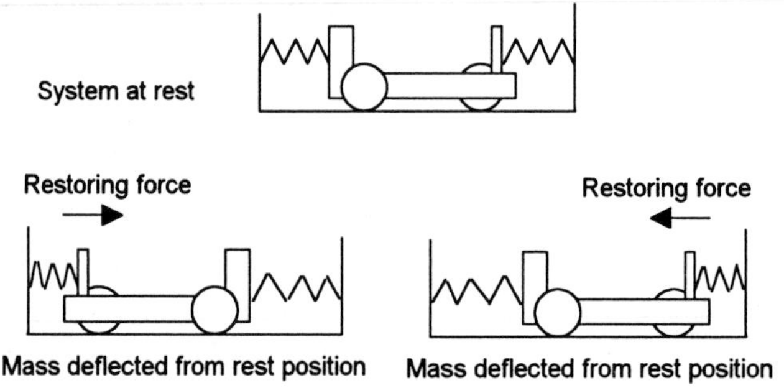

Figure 6.3 *A basic mechanical system*

When the trolley, the mass of the system, is pulled to one side then one of the springs is compressed and the other stretched and this has the effect of providing a restoring force which is directed in such a direction as to endeavour to restore the trolley back to its original position. If the trolley is released from this deflected position, the restoring force causes the trolley to move back towards its original rest position and overshoot that position. The restoring force then reverses its direction to still be directed towards the rest position and so oscillations occur. If the displacement from the rest position is measured as a function of time then the result is as shown in Figure 6.4, the displacement variation with time being described by a cosine graph.

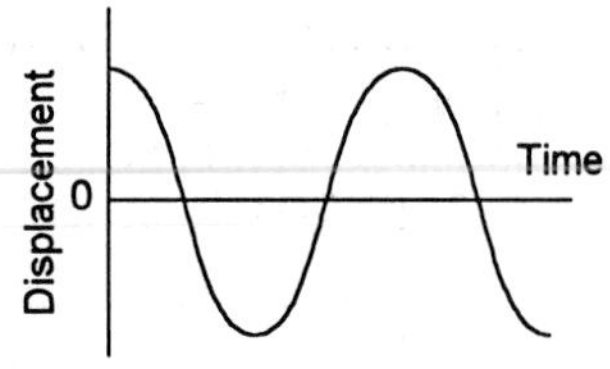

Figure 6.4 *Graph of the displacement with time*

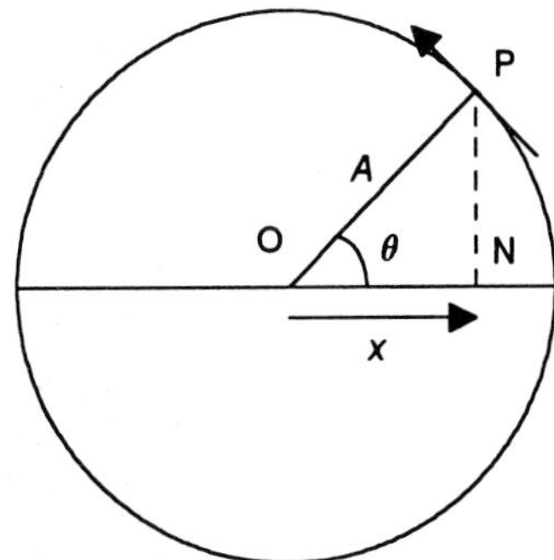

Figure 6.5 *Horizontal displacement for a point moving in a circular path with a constant angular velocity*

Such a type of displacement variation with time can be produced by the horizontal displacement x from the centre of a point P rotating in a circular path with a constant angular velocity ω (Figure 6.5), i.e.

$$x = A \cos \theta$$

where A is the amplitude of the oscillation. Since we have $\theta = \omega t$, then we can describe such oscillations by:

$$x = A \cos \omega t \qquad [1]$$

Note that the above equation has $x = A$ when $t = 0$. If we want to describe the displacement such that $x = 0$ when $t = 0$, then equation [1] becomes:

$$x = A \sin \omega t$$

The frequency f of the oscillations is the number of cycles completed per second and is thus $\omega/2\pi$. The term *angular frequency* is sometimes used for ω since $\omega = 2\pi f$ and it is thus just the frequency multiplied by 2π. Thus equation [1] can be written as:

$$x = A \cos 2\pi ft \qquad [2]$$

The linear velocity v at some instant is the rate of change of displacement dx/dt and thus, differentiating equation [1] gives:

$$v = -A\omega \sin \omega t \qquad [3]$$

Note that the above equation is when we have $x = A$ when $t = 0$ and so $v = 0$ when $x = A$. If we have $x = 0$ when $t = 0$ then we have:

$$v = A \cos \omega t$$

In Figure 5.5 we have $\sin \theta = \text{PN}/A$, and since $\text{PN}^2 = A^2 - x^2$ then $A \sin \theta = \sqrt{(A^2 - x^2)}$ and so equation [3] can be written as:

$$v = -\omega \sqrt{A^2 - x^2} \qquad [4]$$

The maximum velocity is when $x = 0$, i.e. as the mass passes through its rest position, and is:

$$\text{maximum velocity} = -\omega A \qquad [5]$$

The linear acceleration a at an instant is the rate of change of velocity $\text{d}v/\text{d}t$ and thus differentiating equation [3] gives:

$$a = -\omega^2 A \cos \omega t$$

This can be written as:

$$a = -\omega^2 x \qquad [6]$$

The acceleration has a maximum value when x equals the maximum displacement A:

$$\text{maximum acceleration} = -\omega^2 A \qquad [7]$$

The restoring force $F = ma$ and is thus given by equation [6] as:

$$F = -m\omega^2 x \qquad [8]$$

The minus sign indicates that the direction of this restoring force is always in the opposite direction to that for which x increases. Simple harmonic motion is thus defined as:

> *Simple harmonic motion* (SHM) is said to occur when the motion is under the action of a restoring force which is always directed to a fixed point and has a magnitude which is proportional to the displacement from that point.

We can write, using equation [8], the angular frequency as:

$$\omega = \sqrt{\frac{F}{mx}}$$

and thus, since the periodic time $T = 1/f = 2\pi/\omega$:

$$T = 2\pi\sqrt{\frac{\text{mass}}{\text{force per unit displacement}}} \qquad [9]$$

or:

$$f = \frac{1}{2\pi}\sqrt{\frac{\text{force per unit displacement}}{\text{mass}}} \qquad [10]$$

The larger the force needed to produce unit displacement the higher the frequency. Thus a high 'stiffness' mechanical system with its large force per unit displacement will have a high frequency of oscillation.

Example

An object moving with simple harmonic motion has an amplitude of 1.2 m and a periodic time of 3 s. Determine the maximum velocity and maximum acceleration and state at what points in the oscillation they occur.

The angular frequency $\omega = 2\pi f = 2\pi/T = 2\pi/3 = 2.09$ rad/s. Thus, using equation [5]:

$$\text{maximum velocity} = -\omega A = -2.09 \times 1.2 = -2.51 \text{ m/s}$$

The maximum velocity occurs when the displacement from the rest position is zero. The maximum acceleration is given by equation [7] as:

$$\text{maximum acceleration} = -\omega^2 A = -2.09^2 \times 1.2 = -5.24 \text{ m/s}^2$$

The maximum acceleration occurs when the displacement from the rest position is a maximum.

Example

An object moves with simple harmonic motion and has an amplitude of 500 mm and a frequency of 4 Hz. Determine the velocity and acceleration of the object when it is 200 mm from its rest position.

The angular frequency $\omega = 2\pi f = 2\pi \times 4 = 25.1$ rad/s. The velocity is thus given by equation [4] as:

$$v = -\omega\sqrt{A^2 - x^2} = -25.1\sqrt{0.5^2 - 0.2^2} = -11.5 \text{ m/s}$$

The acceleration is given by equation [6] as:

$$a = -\omega^2 x = -25.1^2 \times 0.2 = -126 \text{ m/s}$$

Example

An object starts from its rest position at O and moves with simple harmonic motion about the point O with an amplitude of 0.5 m and a periodic time of ½π s. Determine the time to, and distance, from O to a point will be where the oscillating object has a velocity of 1 m/s.

The angular frequency $\omega = 2\pi f = 2\pi/T = 2\pi/½\pi = 4$ rad/s. Thus we have $x = 0.5 \sin 4t$ and $v = a\omega \cos \omega t = -0.5 \times 4 \cos 4t$. For a velocity of 1 m/s we have $1 = -0.5 \times 4 \cos 4t$ and so $\cos 4t = -0.5$ and so $4t = \pi/3$ and $t = \pi/6$ s. The displacement $x = 0.5 \sin 4\pi/6 = 0.5 \sin 120° = 0.29$ m.

Revision

1 An object is oscillating with simple harmonic motion of amplitude 40 mm and frequency 40 Hz. Determine the velocity and the acceleration when the object is at the maximum displacement from the centre and at the centre of its oscillation.

2 The velocity of an object which is oscillating with simple harmonic motion is 4 m/s when it is 3 m from the central rest position and 3 m/s when it is 4 m. Determine the velocity of the object as it passes through the central rest position.

3 An object of mass 0.1 kg oscillates with simple harmonic motion of frequency 15 Hz. Determine the restoring force acting on the object when it is at a displacement of 30 mm.

4 An object oscillates with simple harmonic motion of amplitude 25 mm. If the period of oscillation is π s, determine the maximum speed and maximum acceleration.

6.2.1 Energy of simple harmonic motion

The velocity of an object when oscillating with simple harmonic motion and at a displacement x from its central rest position is given by equation [4] as:

$$v = -\omega\sqrt{A^2 - x^2}$$

Thus the kinetic energy is:

$$\text{kinetic energy} = ½mv^2 = ½m\omega^2(A^2 - x^2) \qquad [11]$$

Because the restoring force is proportional to the displacement, the work done to move the object from its central rest position to a displacement x is given by the average force acting over that displacement multiplied by the displacement. The force is zero at the central position and given by equation [8] as $m\omega^2 x$ at displacement x. Thus:

$$\text{work done} = \tfrac{1}{2}m\omega^2 x^2$$

Thus the potential energy of the object when displaced by x is:

$$\text{potential energy} = \tfrac{1}{2}m\omega^2 x^2 \qquad [12]$$

The total energy at this displacement is the sum of the potential and kinetic energies (equations [11] and [12]) and thus is:

$$\text{energy} = m\omega^2 A^2 \qquad [13]$$

The total energy is thus constant at all displacements, depending only on the amplitude of the oscillation. What varies at different displacements is the fraction of the energy that is kinetic energy and the fraction that is potential energy. As an object oscillates there is a continual changing of potential energy to kinetic energy and vice versa with, in the absence of losses or inputs to the system, i.e. damping or forcing, the sum remaining constant.

Example

Determine the energy of an object of mass 0.5 kg oscillating with simple harmonic motion of amplitude 100 mm and frequency 5 Hz.

The angular frequency $\omega = 2\pi f = 2\pi \times 5 = 31.4$ rad/s. Thus, using equation [13]:

$$\text{energy} = m\omega^2 A^2 = 0.5 \times 31.4^2 \times 0.1^2 = 4.93 \text{ J}$$

Revision

5 An object of mass 0.1 kg oscillates with simple harmonic motion of frequency 15 Hz and amplitude 50 mm. Determine the total energy of the object.

6.3 Undamped oscillations

The following are derivations of the natural frequencies of a number of mechanical systems executing oscillations when damping is assumed to be negligible. The derivations are based on the technique of determining how the restoring force varies with displacement and then, provided the restoring force is proportional to the displacement and directed towards the rest position, using equation [9] to obtain the frequency.

Figure 6.6 *Mass on a spring*

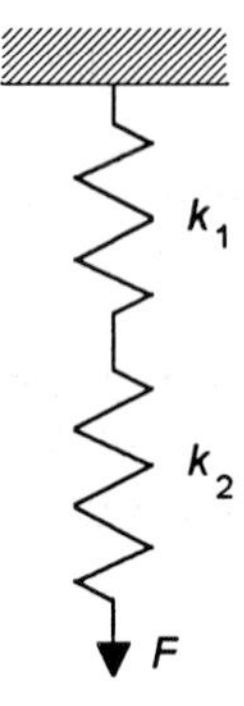

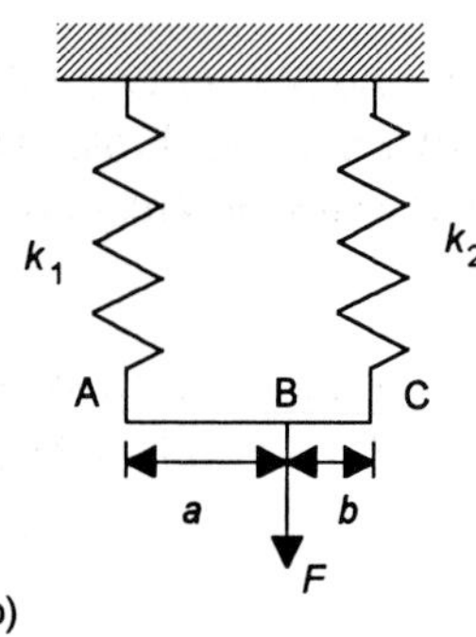

Figure 6.7 *Springs in (a) series, (b) parallel*

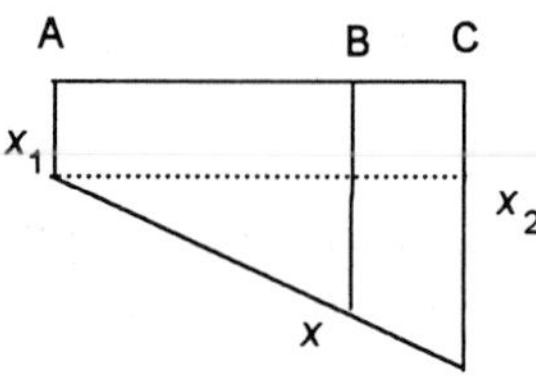

Figure 6.8 *Deflections*

6.3.1 Mass on a spring

Consider a mass suspended from a vertical spring (Figure 6.6), the mass of the spring being assumed to be negligible. It is made to oscillate in the vertical direction by the mass being pulled down, so extending the spring, and then the mass is released. The spring then exerts a restoring force on the mass. Assuming that the spring obeys Hooke's law, then the restoring force F is proportional to the displacement x of the end of the spring from its rest position. The force is always directed towards the rest position and thus we can write:

$$F = -kx$$

where k is the spring stiffness. The motion is simple harmonic because F is proportional to $-x$. The magnitude of the force per unit displacement is k and thus, using equation [9], i.e.

$$f = \frac{1}{2\pi}\sqrt{\frac{\text{force per unit displacement}}{\text{mass}}}$$

then:

$$f = \frac{1}{2\pi}\sqrt{\frac{k}{m}} \qquad [14]$$

For two springs in series (Figure 6.7(a)), one having a stiffness k_1 and the other a stiffness k_2, if a force F is applied then the same force must be applied to each spring and so the series arrangement will stretch by $x = x_1 + x_2$, where $x_1 = F/k_1$ and $x_2 = F/k_2$. Thus the overall system will have a stiffness k of:

$$k = \frac{F}{x} = \frac{1}{\frac{1}{k_1} + \frac{1}{k_2}} = \frac{k_1 k_2}{k_1 + k_2} \qquad [15]$$

For two springs in parallel (Figure 6.7(b)), one having a stiffness k_1 and the other a stiffness k_2, if a force F is applied at point B then if we take moments about A we have $Fa = F_2(a + b)$, where F_2 is the force acting on the spring with stiffness k_2 and if we take moments about C we have $Fb = F_1(a + b)$, where F_1 is the force acting on the spring with stiffness k_1. The deflection of point A is thus $x_1 = F_1/k_1 = Fb/k_1(a + b)$ and the deflection of point C is $x_2 = F_2/k_2 = Fa/k_2(a + b)$. As indicated by Figure 6.8, the deflection x of point B is:

$$x = x_1 + \frac{a}{a+b}(x_2 - x_1)$$

$$= \frac{Fb}{k_1(a+b)} + \frac{a}{a+b}\left(\frac{Fa}{k_2(a+b)} - \frac{Fb}{k_1(a+b)}\right)$$

$$= \frac{F}{(a+b)^2}\left(\frac{a^2}{k_2} + \frac{b^2}{k_1}\right)$$

and thus the effective stiffness k of the spring arrangement is:

$$k = \frac{F}{x} = \frac{(a+b)^2}{\left(\frac{a^2}{k_2} + \frac{b^2}{k_1}\right)} \qquad [16]$$

If $k_1 = k_2$ and $a = b$, then $k = 2k_1$.

Example

A spring supports a carrier of mass 2 kg and when a 10 kg mass is placed on the carrier the spring extends by 50 mm. The carrier and load are then pulled down a further 75 mm and released. Determine the frequency of the oscillations for vertical oscillations.

A force of $10g$ N causes an extension of 50 mm and thus, assuming the spring obeys Hooke's law, the stiffness $k = 10g/0.050 =$ 1962 N/m. The frequency of oscillation is given by equation [14] as:

$$f = \frac{1}{2\pi}\sqrt{\frac{k}{m}} = \frac{1}{2\pi}\sqrt{\frac{1962}{12}} = 2.04 \text{ Hz}$$

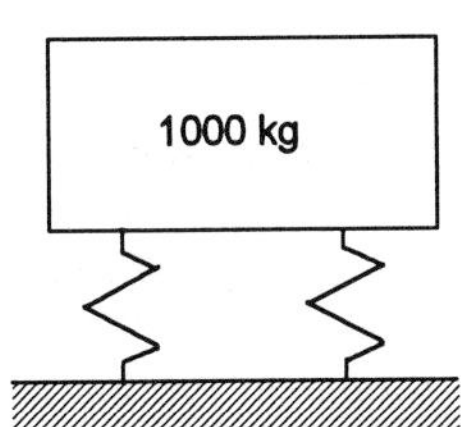

Figure 6.9 *Example*

Example

A machine of mass 1000 kg is mounted centrally on two rubber pads, its centre of mass being central (Figure 6.9). Each pad has a stiffness of 500 kN/m. Determine the natural frequency of oscillation of the system.

The total stiffness of the supports is (see note under equation [16]) $2 \times 500 = 1000$ kN/m. Thus the frequency is given by equation [14] as:

$$f = \frac{1}{2\pi}\sqrt{\frac{k}{m}} = \frac{1}{2\pi}\sqrt{\frac{1000 \times 1000}{1000}} = 5.0 \text{ Hz}$$

Revision

6 A mass of 100 kg is suspended from a vertical spring and stretches it by 50 mm. Determine the natural frequency of the system when it is set oscillating vertically.

7 A mass of 10 kg is suspended from a vertical spring with a spring constant of 2 kN/m. Determine the natural frequency of the system when it is set oscillating vertically.

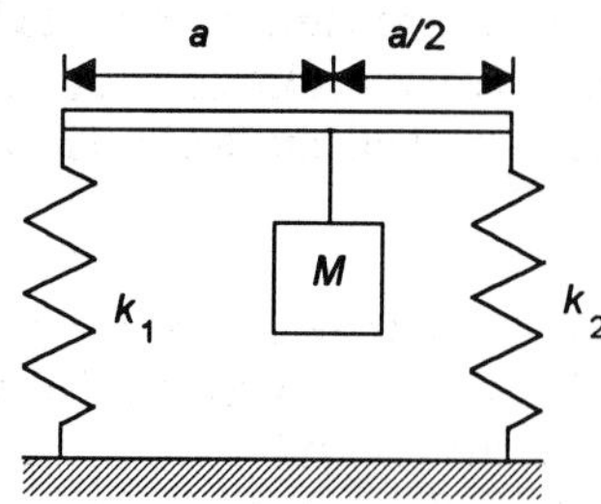

Figure 6.10 *Revision problem 8*

8 Determine the natural frequency for vertical oscillations of the system shown in Figure 6.10.

9 Determine the natural frequency for vertical oscillations of the system shown in Figure 6.11, it being a mass of 3 kg suspended by two series connected springs of stiffness 20 kN/m and 60 kN/m.

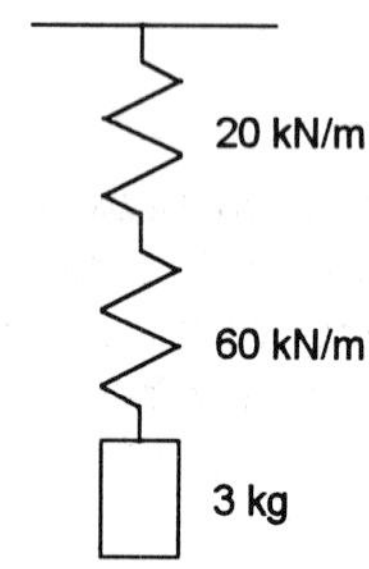

Figure 6.11 *Revision problem 9*

Activity

Suspend a hanger or scale pan from the end of a spiral spring. Put a mass m on the hanger or in the scale pan, then pull it down a small distance and allow it to oscillate vertically. Determine the time taken for a number of oscillations, e.g. 20, and hence obtain a value for the periodic time T. Repeat the experiment for different values of m. Equation [14] can be written as:

$$T^2 = \frac{4\pi^2}{k}M$$

where M is the mass of the hanger/scale pan plus m. Hence, if you plot a graph of T^2 against M, the graph should be a straight line with a slope of $4\pi^2/k$. Use such a graph to obtain a value for the stiffness k. Check the value you obtain by loading the spring and measuring the static deflections resulting from the loads, hence plotting a graph of load against extension and obtaining k from the slope of the graph.

6.3.2 Cantilever

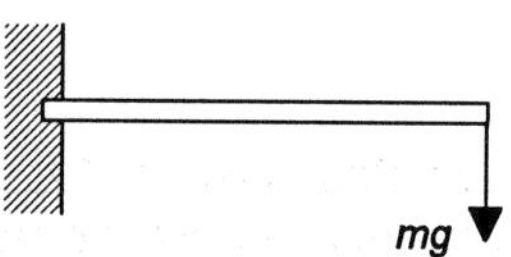

Figure 6.12 *Cantilever*

Consider a cantilever of negligible mass with a point load of mass m at its free end (Figure 6.12), the load resulting in a vertical deflection d from the horizontal. Since the deflection is proportional to the load we have a stiffness $k = mg/d$. If the cantilever is now pulled down a further distance x at the free end, then the restoring force is $F = -kx = -mgx/d$. The restoring force is proportional to the displacement and thus when the cantilever is released it performs oscillations with simple harmonic motion. Hence, using equation [9], i.e.

$$f = \frac{1}{2\pi}\sqrt{\frac{\text{force per unit displacement}}{\text{mass}}}$$

then:

$$f = \frac{1}{2\pi}\sqrt{\frac{mg/d}{m}} = \frac{1}{2\pi}\sqrt{\frac{g}{d}} \qquad [17]$$

Example

A machine is placed on a horizontal steel girder which is fixed at both its ends. As a result of placing the machine on the girder it deflects by 2 mm at the point where the machine is mounted. Estimate the natural frequency of the transverse oscillations that can occur if the mass of the girder is neglected and the deflection is proportional to the load.

As with the cantilever, the deflection d is proportional to the load and so we have a stiffness $k = mg/d$. This is the same as the theory derived above for a load at the end of a cantilever and so we can use the same equation, namely equation [17], for the estimate. Thus:

$$f = \frac{1}{2\pi}\sqrt{\frac{g}{d}} = \frac{1}{2\pi}\sqrt{\frac{9.81}{0.002}} = 11.1 \text{ Hz}$$

Revision

10 What will be the frequency of transverse oscillations of the cantilever when a mass of 2 kg is attached to the free end? The stiffness of the cantilever is such that when a force of 9 N is applied to the free end of a cantilever a deflection of 75 mm is produced.

Activity

Clamp one end of a metre rule to the edge of a bench. Suspend a scale pan or hanger from the free end. Place a mass m in the scale pan or on the hanger and determine the amount d by which the free end moves downwards. Then deflect the free end through a small vertical distance and then allow it to freely oscillate. Determine the time for, say, 20 oscillations and hence obtain a value for the periodic time T. Repeat this for a number of different loads. Equation [17] can be written as:

$$T^2 = \frac{4\pi^2}{g} d$$

A graph of T^2 against d should be a straight line with a slope of $4\pi^2/g$. From a measurement of the slope, obtain a value for the acceleration due to gravity.

6.3.3 Simple pendulum

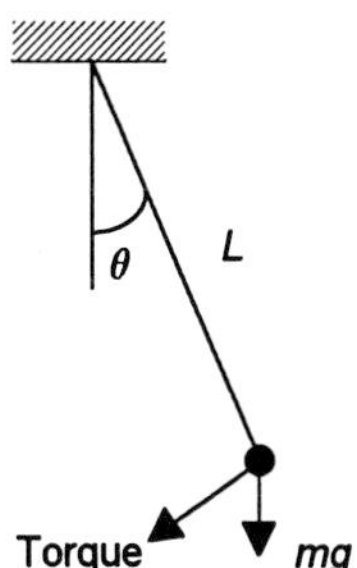

Figure 6.13 *Pendulum*

Consider a simple pendulum in which the mass m of the pendulum acts at a distance L from the point of suspension (Figure 6.13). Note the use of the term 'simple', this is because we will make the assumption that all the mass of the pendulum is concentrated in a small bob on the end of a string of negligible mass; if these assumptions cannot be made the pendulum is termed compound and the analysis of the motion of the pendulum has to be modified from that which follows. When the pendulum is pulled through a small angle θ from the vertical, there is a resultant torque acting on the pendulum bob which arises from the component of the weight mg which is tangential to the arc of motion of the bob, i.e. $mg \sin \theta$. The torque is thus $-mgL \sin \theta$. Since we are only considering small angles $\sin\theta \approx \theta$ and thus the torque is approximately $-mgL\theta$. This torque produces an angular acceleration α, where:

$$\text{torque} = -mgL\theta = I\alpha$$

where I is the moment of inertia of the system. Since the pendulum is a mass concentrated at a distance L from the axis of rotation, the moment of inertia is mL^2. Thus:

$$-mgL\theta = mL^2\alpha$$

The angular acceleration is thus proportional to the angular displacement $-\theta$. This compares with the statement already used for simple harmonic motion where the linear acceleration is proportional to the displacement and we consider angular simple harmonic motion to be occurring. Note that we only obtain simple harmonic motion when we can make the approximation $\sin\theta \approx \theta$. The above equation gives:

$$\text{angular acceleration} = -\frac{g}{L}\theta \qquad [18]$$

This is the angular equivalent of equation [6], i.e.

$$\alpha = -\omega^2\theta \qquad [19]$$

and thus, since $\omega = 2\pi f$:

$$f = \frac{1}{2\pi}\sqrt{\frac{g}{L}} \qquad [20]$$

Note that equation [19] gives $T = I\alpha = -I\omega^2\theta$ and so the equivalent of equation [9] for angular simple harmonic motion as:

$$f = \frac{1}{2\pi}\sqrt{\frac{\text{torque/unit angular displacement}}{\text{moment of inertia}}} \qquad [21]$$

Example

How does the periodic time of a simple pendulum with a length of 1.21 m compare with that for a pendulum with length 1.00 m?

Using equation [20], since the periodic time is the reciprocal of the frequency, we have:

$$\text{periodic time} \propto \sqrt{L}$$

and thus:

$$\text{ratio of the periodic times} = \sqrt{\frac{1.21}{1.00}} = 1.10$$

Revision

11 A pendulum which has a periodic time of 1.000 s when its support is stationary is placed in a lift which ascends with an acceleration of 0.02 m/s. What effect will this have on the periodic time?

Activity

Construct a simple pendulum by suspending a small ball of metal from the end of a light thread. Measure the length L of the pendulum from its point of support to the centre of gravity of the bob. Determine the time taken for, say, 20 oscillations when the pendulum bob is given a small displacement. Hence, obtain a value for the periodic time T. Repeat the experiment for a number of different values of L. Equation [20] can be written as:

$$T^2 = \frac{4\pi^2}{g}L$$

Thus a graph of T^2 against L will be a straight line with a slope of $4\pi^2/g$. Hence, from a measurement of the slope of such a graph, obtain a value for the acceleration due to gravity g.

6.3.4 Torsional oscillations

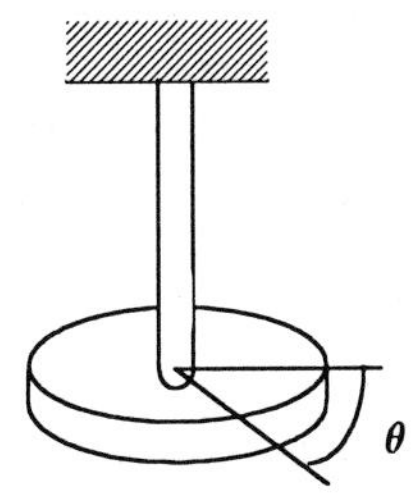

Figure 6.14 *Torsional oscillations*

Consider a disc, with moment of inertia I about an axis at right angles to its plane and through its centre, which is mounted on a central axial slender shaft (Figure 6.14). When the disk is rotated through some angle θ, the shaft is twisted. The restoring torque T produced by this twist is proportional to θ (see Chapter 3 and equation [10], i.e. $T = G\theta J/L$ with G = shear modulus, J = polar second moment of area and L = length), thus:

$$\text{restoring torque} = -k\theta$$

where k is the torque per radian twist or torsional stiffness (see note above, it is GJ/L). When the system is released and allowed to oscillate then:

$$\text{restoring torque} = I\alpha = -k\theta$$

The oscillation is thus angular simple harmonic and so, using equation [21]:

$$f = \frac{1}{2\pi}\sqrt{\frac{k}{I}} \qquad [22]$$

Example

A flywheel of mass 30 kg and radius of gyration 150 mm about its central axis is attached centrally to one end of a slender shaft, the other end being effectively clamped. The shaft has a torsional stiffness of 1.6 kN m/rad. Determine its frequency of oscillations when the flywheel is given an initial angular displacement and then released.

The moment of inertia of the flywheel is:

$$I = mk^2 = 30 \times 0.150^2 = 0.675 \text{ kg m}^2$$

Thus, using equation [22]:

$$f = \frac{1}{2\pi}\sqrt{\frac{1600}{0.675}} = 7.75 \text{ Hz}$$

Revision

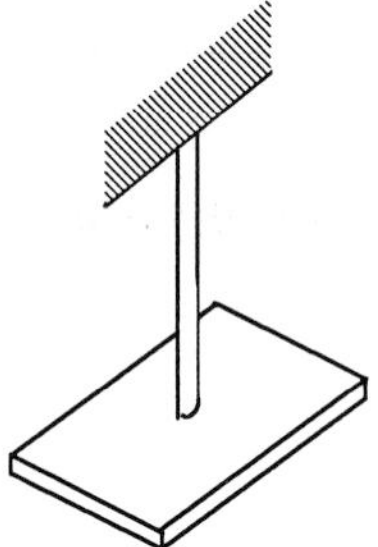

Figure 6.15 *Revision problem 12*

12 A rectangular plate 200 mm by 300 mm and having a mass of 10 kg is fixed by its centre to the end of a slender rod having a torsional stiffness of 1.5 N m/rad (Figure 6.15). Determine the natural frequency of oscillation when the plate is given an angular displacement and then released.

6.4 Damped oscillations

So far in this chapter in the discussion of oscillations, it has been assumed that the only force acting on an object performing oscillations is the restoring force. However, there are other forces, since if there was only the restoring force an object would continue oscillating for ever. Thus if the oscillations of a mass suspended from a fixed support are considered; when the mass is pulled down and then released it oscillates with the amplitude of each successive oscillation becoming smaller (Figure 6.16), eventually stopping as a result of frictional forces. When damping occurs there is a loss of energy from the oscillating system as work is done against such forces as friction or air resistance. Since the total energy of an oscillating system is (equation [13]):

$$\text{energy} = m\omega^2A^2$$

then diminishing this energy as a result of energy loss means the amplitude decreases. The oscillation is said to be *damped*, the amount of damping determining how rapidly the amplitude diminishes.

Figure 6.16 shows how the displacement varies with time with moderate damping. If the damping is increased there comes a point at which the oscillations just fail to materialise and the displacement becomes of the form shown in Figure 6.17. Such damping is termed *critical damping*. Critical damping is defined as the value of damping needed for there to be no oscillations and the displacement reach zero in the minimum time. If the damping is increased beyond this, the time to

Application

Car shock absorbers have damping mechanisms deliberately introduced so that the vibrations of the suspension relative to the car frame are rapidly damped out. Without this, passengers in their seat would oscillate for many oscillations every time the car hits a bump in the road. With an instrument pointer of a moving galvanometer, when a current occurs the coil of the instrument is given a deflection and can start oscillating. Such oscillations are generally undesirable since we want the instrument pointer to quickly settle down to the steady-state reading. Thus damping is introduced. This typically involves using a metal former for the galvanometer coil, its motion in the magnetic field of the instrument resulting in eddy currents being produced which oppose the coil motion.

reach zero displacement is increased; the system is said to be *over damped*. The system with less damping than critical damping gives oscillations and is said to be *under damped*.

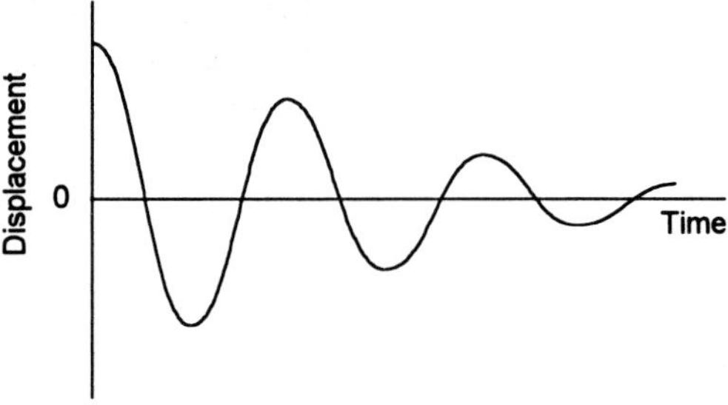

Figure 6.16 *Damping reducing the amplitude of the oscillation*

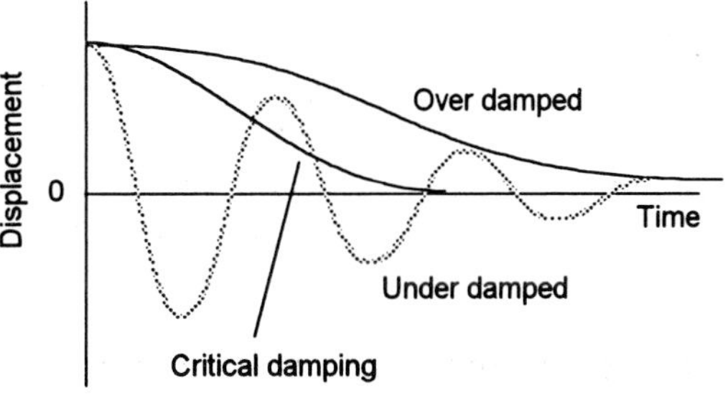

Figure 6.17 *Effect of different degrees of damping*

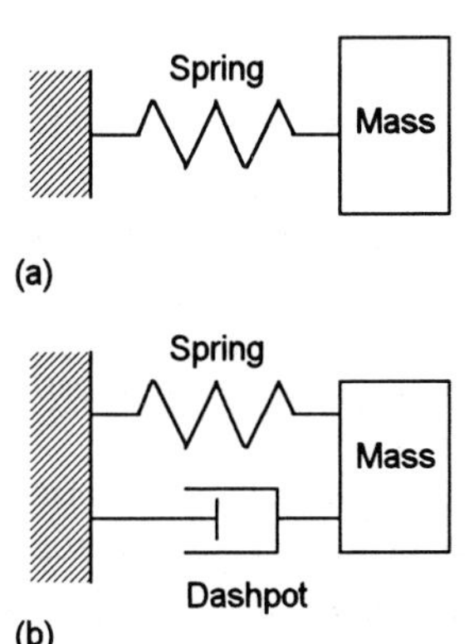

Figure 6.18 *Models for (a) undamped, (b) damped systems*

A basic model for an undamped oscillating system is one which has just mass and elasticity (Figure 6.18(a)). We can modify the model to include damping by introducing a dashpot (Figure 6.18(b)). When the piston of the dashpot moves, air has to either escape from or get into the trapped air space. The result is a resistance to motion which depends on the velocity with which the piston moves, the faster it moves the greater the resistance to motion. With the undamped system, the restoring force acting on the mass is $-kx$ and the free-body diagram is as shown in Figure 6.19(a). Thus, for the acceleration a of the mass m, we have:

$$-kx = ma$$

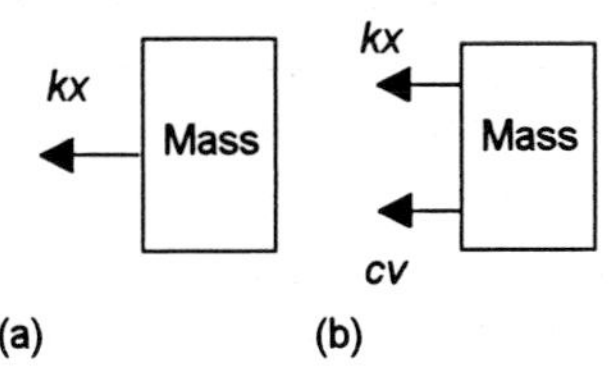

Figure 6.19 *Free-body diagrams for (a) undamped, (b) damped systems*

With the damped system, there is a damping force acting on the mass which is typically proportional to the velocity v (this is normally the case where damping is due to air resistance, the viscosity of a liquid or electromagnetic induction) and written as cv, where c is a constant called the damping constant. The free-body diagram is as shown in Figure 6.19(b). Thus we have:

$$-kx - cv = ma$$

Example

A mass oscillating on the end of a spring is damped so that it loses 16/25 of its energy in each cycle. How does the amplitude vary with time?

Equation [13] gives the energy as $m\omega^2A^2$. Thus the amplitude is proportional to the square root of the energy. The energy in successive cycles is E and $9/25E$. Thus the ratio of the amplitude in successive cycles is $\sqrt{(9/25)} = 3/5$. Each successive amplitude is three-fifths that of the previous oscillation.

6.5 Forced oscillations

If you give a child on a swing an initial push, the oscillations will usually have a periodic time of about two or three seconds and will gradually diminish in amplitude unless further pushes are given. To build up large amplitude oscillations it is necessary for the pushes applied to the swing to be at just the right time intervals. If we tried to push several times a second then the amplitude of the oscillation would be quite small; similarly if we pushed just once every ten seconds. However, it we push each time the swing is at its maximum displacement, i.e. with the same periodic time as that of the swing, then very large amplitude oscillations can be built up.

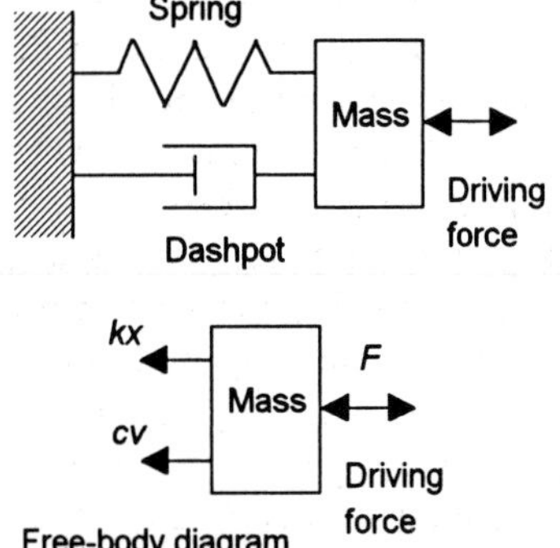

Figure 6.20 *Model for forced oscillation system*

The above represents a situation where an object with a natural frequency is supplied with an externally applied periodic force. The resulting oscillations are said to be *forced oscillations*. The frequency of the applied external force is called the *driver frequency*. Figure 6.20 shows the basic model for such a system.

With the child on the swing being pushed, the maximum amplitude oscillation occurs when the applied frequency of the 'driver' is exactly equal to the natural frequency with which the system freely oscillates. This special condition is termed *resonance*.

Figure 6.21 shows how the amplitude of the oscillation depends on the driver frequency for different values of damping in the system. When the largest amplitude oscillations occur the system is in resonance. The less the damping the larger the amplitude at the resonance frequency and the 'sharper' the graph, i.e. the large response over a narrow range of frequencies. With more damping there is a smaller response and the graph is less 'sharp'. With very low damping, resonance occurs when the driver frequency is equal to the natural frequency of the oscillator. At resonance with low damping, relatively small forces can cause very large amplitude oscillations.

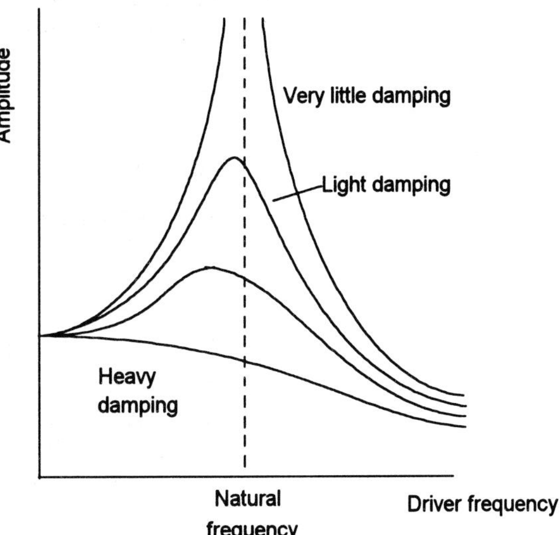

Figure 6.21 *The amplitude of oscillations as a function of driver frequency and damping*

Application
A notable example of resonance was provided by the failure of the Tacoma Narrows Bridge in 1940. The bridge was set into oscillation by relatively small forces produced by winds of about 60 km/h. Because the structure had very little damping and the natural frequency of the bridge was close to frequency of the oscillations brought about by the wind, the bridge was set into large amplitude oscillations and collapsed.

Resonance is a phenomenon that occurs in many situations. For example, if the wheel of a car is out of balance it can oscillate like a mass on a spring, i.e. the car suspension, and so has a natural frequency. At certain speeds there might be a forcing frequency applied which is close enough to the natural frequency to cause relatively large amplitude oscillations and the driver of the car can more readily perceive the oscillations, even though the road surface over which the car was being driven was not particularly rough. Aeroplanes are tested, by shaking them at a wide range of frequencies, to determine their resonant frequencies and so ensure that they are unlikely to occur when the plane is in service. If they did occur, the resulting large amplitude oscillations could result in damage to the plane. Soldiers marching in step over a bridge apply a forcing frequency and, though the force applied is small, if it is near to the natural frequency it can result in large amplitude oscillations which destroy the bridge. This happened in 1831 in Manchester when 60 marching soldiers broke the Brough suspension bridge over the River Irwell. In 1850 a French infantry battalion marched in step across the Angers suspension bridge and the resulting oscillations destroyed it with, as a consequence, the deaths of 226 of the soldiers. Soldiers marching across bridges have to break step to avoid the possibility of such an event occurring.

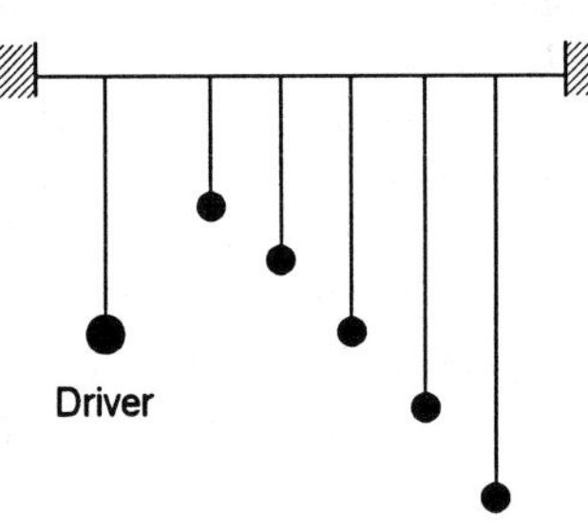

Figure 6.22 *Activity*

Activity

Suspend a number of pendulums with light bobs and of different lengths from a cord, as shown in Figure 6.22. Then suspend a pendulum with a more massive bob from the cord, this being the driver pendulum. Set the driver pendulum in oscillation and observe what happens to the driven pendulums. Which of them has the greatest amplitude? Why?

Problems

1 An object oscillates with simple harmonic motion of frequency 4 Hz and amplitude 150 mm. Determine the maximum velocity of the object.

2 An object oscillates with simple harmonic motion. At a displacement of 1 m from its central rest position the acceleration is 200 m/s^2. Determine the frequency of the oscillation.

3 An object oscillates with simple harmonic motion of frequency 4 Hz and amplitude 150 mm. Determine the velocity and acceleration of the particle at a displacement of 90 mm.

4 An object oscillates with simple harmonic motion. Determine the periodic time if the acceleration is 4 m/s^2 when the displacement from the centre of the oscillation is 2 m.

5 An object oscillates with simple harmonic motion and has a periodic time of 8 s and amplitude 4 m. Determine the maximum velocity and the velocity when the object is 2 m from its central rest position.

6 An object of mass 0.2 kg oscillates with simple harmonic motion of frequency 10 Hz and amplitude 70 mm. Determine the energy of the oscillating object.

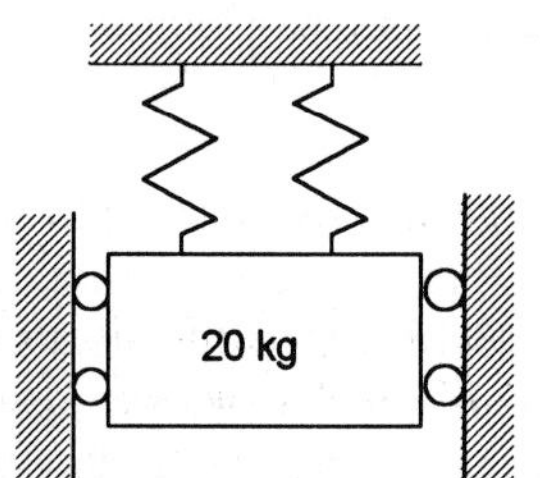

Figure 6.23 *Problem 8*

7 A mass of 5 kg is suspended from a fixed support by a vertical spring. If the natural frequency of oscillation is 2 Hz for vertical oscillations, what is the stiffness of the spring?

8 Determine the natural frequency of oscillation for the system shown in Figure 6.23 of a mass of 20 kg suspended by two identical symmetrically spaced springs, each having a stiffness of 500 N/m. The arrangement is constrained so that only vertical oscillations can occur.

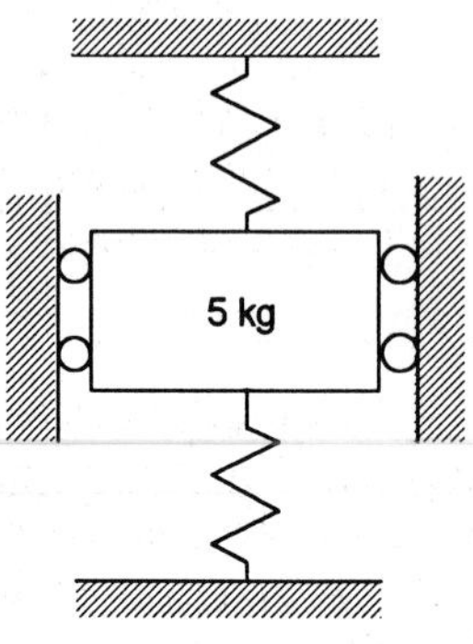

Figure 6.24 *Problem 9*

9 Determine the natural frequency of oscillation for the system shown in Figure 6.24 of a mass of 5 kg suspended by two identical springs, each having a stiffness of 500 N/m. The arrangement is constrained so that only vertical oscillations can occur.

10 A horizontal platform of mass 2 kg rests symmetrically on four vertical springs, each of stiffness 15 kN/m. Determine the natural frequency for vertical oscillations.

11 When four people with a combined mass of 300 kg get into a car, the springs are observed to compress by 50 mm. If the total load

supported by the springs is 900 kg, determine the natural frequency of the system for vertical oscillations.

12 A simple pendulum has a periodic time of 1.000 s where the acceleration due to gravity is 9.81 m/s^2. What will be its periodic time when it is taken to a place where the acceleration due to gravity is 9.85 m/s^2?

13 A flywheel of moment of inertia 5 kg m^2 about an axis perpendicular to its face and through its centre is attached centrally to one end of a slender shaft, the other end being effectively clamped. The shaft has a torsional stiffness of 500 N m/rad. Determine its frequency of oscillations when the flywheel is given angular displacement and then released.

14 A disc with a moment of inertia of 0.06 kg m^2 about an axis through its centre and at right angles to its face is attached at its centre to the lower end of a vertical rod, the other end of the rod being rigidly fixed. The rod has a uniform circular cross-section of radius 4.5 mm, a length of 1.75 m and modulus of rigidity of 80 GPa. When the disc is given an angular displacement and released, what will be the frequency of the oscillations?

15 A U-tube is partially filled with a liquid so that the total length of the liquid in the tube is L. When the liquid is set oscillating in the tube, show that the frequency of the oscillations is given by:

$$f = \frac{1}{2\pi}\sqrt{\frac{L}{2g}}$$

16 A cylinder with a uniform cross-section is floating vertically in a liquid with a depth h immersed. When the cylinder is pushed down a little way and released, oscillations occur. Show that the frequency of the oscillations is given by:

$$f = \frac{1}{2\pi}\sqrt{\frac{g}{h}}$$

17 A small coin rests on a horizontal platform which is oscillating with simple harmonic motion of frequency 2.5 Hz. What is the maximum amplitude of this motion which will allow the coin to remain in contact with the platform at all times?

18 Figure 6.3 shows a trolley of mass m on a horizontal table and tethered by springs at each end to fixed supports. For the springs initially taut and with identical spring stiffness k, and neglecting frictional effects, determine the frequency of the oscillation when the trolley oscillates back-and-forth along the line of the springs.

19 A mountaineer of mass 80 kg is on the end of a rope of length 35 m when he loses his hold and becomes suspended by the rope. If the rope stretches by 1.6 m under his weight, what will be the periodic time of his oscillations if (a) they are vertical oscillations, (b) he swings like a pendulum bob?

20 A mass oscillating on the end of a spring is damped so that it loses half of its energy in each cycle. How does the amplitude vary with time?

21 An oscillating system has a damping mechanism which dissipates 36% of the oscillator energy each complete oscillation. If the initial amplitude was 30 mm, what will be the amplitude after 1, 2 and 3 oscillations?

7 D.c. theory

7.1 Introduction

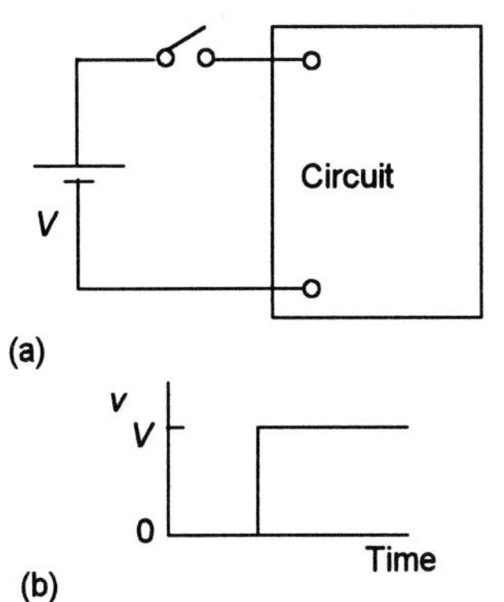

Figure 7.1 *Step voltage input*

This chapter is a review of the fundamental principles of d.c. circuits, the behaviour of the components used in such circuits, namely resistors, capacitors and inductors, and a brief consideration of d.c. motors.

It starts with a consideration of basic circuits involving resistors and the use of Kirchhoff's laws in their analysis. The circuits used as voltage and current dividers are discussed.

The chapter then takes a look at fundamental principals involved with capacitors and inductors and their behaviour in d.c. circuits. It is concerned with the current and voltage changes that occur in a circuit when there is a change in applied voltage as a result of perhaps when a d.c. voltage V applied to some circuit is switched on (Figure 7.1(a)), the voltage input to the circuit rising abruptly from 0 to V when the switch is closed. Such an input voltage is termed a *step voltage* and is represented by Figure 7.1(b). The current and voltages that occur in the circuit while they are reacting to such an input and their changing to their steady state values are termed *transients*. Such transients occur whenever the applied voltage changes and thus can occur, for example, when we apply a constantly changing input voltage or switch on a sinusoidal alternating voltage. In this chapter we restrict the consideration to the transients produced in circuits when subject to a step input voltage.

Finally the basic principles and characteristics of d.c. motors are discussed.

7.1.1 Basic terms

The following are basic terms used in connection with electrical circuits:

1 *Current*

Current is defined as the rate of movement of charge in a circuit. When there is a current I then the charge Q being moved in a time t is given by $I = Q/t$. Current has the unit of ampere or amp (A), the unit being defined as that constant current which, when flowing in two straight parallel conductors of infinite length, of negligible cross-section, and placed 1 m apart in a vacuum, produces between the conductors a force of 2×10^{-7} N per metre length.

2 *Charge*

The unit of charge is the coulomb (C) and is defined as the quantity of electricity passing a point in a circuit when a current of 1 A flows for 1 s. In an electric circuit the charge carriers through electrical

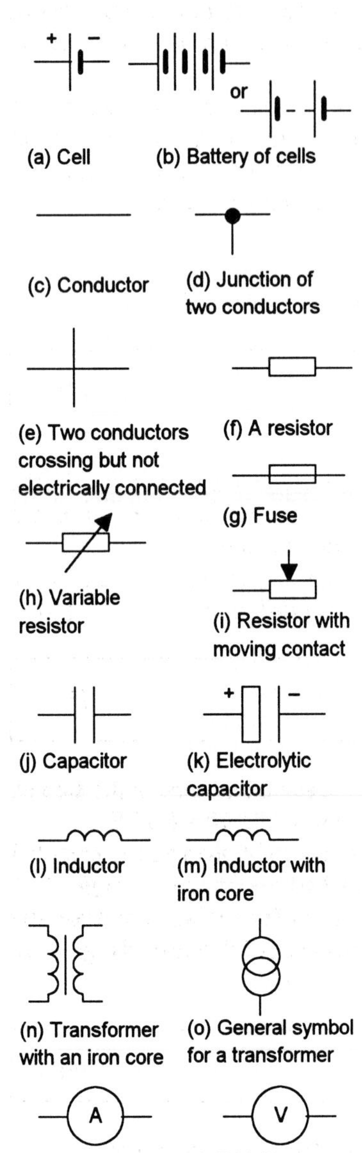

Figure 7.2 *Basic symbols*

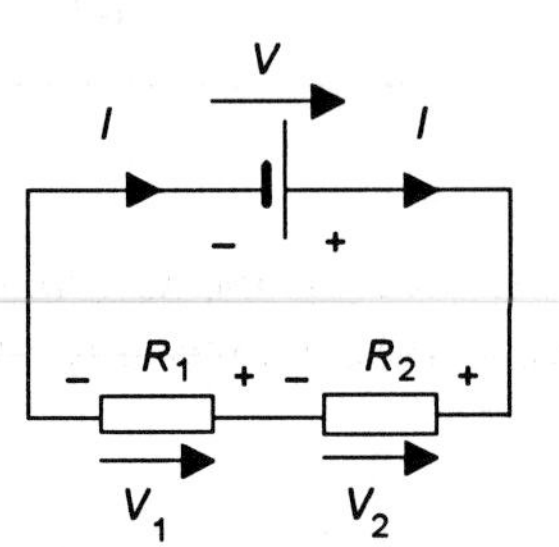

Figure 7.3 *Circuit notation*

conductors are electrons. One coulomb of charge is the total charge carried by 6.24×10^{18} electrons.

3 *Electrical power*
When there is a current through a circuit then charge is continuously being moved through it. Energy has to be continually supplied to keep the charge moving. The rate at which this energy is required is called the *power P*. Thus if an energy W is required over a time t then $P = W/t$.

4 *E.m.f*
To establish a flow of charge through a circuit, it is necessary to exert some sort of force on the charge carriers to cause them to move. This is termed the *electromotive force* (e.m.f.). A battery is a source of e.m.f.; chemical energy is transformed in the battery into energy which is used to move the charge carriers round the circuit. The e.m.f. is a measure of the potential energy per charge that is given by the battery. The unit of e.m.f. is the volt (V) and an e.m.f. of 1 V is when the battery gives 1 J to each coulomb of charge.

5 *Potential difference*
Energy is required to establish a flow of charge through a component. The term *voltage* or *potential difference* is used for the energy required to move a unit charge between two points in a circuit and has the unit of volt (V). Thus if V is the potential difference between two points in a circuit then the energy W required to move a charge Q between the points is given by $V = W/Q$. Thus we can write:

$$P = \frac{W}{t} = \frac{Q}{t} \times \frac{W}{Q} = IV \qquad [1]$$

A potential difference of 1 V is said to exist between two points of a conducting wire carrying a constant current of 1 A when the power dissipated between these points is equal to 1 W.

7.1.2 Circuit symbols and sign conventions

Figure 7.2 shows basic symbols used in electrical circuit diagrams. In circuit diagrams (Figure 7.3), the convention adopted is to indicate the direction of the current in a circuit conductor by an arrow with the direction of the current being taken as flowing out of the positive plate of a cell (the long line on the cell symbol), round the circuit and back into the negative plate (the short fat line on the symbol). This is the convention that was originally adopted by those who first investigated electrical currents and who thought that current flow was caused by the motion of positive charge carriers rather than negative charge. Thus they assumed that the current flowed from the positive terminal to the negative terminal and this convention for labelling current direction is still used. Now we know that in conductors the current flow is by electrons, negative charge carriers, and so the direction of the flow of electrons is in the opposite direction to the direction used to label the current in an electrical circuit.

The direction of the potential difference V between two points in a circuit is indicated by an arrow between the points and which is parallel to the conductor (Figure 7.3). The arrow for the potential difference points towards the point which is taken to be more positive.

7.2 Resistors in d.c. circuits

Application
There are a number of types of fixed value resistors.
Carbon composition resistors are made by mixing finely ground carbon with a resin binder and insulating filler, compressing it into a cylindrical shape, and then firing it. Copper leads are provided at each end of the cylindrical shape and the resistor coated with an insulating plastic film. The resistance of the resistor depends on the ratio of carbon to insulating filler.
Film resistors are made by depositing an even film of a resistive material on a ceramic rod. The materials used may be carbon (carbon-film resistors), nickel chromium (metal film resistors), a mixture of metals and glass (metal glaze resistors), or a mixture of a metal and an insulating oxide (metal oxide resistors). The required resistance value is obtained by cutting a spiral in the film.
Wire-wound resistors are made by winding resistor wire, such as nickel chromium or a copper-nickel alloy, onto a ceramic tube. The whole resistor is then coated with an insulator. Wire-wound resistors have power ratings from about 1 W to 25 W, the other forms tending to have ratings of about 0.25 W.

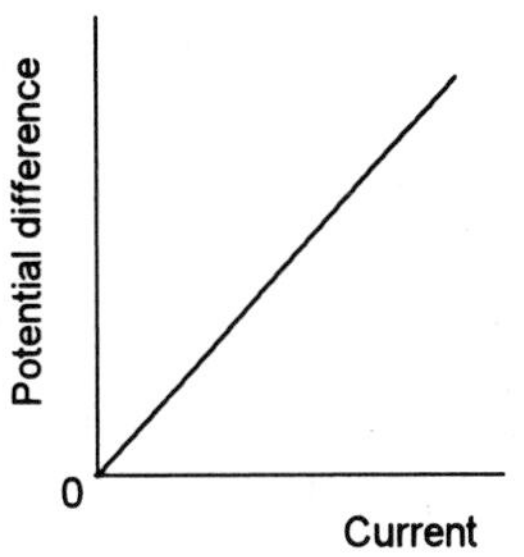

Figure 7.4 *Circuit element obeying Ohm's law*

The electrical *resistance* R of a circuit element is defined by:

$$R = \frac{V}{I} \qquad [2]$$

where V is the potential difference across an element when I is the current through it. The unit of resistance is the ohm (Ω) when the potential difference is in volts and the current in amps.

It is sometimes more convenient to use the reciprocal value of a resistance, i.e. $1/R$. This reciprocal value is called the *conductance* (G).

$$G = \frac{1}{R} = \frac{I}{V} \qquad [3]$$

With the resistance in ohms, the unit for the conductance is the siemen (S). Thus a resistance of 10 Ω is a conductance of 1/10 = 0.10 S.

For many resistors, if the temperature does not change, the potential difference across a resistor is proportional to the current through it. A graph of potential difference V against current I is a straight line passing through the origin (Figure 7.4) with a constant of V/I. Hence the gradient is the resistance R. Thus we can write:

$$V = RI \qquad [4]$$

This is called *Ohm's law*. Since the power $P = IV$, we can write:

$$P = IV = I^2R = \frac{V^2}{R} \qquad [5]$$

Example

A 200 Ω resistor has a power rating of 2 W. What is the maximum current that can be used with the resistor without exceeding the power rating?

Using $P = I^2R$:

$$I = \sqrt{\frac{P}{R}} = \sqrt{\frac{2}{200}} = 0.1 \text{ A}$$

The maximum current is thus 0.1 A or 100 mA.

Revision

1 An electric lamp dissipates 2 W when the current through it is 100 mA. What is (a) the potential difference across the lamp, (b) the lamp resistance?

2 A 200 Ω resistor has a power rating of 250 mW. What should be the maximum current used with it?

7.2.1 Resistors in series

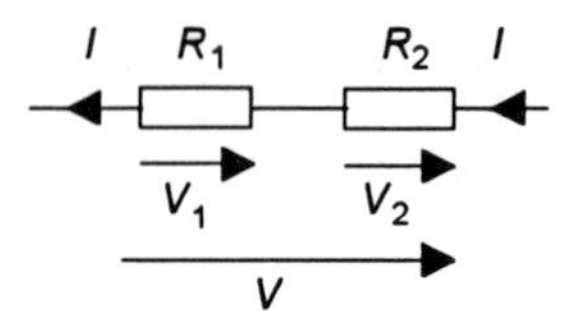

Figure 7.5 *Resistors in series*

When resistors are in *series* (Figure 7.5) the same current I flows through each resistor. Thus the potential difference V_1 across resistance R_1 is IR_1. The potential difference V_2 across R_2 is IR_2. The potential difference V across the series arrangement is the sum of the potential differences and so:

$$V = V_1 + V_2 = IR_1 + IR_2 = I(R_1 + R_2)$$

We could replace the two resistors by a single, equivalent, resistor R if we have $V = IR$. Hence:

$$R = R_1 + R_2 \qquad [6]$$

Application
A common form of indicator used with microprocessor systems is the light-emitting diode (LED). A series resistor has generally to be used with the LED, some indeed come with built-in series resistors, as typically a LED is limited to a current of about 20 mA and has a resistance such that there is then a voltage drop of about 2.1 V across it. This means that when used with a microprocessor giving a 5 V output, 2.9 V has to be dropped across a series resistor. This means a resistance of about 2.9/0.020 = 145 Ω is used.

Example

A circuit has resistors of 10 Ω, 20 Ω and 40 Ω connected in series with a current of 0.1 A flowing through them. What is (a) the total circuit resistance, (b) the potential difference across each resistor?

(a) The total resistance is 10 + 20 + 40 = 70 Ω.
(b) The potential differences across each of the resistors are 10 × 0.1 = 1.0 V, 20 × 0.1 = 2.0 V, 40 × 0.1 = 4.0 V.

Revision

3 A circuit has resistors of 100 Ω, 200 Ω and 500 Ω connected in series. If the potential difference across the circuit is 10 V, what will be (a) the circuit current, (b) the potential differences across each resistor?

4 The current through a light-emitting diode (LED) has to be limited to 5 mA. If the LED has a resistance of 250 Ω, what resistance should be connected in series with it when it is used with a 5 V supply?

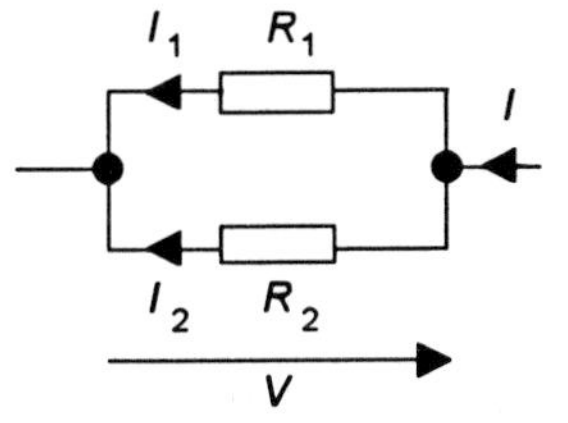

Figure 7.6 *Resistors in parallel*

7.2.2 Resistors in parallel

For circuit elements in *parallel* (Figure 7.6), the same potential difference occurs across each element and, since charge does not accumulate at a junction, the current entering a junction must equal the sum of the current leaving it, i.e. $I = I_1 + I_2$. For resistance R_1 we have $V = I_1R_1$ and for resistance R_2 we have $V = I_2R_2$, thus:

$$I = I_1 + I_2 = \frac{V}{R_1} + \frac{V}{R_2} = V\left(\frac{1}{R_1} + \frac{1}{R_2}\right)$$

We could replace the two resistors by a single, equivalent, resistor R if we have $I = V/R$. Hence:

$$\frac{1}{R} = \frac{1}{R_1} + \frac{1}{R_2} \qquad [7]$$

We can write this equation in terms of conductances as:

$$G = G_1 + G_2 \qquad [8]$$

Example

A circuit consists of two resistors of 5 Ω and 10 Ω in parallel with a voltage of 2 V applied across them. Determine (a) the total circuit resistance, (b) the current through each resistor.

(a) The total resistance R is given by:

$$\frac{1}{R} = \frac{1}{R_1} + \frac{1}{R_2} = \frac{1}{5} + \frac{1}{10} = \frac{3}{10}$$

Hence $R = 3.3$ Ω.
(b) The same potential difference will be across each resistor. Thus the currents are $I_1 = V/R_1 = 2/5 = 0.4$ A = 400 mA and $I_2 = V/R_2 = 2/10 = 0.2$ A = 200 mA.
(c) The total current entering the arrangement can be obtained by either adding the currents through each of the branches or determining the current taken by the equivalent resistance. Thus, $I = 0.4 + 0.2 = 0.6$ A or 600 mA, or $I = V/R = 2/3.3 = 0.61$ A or 610 mA. Rounding errors account for the differences in answers.

Revision

5 A circuit consists of two resistors of 10 Ω and 30 Ω in parallel. If the potential difference across the parallel arrangement is 12 V, what will be the current through each resistor?

6 What is the total resistance of a circuit which consists of resistances of 50 Ω, 100 Ω and 200 Ω in parallel?

7.2.3 Circuits with both series and parallel resistors

For a circuit consisting of both series and parallel components, we can use the rules for combining series resistors and for combining parallel resistors to simplify each part of the circuit in turn and so obtain a simpler equivalent circuit. As an illustration, consider the circuit shown in Figure 7.7(a). As a first step we can reduce the two parallel resistors to their equivalent, thus obtaining circuit (b) with $R_p = R_2R_3/(R_2 + R_3)$. We then have R_1 in series with R_p and so can obtain the equivalent resistance $R_e = R_1 + R_2R_3/(R_2 + R_3)$ and circuit (c).

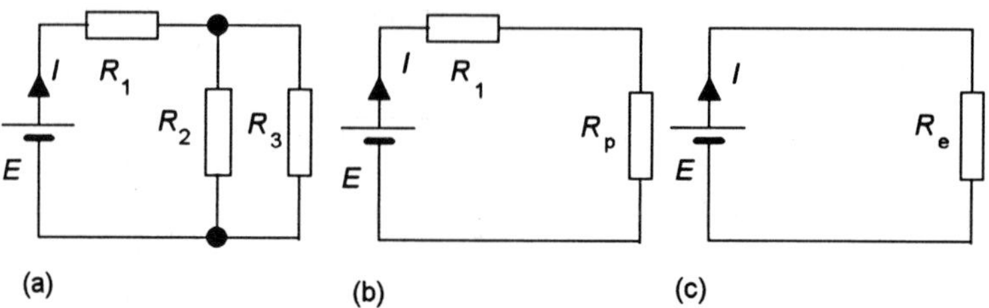

Figure 7.7 *(a) Circuit, (b) first reduction, (c) second reduction*

Figure 7.8(a) gives another illustration. As a first step we can reduce the two series resistors to their equivalent, thus obtaining circuit (b) with $R_s = R_1 + R_2$. We then have R_s in parallel with R_3 and so can obtain the equivalent resistance $R_e = R_sR_3/(R_s + R_3) = (R_1 + R_2)R_3/(R_1 + R_2 + R_3)$.

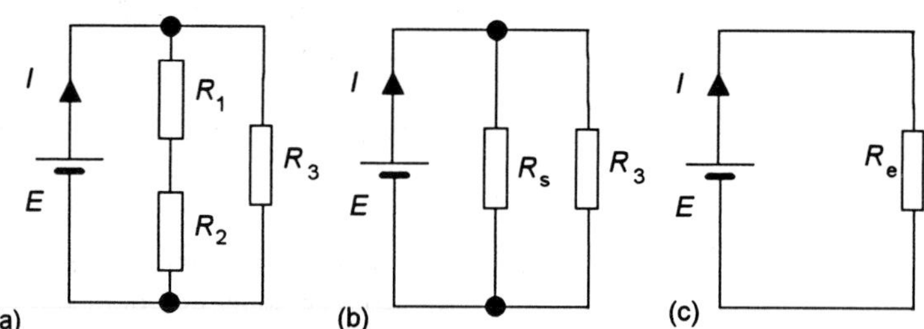

Figure 7.8 *(a) Circuit, (b) first reduction, (c) second reduction*

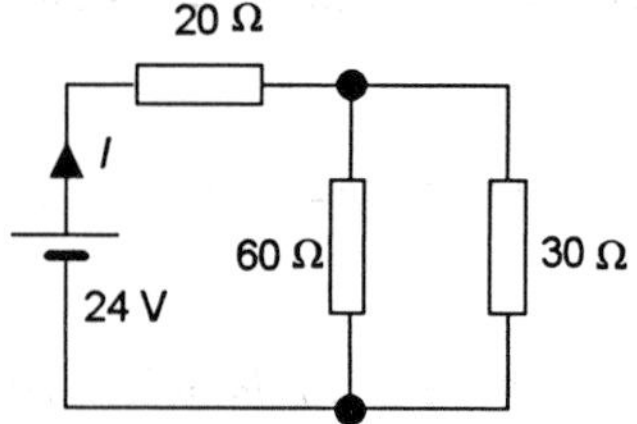

Figure 7.9 *Example*

Example

Determine the current I taken from the voltage source in the circuit given by Figure 7.9.

For the two resistors in parallel we have:

$$\frac{1}{R_p} = \frac{1}{60} + \frac{1}{30} = \frac{1+2}{60} = \frac{1}{20}$$

Hence the circuit reduces to that in Figure 7.10(a). For the two resistors in series, the equivalent resistance is 20 + 20 = 40 Ω. We now have the circuit shown in (b) and so I is $E/R_e = 24/40 = 0.6$ A.

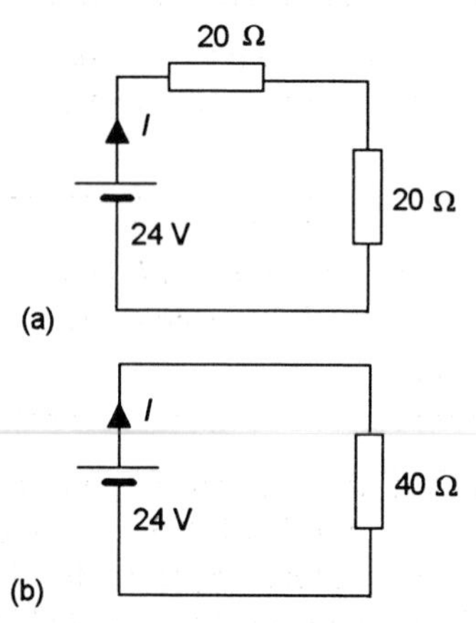

Figure 7.10 *Example*

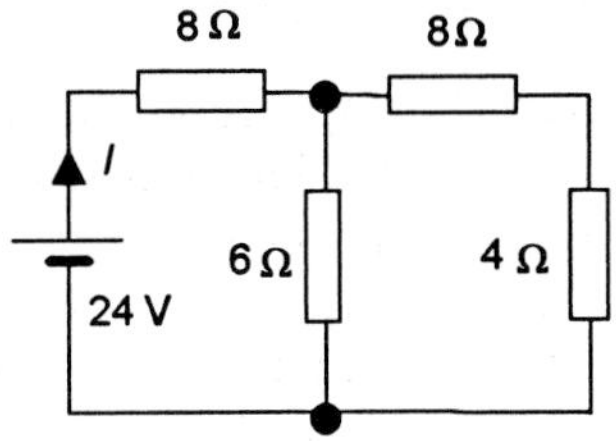

Figure 7.11 *Example*

Example

Determine the current I taken from the voltage source in the circuit given by Figure 7.11.

The equivalent resistance for the two series resistors is 8 + 4 = 12 Ω. This gives the simpler circuit shown in Figure 7.12(a). The equivalent resistance for the two parallel resistors is 12 × 6/(12 + 6) = 4 Ω. This gives Figure 7.12(b). The equivalent resistance for the two series resistors is 8 + 4 = 12 Ω. Hence we end up with the equivalent circuit shown in Figure 7.12(c). The current I is thus 24/12 = 2 A.

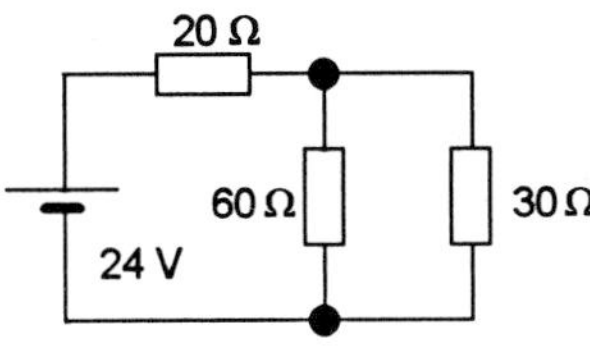

Figure 7.13 *Revision problem 7*

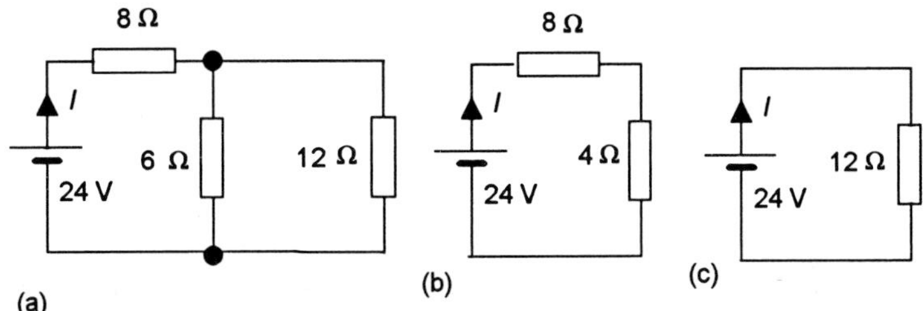

Figure 7.12 *Example*

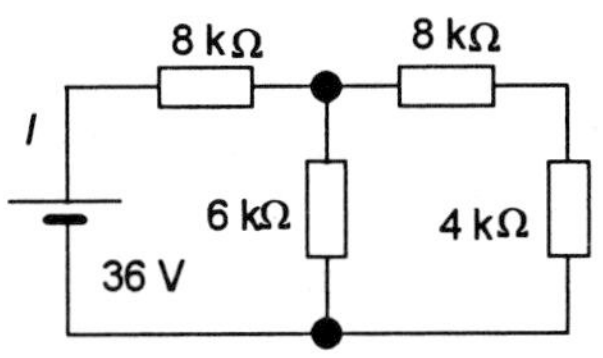

Figure 7.14 *Revision problem 8*

Revision

7 Determine the voltage across the 60 Ω resistor in the circuit shown in Figure 7.13.

8 Determine the current I in the circuit shown in Figure 7.14.

7.3 Kirchhoff's laws

The fundamental laws we use in circuit analysis are *Kirchhoff's laws*. The laws can be stated as:

> *Kirchhoff's current law states that at any junction in an electrical circuit, the current entering it equals the current leaving it.*

> *Kirchhoff's voltage law states that around any closed path in a circuit, the sum of the voltage drops across all the components is equal to the sum of the applied voltage rises.*

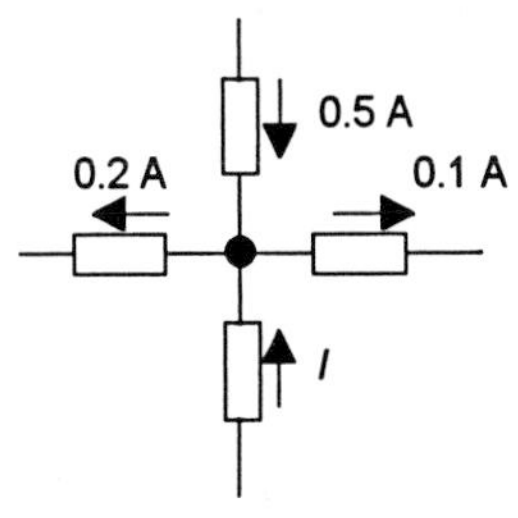

Figure 7.15 *Currents at a junction*

As an illustration of the current law, consider the circuit junction shown in Figure 7.15. The current entering the junction is 0.5 + I A and the current leaving it is 0.2 + 0.1 A and so the current I must be –0.2 A. This means the current I is in the opposite direction to that indicated in the figure.

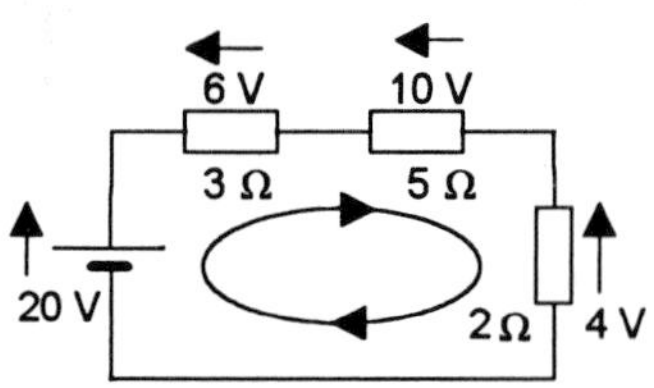

Figure 7.16 *Voltages round a closed loop*

As an illustration of the voltage law, consider the circuit shown in Figure 7.16. A closed loop is any path that begins and ends at the same point. Thus, proceeding round the loop shown in the figure we have a voltage rise of 20 V and voltage drops of 6 + 10 + 4 V. The sum of the voltage drops equals the sum of the voltage rises.

7.3.1 Voltage and current division

Resistors in series can be used as a simple method of voltage division. Consider two resistors R_1 and R_2 in series, as illustrated in Figure 7.17 with a voltage V applied across them. Kirchhoff's voltage law tells us that the voltage drops across the two resistors will be equal to the supply voltage V and so we have the voltage V divided between the two resistors.

The total resistance across which the voltage is applied is $R_1 + R_2$. Thus, the current I through the series resistors is:

$$I = \frac{V}{R_1 + R_2}$$

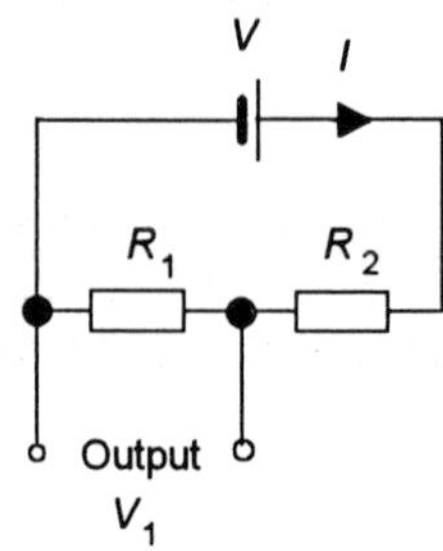

Figure 7.17 *Voltage divider circuit*

The voltage across resistor R_1 is IR_1. Hence if this voltage is taken as the output voltage, we have:

$$V_1 = \frac{R_1}{R_1 + R_2} V \qquad [9]$$

Thus the output voltage is the fraction $R_1/(R_1 + R_2)$ of the input voltage. The result is thus a *voltage-divider circuit.*

It is often necessary to obtain a continuously variable voltage from a fixed supply voltage. For this a resistor with a sliding contact is used, such a circuit element being known as a *potentiometer* (Figure 7.18). The position of the sliding contact determines the ratio of the resistances R_1 and R_2 and hence the output voltage.

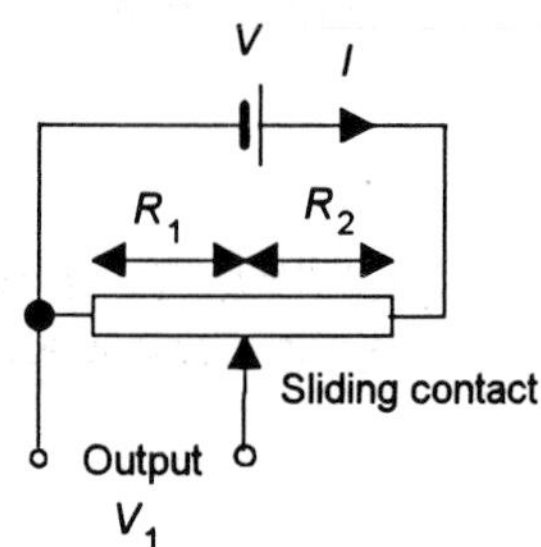

Figure 7.18 *Potentiometer*

The voltage division rule can be stated as:

> *The voltage across any resistance in a series circuit is the source voltage multiplied by the ratio of that resistance to the total resistance of the circuit.*

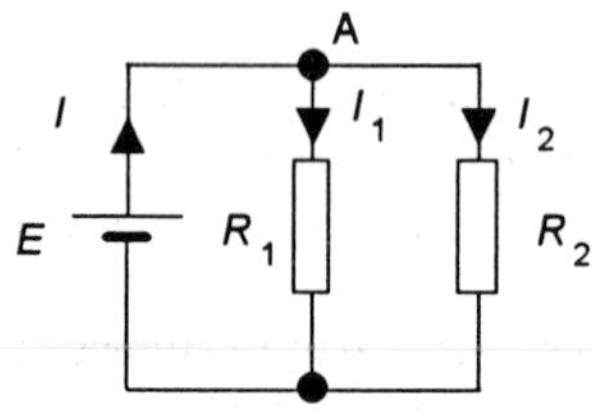

Figure 7.19 *Current division with parallel circuit*

For a parallel circuit we have a *current division* rule. Consider the circuit shown in Figure 7.19. Kirchhoff's current law tells us that the current entering junction A will equal the current leaving it.

$$I = I_1 + I_2$$

Thus the current input to the junction has been divided between the two exit paths from the junction.

The equivalent resistance for the two parallel resistors is $R_e = R_1R_2/(R_1 + R_2)$ and so $E = IR_e = IR_1R_2/(R_1 + R_2)$. The potential drop across each resistor is the same, namely E. Thus the current I_1 through resistor R_1 is:

$$I_1 = \frac{E}{R_1} = \frac{IR_e}{R_1} = I\frac{R_1R_2}{R_1(R_1+R_2)} = I\frac{R_2}{R_1+R_2} \quad [10]$$

The current I_2 through resistor R_2 is:

$$I_2 = \frac{E}{R_2} = \frac{IR_e}{R_2} = I\frac{R_1R_2}{R_2(R_1+R_2)} = I\frac{R_1}{R_1+R_2} \quad [11]$$

Thus we can state a current division rule as:

> *With two resistors in parallel, the current in each resistor is the total current multiplied by the fraction of the resistance of the opposite resistor divided by the sum of the two resistances.*

When we have a series–parallel circuit then we can use the voltage division rule to determine the voltage drop across each group of series elements and the current division rule to determine the current through each branch of parallel elements.

Example

Determine the voltage drops across, and the current through, each of the resistors in the circuit given in Figure 7.20.

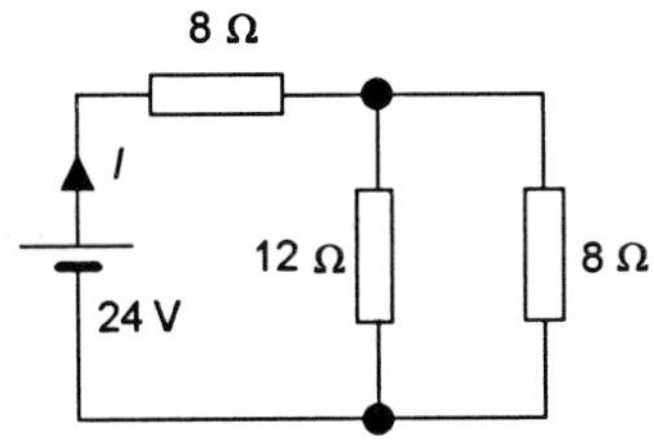

Figure 7.20 *Example*

The equivalent resistance R_p of the pair of parallel resistors is:

$$R_p = \frac{12\times 8}{12+8} = 4.8\ \Omega$$

We can thus draw the equivalent circuit of Figure 7.21. The voltage drop across the 8 Ω resistor is given by the voltage divider rule as:

$$V_8 = 24\frac{8}{8+4.8} = 15\ \text{V}$$

The voltage drop across the parallel resistors is given by the voltage divider rule as:

$$V_p = 24\frac{4.8}{8+4.8} = 9\ \text{V}$$

We can check these values by the use of Kirchhoff 's voltage law which states that: around any closed path in a circuit that the sum of the voltage drops and voltages rises from sources is zero. Thus we have 15 + 9 = 24 V.

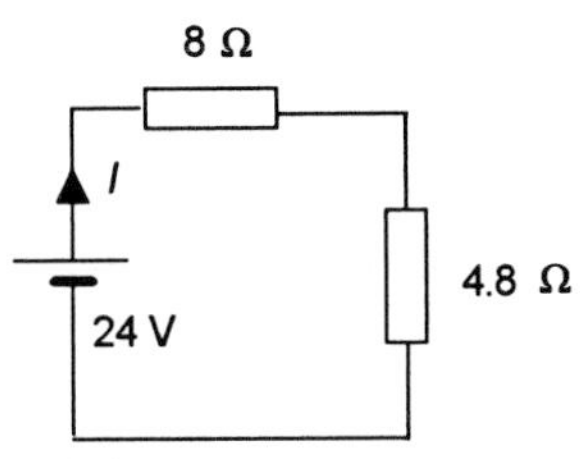

Figure 7.21 *Example*

The total circuit resistance is the sum of the series resistances in the equivalent circuit in Figure 7.21, i.e. 8 + 4.8 = 12.8 Ω. Thus the circuit current I is 24/12.8 = 1.875 A. This is the current through the 8 Ω resistor. It is the current entering the parallel arrangement and being divided between the two resistors. Using the current divider rule, the current through the 12 Ω resistor is:

$$I_{12} = 1.875\frac{8}{12+8} = 0.75 \text{ A}$$

and the current through the 8 Ω resistor is:

$$I_8 = 1.875\frac{12}{12+8} = 1.125 \text{ A}$$

We can check these values by using Kirchhoff 's current law: the current entering a junction equals the current leaving it. The current entering the parallel arrangement is 1.875 A and the current leaving it is 0.75 + 1.125 = 1.875 A.

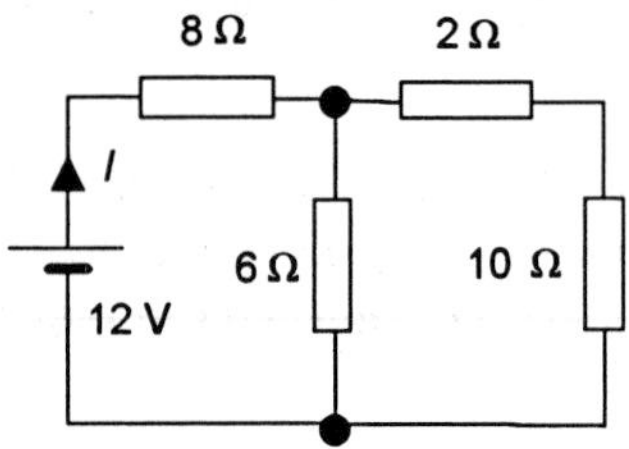

Figure 7.22 *Example*

Example

Determine the voltage drops across, and the current through, each of the resistors in the circuit given in Figure 7.22.

For the series arrangement of the 2 Ω and 10 Ω, the equivalent resistance is 12 Ω. This gives the simplified circuit of Figure 7.23(a). For the parallel arrangement of the 12 Ω and 6 Ω, the equivalent resistance is 12 × 6/(12 + 6) = 4 Ω. This gives the simplified circuit of Figure 7.23(b). The voltage drop across the 8 Ω resistor is thus given by the voltage division rule as:

$$V_8 = 12\frac{8}{8+4} = 8 \text{ V}$$

The voltage drop across the parallel arrangement is:

$$V_p = 12\frac{4}{8+4} = 4 \text{ V}$$

We can check this with Kirchhoff's voltage law: 8 + 4 = 12 V. For the series element of the parallel arrangement we have 4 V across 2 Ω in series with 10 Ω. Hence, using the voltage divider rule:

$$V_2 = 4\frac{2}{2+10} = 0.67 \text{ A}$$

$$V_{10} = 4\frac{10}{2+10} = 3.33 \text{ A}$$

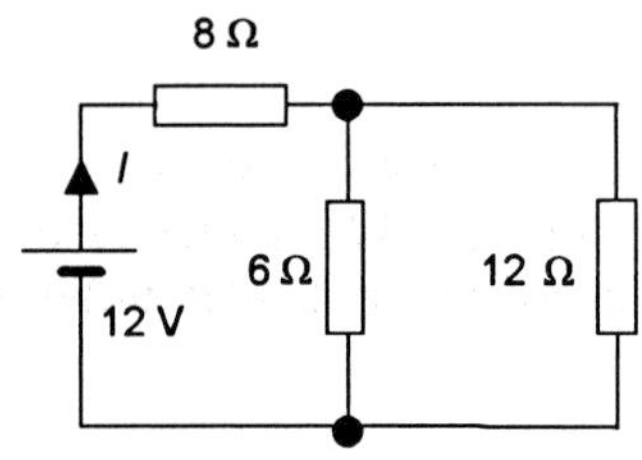

(a)

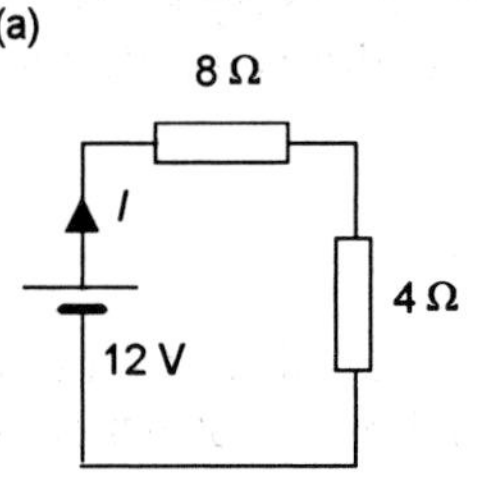

(b)

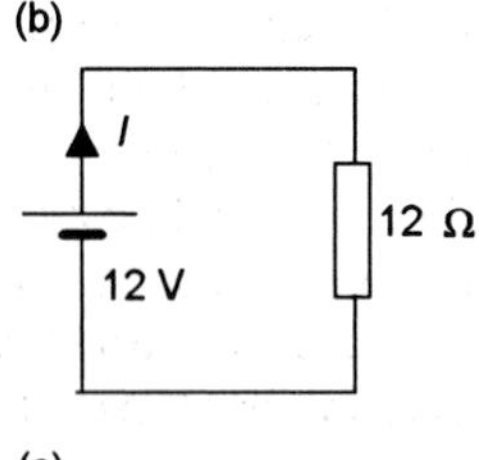

(c)

Figure 7.23 *Example*

We can check this with Kirchhoff's voltage law: 0.67 + 3.33 = 4 V.

For the currents through the resistors we need the total circuit current *I*. The simplified circuit of Figure 7.23(b) has 8 Ω and 4 Ω in series and so we can obtain the simplified circuit of Figure 7.23(c). The circuit current is thus I = 12/(8 + 4) = 1.0 A. This is the current through the 8 Ω resistor. It is also the current entering the parallel arrangement. Using the current divider rule for Figure 7.23(a):

$$I_6 = 1.0\frac{12}{6+12} = 0.67 \text{ A}$$

$$I_{12} = 1.0\frac{6}{6+12} = 0.33\text{ A}$$

Example

A potentiometer has a total track resistance of 50 Ω and is connected across a 20 V supply. What will be the resistance between the slider and the track end when the output voltage taken off between these terminals is 5.0 V?

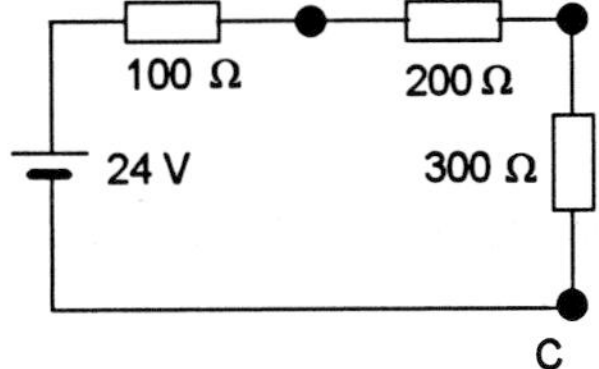

Figure 7.24 *Revision problem 9*

The fraction of the output voltage required is 5.0/20 and thus this must be the ratio of the resistance between one end and the slider to the sum of the two resistances, i.e. the total potentiometer resistance. Thus $R/50 = 5.0/20$ and so $R = 12.5$ Ω.

Revision

9 Use the voltage–divider rule to determine the voltages between points (a) A and B and (b) A and C in the circuit shown in Figure 7.24.

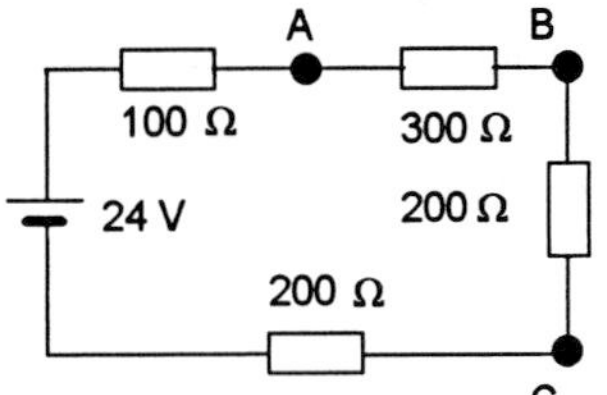

Figure 7.25 *Revision problem 11*

10 Use the current–divider rule to determine the resistance of an ammeter shunt if the meter has a resistance of 50 Ω and indicates a current of 1 mA when the current supplied to the shunted meter is 100 mA.

11 Use the voltage–divider rule to determine the voltages between points (a) A and B, (b) A and C, (c) B and C in the circuit shown in Figure 7.25.

7.3.2 Node and mesh analysis of circuits

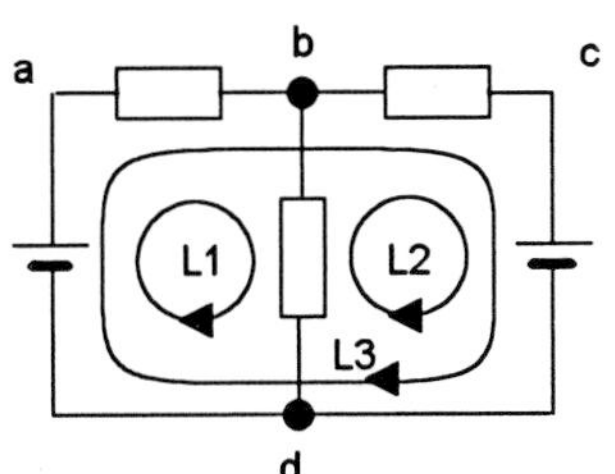

Figure 7.26 *Nodes and Loops*

For circuits, the term *node* is used for a point in a circuit where two or more devices are connected together, i.e. it is a junction at which we have current entering and current leaving. The term *loop* is a sequence of circuit elements that form a closed path. For the circuit shown in Figure 7.26, there are four different nodes a, b, c and d and three loops L1, L2 and L3. Loop 1 is through a, b and d, loop 2 is through b, c and d and loop 3 is round the outer elements of the circuit, i.e. a, b, c and d. One way we can use Kirchhoff's laws to analyse circuits is to write equations for the currents at every node in the circuit and write equations for every loop in the circuit. We then have to solve the resulting set of simultaneous equations.

There are two methods that can be used to reduce the number of simultaneous equations that have to be solved, these being *node analysis* and *mesh analysis*.

Node analysis uses Kirchhoff's current law to evaluate the voltage at each principal node in a circuit. A *principal node* is a point where three or more elements are connected together. Thus in Figure 7.26, just b and c are principal nodes. One of the principal nodes is chosen to be a reference node so that the potential differences at the other nodes are with reference to it; in Figure 7.26 we might choose d to be the reference node. Kirchhoff's current law is then applied to each non-reference node. The procedure is thus:

1 Draw a labelled circuit diagram and mark the principal nodes.

2 Select one of the principal nodes as a reference node.

3 Apply Kirchhoff's current law to each of the non-reference nodes, using Ohm's law to express the currents through resistors in terms of node voltages.

4 Solve the resulting simultaneous equations. If there are n principal nodes there will be $(n - 1)$ equations.

5 Use the derived values of the node voltages to determine the currents in each branch of the circuit.

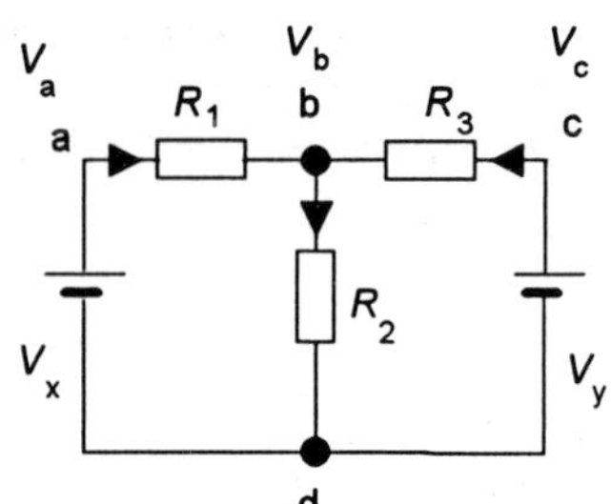

Figure 7.27 *Node analysis*

As an illustration of the application of the above method of circuit analysis, consider the circuit shown in Figure 7.27. There are four nodes a, b, c and d, of which b and d are principal nodes. If we take node d as the reference node, then the voltages V_a, V_b and V_c are the node voltages relative to node d. This means that the potential difference across resistor R_1 is $(V_a - V_b)$, across resistor R_2 is V_b and across R_3 is $(V_c - V_b)$. Thus the current through R_1 is $(V_a - V_b)/R_1$, through resistor R_2 is V_b/R_2 and through R_3 is $(V_c - V_b)/R_3$. Thus, applying Kirchhoff's current law to node b gives:

$$\frac{V_a - V_b}{R_1} + \frac{V_c - V_b}{R_3} = \frac{V_b}{R_2}$$

But $V_a = V_x$ and $V_c = V_y$ and so:

$$\frac{V_x - V_b}{R_1} + \frac{V_y - V_b}{R_3} = \frac{V_b}{R_2}$$

Hence the voltage at node b can be determined and hence the currents in each branch of the circuit.

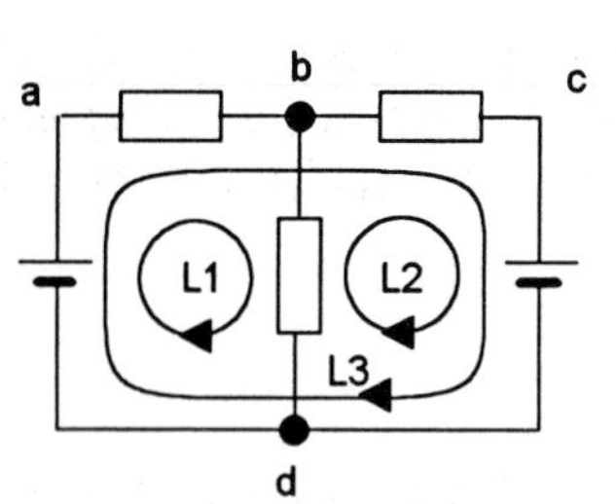

Figure 7.28 *Meshes and Loops*

The term *mesh* is used for a loop which does not contain any other loops within it. Thus for Figure 7.28 loops 1 and 2 are meshes but loop 3 is not. Mesh analysis involves defining a current as circulating round each mesh. The same direction must be chosen for each mesh current and the usual convention is to make all the mesh currents circulate in a clockwise direction. Thus for Figure 7.28 we would define a current I_1 as circulating round mesh 1 and a current I_2 round mesh 2. Having specified mesh currents, Kirchhoff's voltage law is then applied to each mesh. The procedure is thus:

1 Label each of the meshes with clockwise mesh currents.

2 Apply Kirchhoff's voltage law to each of the meshes, the potential differences across each resistor being given by Ohm's law in terms

of the current through it and in the opposite direction to the current. The current through a resistor which borders just one mesh is the mesh current; the current through a resistor bordering two meshes is the algebraic sum of the mesh currents through the two meshes.

3 Solve the resulting simultaneous equations to obtain the mesh currents. If there are n meshes there will be n equations.

4 Use the results for the mesh currents to determine the currents in each branch of the circuit.

Note that mesh analysis can only be applied to planar circuits, these being circuits that can be drawn on a plane so that no branches cross over each other.

As an illustration of the above method of circuit analysis, consider the circuit shown in Figure 7.29. There are three loops ABCF, CDEF and ABCDEF but only the first two are meshes. We define currents I_1 and I_2 as circulating in a clockwise direction in these meshes.

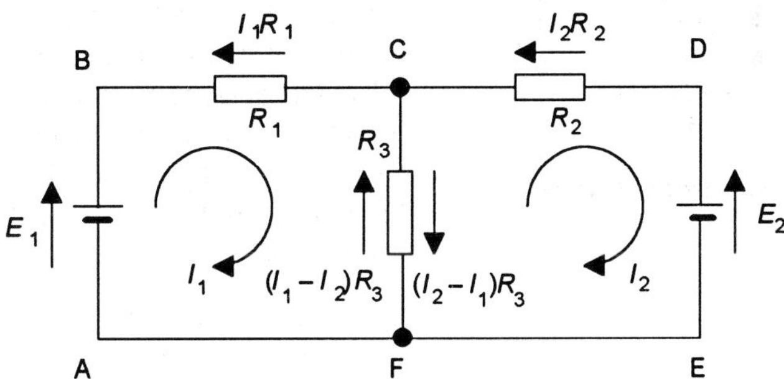

Figure 7.29 *Mesh currents*

Consider mesh 1. The current through R_1 is the mesh current I_1 . The current through R_2, which is common to the two meshes, is the algebraic sum of the two mesh currents, i.e. $I_1 - I_2$. Thus, applying Kirchhoff's voltage law to the mesh gives

$$E_1 - I_1R_1 - (I_1 - I_2)R_2 = 0$$

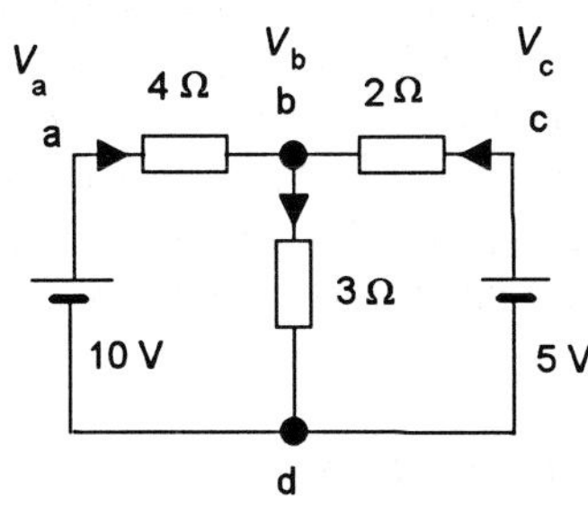

Figure 7.30 *Example*

For mesh 2 we have a current of I_2 through R_3 and a current of $(I_2 - I_1)$ through R_2. Thus applying Kirchhoff's voltage law to this mesh gives:

$$-E_2 - I_2R_2 - (I_2 - I_1)R_2 = 0$$

We thus have the two simultaneous equations for the two meshes.

Example

Use node analysis to determine the currents in each branch of the circuit shown in Figure 7.30.

The nodes are a, b, c and d with nodes b and d being principal nodes. Node d is taken as the reference node. If V_a, V_b and V_c are the node voltages relative to node d then the potential difference across the 4 Ω resistor is $(V_a - V_b)$, across the 3 Ω resistor is V_b and across the 2 Ω resistor is $(V_c - V_b)$. Thus the current through the 4 Ω is $(V_a - V_b)/4$, through the 3 Ω resistor is $V_b/3$ and through the 2 Ω resistor is $(V_c - V_b)/2$. Thus, applying Kirchhoff's current law to node b gives:

$$\frac{V_a - V_b}{4} + \frac{V_c - V_b}{2} = \frac{V_b}{3}$$

But $V_a = 10$ V and $V_c = 5$ V and so:

$$\frac{10 - V_b}{4} + \frac{5 - V_b}{2} = \frac{V_b}{3}$$

$$\frac{2(10 - V_b) + 4(5 - V_b)}{8} = \frac{V_b}{3}$$

$$60 - 6V_b + 60 - 12V_b = 8V_b$$

Thus $V_b = 4.62$ V. The potential difference across the 4 Ω resistor is thus 10 – 4.62 = 5.38 V and so the current through it is 5.38/4 = 1.35 A. The potential difference across the 3 Ω resistor is 4.62 V and so the current is 4.62/3 = 1.54 A. The potential difference across the 2 Ω resistor is 5 – 4.62 = 0.38 V and so the current is 0.38/2 = 0.19 A.

Example

Determine, using mesh analysis, the current through the 20 Ω resistor in the circuit shown in Figure 7.31.

There are two meshes and we define mesh currents of I_1 and I_2 as circulating round them. For mesh 1, Kirchhoff's voltage law, gives:

$$5 - 5I_1 - 20(I_1 - I_2) = 0$$

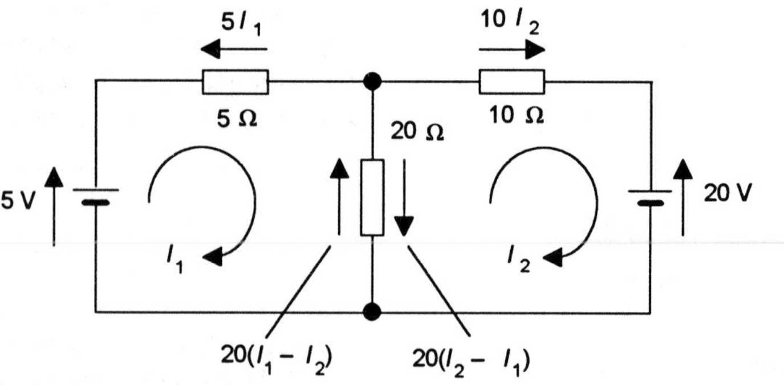

Figure 7.31 *Example*

This can be rewritten as:

$$5 = 25I_1 - 20I_2$$

For mesh 2, applying Kirchhoff's voltage law:

$$-10I_2 - 20 - 20(I_2 - I_1) = 0$$

This can be rewritten as:

$$20 = 20I_1 - 30I_2$$

We now have a pair of simultaneous equations. Multiplying the equation for mesh 1 by 4 and subtracting from it five times the equation for mesh 2 gives:

$$\begin{array}{rrcl} & 20 = & 100I_1 & - \ 80I_2 \\ \text{minus} & \underline{100 =} & \underline{100I_1} & \underline{- \ 150I_2} \\ & -80 = & 0 & + \ 70I_2 \end{array}$$

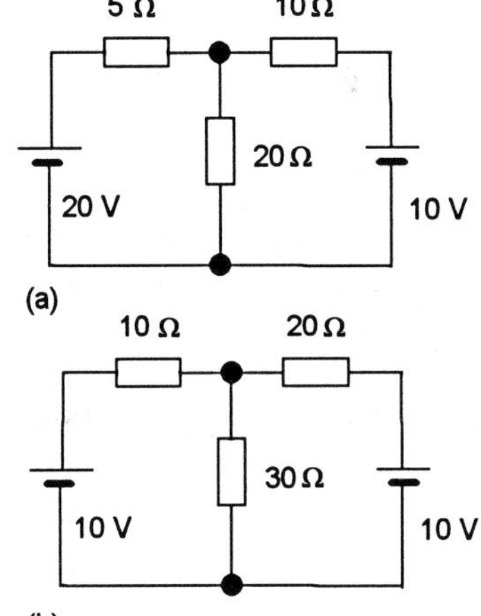

Figure 7.32 *Revision problems 12 and 13*

Thus $I_2 = -1.14$ A and, back substituting this value in one of the mesh equations, $I_1 = -0.71$ A. The minus signs indicate that the currents are in the opposite directions to those indicated in the figure. The current through the 20 Ω resistor is thus, in the direction of I_1, $-0.71 + 1.14 = 0.43$ A.

Revision

12 Use node analysis to determine the currents through each of the resistors in the circuit shown in Figure 7.32.

13 Use mesh analysis to determine the currents through each of the resistors in the circuits shown in Figure 7.32.

7.4 Capacitors

A *capacitor* is basically a pair of parallel conducting plates with a dielectric between them. When they are connected to a d.c. supply and a potential difference V produced between them, one of the plates becomes positively charged and the other negatively charged (Figure 7.33). The amount of charge Q on a plate is found to be directly proportional to V. The constant of proportionality is called the *capacitance* C. Thus:

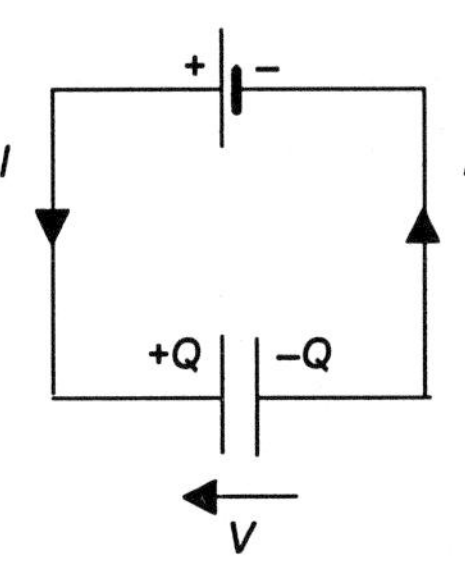

Figure 7.33 *Charging a capacitor*

$$Q = CV \qquad [12]$$

The unit of capacitance is the farad (F), when V is in volts and Q in coulombs. Note that a capacitance of 1 F is a very large capacitance and more usually capacitances will be microfarads (μF), i.e. 10^{-6} F, or nanofarads (nF), i.e. 10^{-9} F, or picofarads (pF), i.e. 10^{-12} F.

For three capacitors in *series* (Figure 7.34), the potential difference V across the arrangement will be the sum of the potential differences across each capacitor. Thus $V = V_1 + V_2 + V_3$. In order to have the same current through all parts of the series circuit we must have the same charge flowing onto and off the plates of each capacitor. Thus each capacitor will have the same charges of $+Q$ and $-Q$. Dividing the above equation throughout by Q gives:

$$\frac{V}{Q} = \frac{V_1}{Q} + \frac{V_2}{Q} + \frac{V_3}{Q}$$

But $C_1 = Q/V_1$, $C_2 = Q/V_2$ and $C_3 = Q/V_3$. Hence, if we replaced the three series-connected capacitors by a single equivalent capacitor with a capacitance given by $C = Q/V$, we must have:

$$\frac{1}{C} = \frac{1}{C_1} + \frac{1}{C_2} + \frac{1}{C_3} \qquad [13]$$

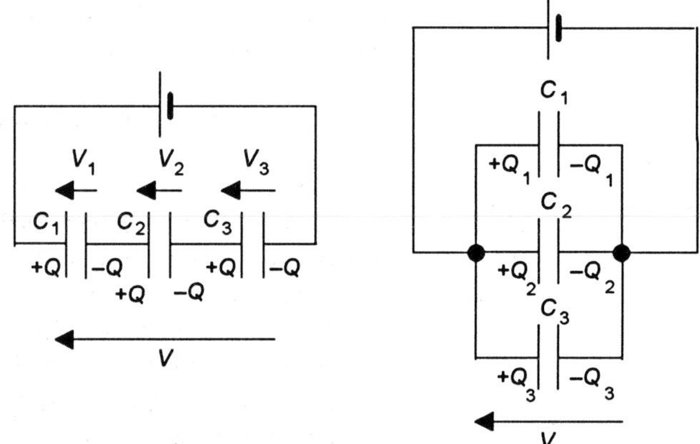

Figure 7.34 *Capacitors in series*

Figure 7.35 *Capacitors in parallel*

For three capacitors in *parallel* (Figure 7.35), the potential difference V across each capacitor will be the same. The charges on each capacitor will depend on their capacitances. If the total charge shared between the capacitors is Q, then $Q = Q_1 + Q_2 + Q_3$ and so dividing by V gives:

$$\frac{Q}{V} = \frac{Q_1}{V} + \frac{Q_2}{V} + \frac{Q_3}{V}$$

But $C_1 = Q_1/V$, $C_2 = Q_2/V$ and $C_3 = Q_3/V$. Hence, if we replaced the three parallel-connected capacitors by a single equivalent capacitor with a capacitance given by $C = Q/V$, we must have:

$$C = C_1 + C_2 + C_3 \qquad [14]$$

Application

Common types of capacitors used in electrical circuits are:

Paper capacitors with a layer of waxed paper as the dielectric sandwiched between two layers of metal foil, the whole being wound into a roll and either sealed in a metal can or encapsulated in resin. They have capacitances between about 10 nF and 10 mF. Working voltages are up to about 600 V.

Plastic capacitors are similar to paper capacitors and consist of layers of plastic film, e.g. polystyrene, between layers of metal foil or metallised plastic film with metal films being deposited on both sides of a sheet of plastic, e.g. polyester. Polystyrene film capacitors have capacitances from about 50 pF to 0.5 mF and working voltages up to about 500 V, metallised polyester about 50 pF to 0.5 mF and up to about 400 V.

Ceramic capacitors in tube, disc or rectangular plate forms, essentially being a plate of ceramic silvered on both sides. They have capacitances in the range 5 pF to 1 mF, or more, and working voltages up to about 1 kV.

Electrolytic capacitors, one form being two sheets of aluminium foil separated by a thick absorbent material, e.g. paper, impregnated with an electrolyte such as ammonium borate, with the whole arrangement being rolled up and put in an aluminium can. Electrolytic action occurs when a potential difference is connected between the plates and results in a thin layer of aluminium oxide being formed on the positive plate. This layer forms the dielectric. Another form uses tantalum instead of aluminium, with tantalum oxide forming the dielectric. The electrolytic capacitor must always be used with a d.c. supply and always connected with the correct polarity. This is because if a reverse voltage is used, the dielectric layer will be removed and a large current can occur with damage to the capacitor. Electrolytic capacitors, because of the thinness of the dielectric, have very high capacitances. Aluminium electrolytic capacitors have capacitances from about 1 mF to 100 000 mF, tantalum ones 1 mF to 2000 mF. Working voltages can be as low as 6 V.

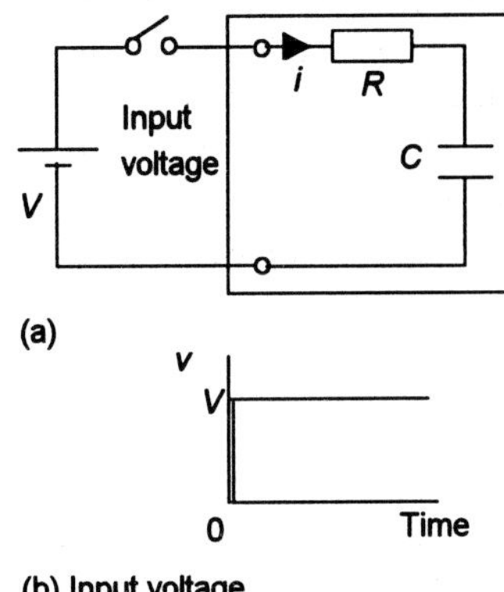

Figure 7.36 *Step voltage applied to RC circuit*

Example

What is the capacitance of a 2 μF capacitor and 4 μF capacitor when connected in (a) series, (b) parallel?

(a) $1/C = 1/2 + 1/4 = 3/4$, hence $C = 1.33$ μF.
(b) $C = 2 + 4 = 6$ μF.

Revision

14 Determine the total equivalent capacitance for three capacitors 60 μF, 30 μF and 20 μF when they are (a) connected in series, (b) connected in parallel.

15 What is the charge on the plates of capacitors of 50 pF, 100 pF and 200 pF when they are connected in parallel?

7.4.1 *RC* circuit: charging

Consider a circuit with series resistance and capacitance and there is a step voltage input to the circuit at time $t = 0$ (Figure 7.36(a)). The voltage is applied across two series components and so we must have:

$$V = \text{voltage across resistance } v_R + \text{voltage across capacitor } v_C$$

The charge q on a capacitor is related to the voltage v_C between its plates by $q = Cv_C$, where C is the capacitance. Initially there is no charge on the capacitor because there has been no current to move charge onto and off its plates. Thus, at the instant we close the switch, there is no voltage across the capacitor and the voltage V must be entirely across the resistance. The current i through the resistor, and hence the initial circuit current, is v_R/R and so initially is V/R. When the capacitor begins to acquire charge then the voltage across it increases. Since $V = v_R + v_C$, this must result in a decrease in the voltage across the resistor. Thus, since $i = v_R/R$, the circuit current i must decrease. When the capacitor is fully charged the circuit current has dropped to zero. There is then no voltage across the resistor and the entire input voltage is across the capacitor. Figure 7.37(a) shows how the circuit current changes with time, Figures 7.37(b) and (c) how the voltages across the resistance and capacitance change with time. The graphs are exponentials.

At any instant of time $V = v_C + v_R$. But $v_R = iR$ and so we can write:

$$V = v_C + iR$$

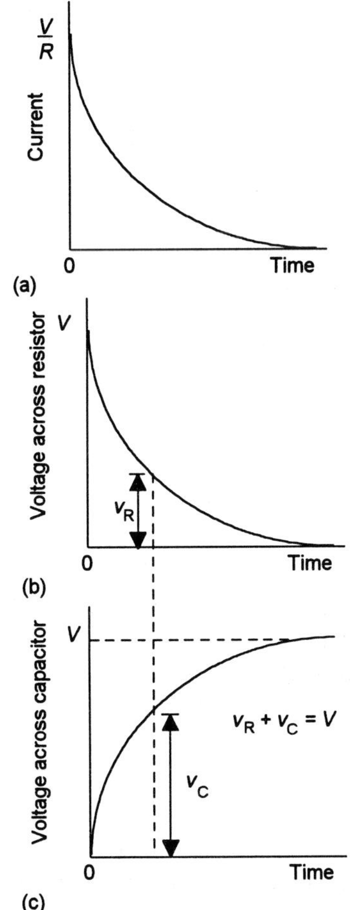

Figure 7.37 *Series RC circuit*

The circuit current i is the rate of movement of charge through the circuit. But each bit of charge moved through the circuit changes the voltage across the capacitor; $q = Cv_C$ and so:

$$i = \text{rate of movement of charge} = \text{rate of change of } Cv_C$$

$$= C \times \text{rate of change of } v_C$$

We can write this, in calculus notation, as:

$$i = \frac{dq}{dt} = \frac{d(Cv_C)}{dt} = C\frac{dv_C}{dt}$$

The current is thus proportional to the rate of change of the voltage across the capacitor and so:

$$V = v_C + iR = v_C + RC\frac{dv_C}{dt}$$

$$V - v_C = RC\frac{dv_C}{dt} \qquad [15]$$

Initially, $v_C = 0$ and so:

initial rate of change of voltage across the capacitor = $V/RC = V/\tau$

The product RC has the unit of time and is termed the *time constant* $\tau = RC$. The significance of the time constant is that, the bigger the time constant the smaller the initial rate of charging of the capacitor and the longer it will take to reach the voltage V.

We can obtain the equation of the graph of v_C with time by integrating equation [15]. Separation of variables gives:

$$\frac{dv_C}{V - v_C} = \frac{dt}{RC}$$

Integration then gives:

$$\int_0^V \frac{dv_C}{V - v_C} = \int_0^t \frac{dt}{RC}$$

$$-\ln(V - v_C) + \ln V = \frac{t}{RC}$$

$$\frac{V}{V - v_C} = e^{t/RC}$$

$$\frac{V - v_C}{V} = e^{-t/RC}$$

$$v_C = V(1 - e^{-t/RC}) \qquad [16]$$

Since $V = v_R + v_C$ then:

$$v_R = V - V(1 - e^{-t/RC}) = V e^{-t/RC} \qquad [17]$$

Since $i = v_R/R$ then:

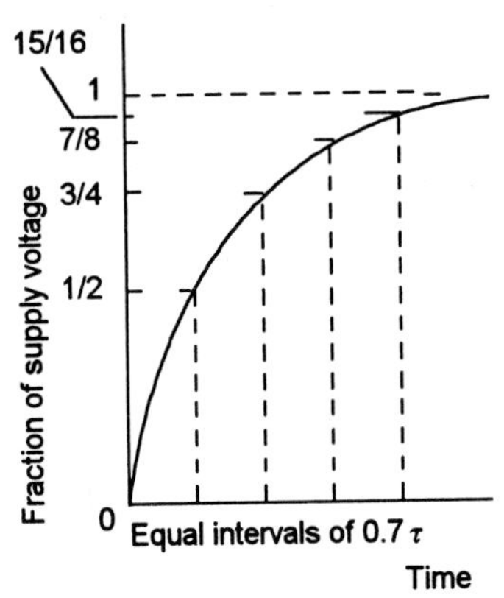

Figure 7.38 *Voltage across the capacitor*

$$i = \frac{V}{R} e^{-t/RC} \qquad [18]$$

The above three equations describe the graphs in Figure 7.37.

The time constant τ is RC. Thus:

$$v_C = V(1 - e^{-t/RC}) = V(1 - e^{-t/\tau})$$

What time will be required for v_C to reach $0.5V$? This gives:

$$0.5V = V(1 - e^{-t/\tau})$$

$$e^{-t/\tau} = 0.5$$

$$-\frac{t}{\tau} = \ln 0.5 = -0.693$$

Thus in a time of 0.693τ the voltage will reach half its steady state voltage. The time taken to reach $0.75V$ is given by:

$$0.75V = V(1 - e^{-t/\tau})$$

$$e^{-t/\tau} = 0.25$$

$$-\frac{t}{\tau} = \ln 0.25 = -1.386$$

Thus in a time of 1.386τ the voltage will reach three-quarters of its steady state value. This is twice the time taken to reach half the steady state voltage. This is a characteristic of exponential graphs: if t is the time taken to reach half the steady state value, then in $2t$ it will reach three-quarters, in $3t$ it will reach seven-eighths, etc. *In each successive time interval of 0.7τ the p.d. across the capacitor reduces its value by a half* (Figure 7.38) (Table 7.1).

Table 7.1 *Growth of the p.d. across the capacitor*

Time	v_C
0	0
$0.7T$	$0.5V$
$1.4T$	$0.75V$
$2.1T$	$0.875V$
$2.8T$	$0.938V$
$3.5T$	$0.969V$

When $t = 1\tau$ then $v_C = V(1 - e^{-1}) = 0.632V$. Thus in a time equal to the time constant the voltage across the capacitor rises to 63.2% of the steady state voltage. When $t = 2\tau$ then $v_C = V(1 - e^{-2}) = 0.865V$. Thus the voltage across the capacitor rises to 86.5% of the steady state voltage. When $t = 3\tau$ then $v_C = V(1 - e^{-3}) = 0.950V$. Thus the voltage across the capacitor rises to 95.0% of the steady state voltage. Table 7.2 summarises this data.

Table 7.2 *Growth of the p.d. across the capacitor*

Time	v_C
0	0
1*T*	0.632*V*
2*T*	0.865*V*
3*T*	0.950*V*

Example

What is the initial rate at which the voltage across a 100 μF capacitor will change with time when it is charged through a resistance of 5 kΩ by a voltage being switched from 0 to 6 V?

$$\text{Initial rate} = \frac{V}{RC} = \frac{6}{5 \times 10^3 \times 100 \times 10^{-6}} = 12 \text{ V/s}$$

Example

What is the time constant for a series *RC* circuit which has a resistance of 2 MΩ and capacitance 10 μF?

The time constant $\tau = RC = 2 \times 10^6 \times 10 \times 10^{-6} = 20$ s.

Example

Determine the equation describing the voltage across a 15 μF capacitor in a series *RC* circuit if the resistance is 40 kΩ and the voltage 1 s after if it is connected to a step voltage of 12 V.

$RC = 40 \times 10^3 \times 15 \times 10^{-6} = 0.6$ s and so the voltage is described by the equation:

$$v_C = V(1 - e^{-t/RC}) = 12(1 - e^{-t/0.6})$$

After 1 s, $v_C = 12(1 - e^{-1/0.6}) = 9.7$ V.

Example

A series *RC* circuit has a resistance of 10 kΩ and a capacitance of 0.1 μF. Determine the current (a) 0.001 s, (b) 0.005 s after a steady voltage of 20 V is connected to the circuit.

$RC = 10 \times 10^3 \times 0.1 \times 10^{-6} = 0.001$ s. Thus:

$$i = \frac{V}{R} e^{-t/RC} = \frac{20}{10 \times 10^{-3}} e^{-t/0.001} = 2\, e^{-t/0.001} \text{ mA}$$

(a) After 0.01 s, $i = 2\ e^{-0.001/0.001} =$ 0.736 mA.
(b) After 0.005 s, $i = 2\ e^{-0.005/0.001} =$ 0.0135 mA.

Revision

16 What is the time constant for a circuit having a capacitance of 8 μF in series with a resistance of 1 MΩ?

17 A 1 μF capacitor is connected in series with a 100 kΩ resistor. Following the connection of a 20 V d.c. supply, what will be the current in the circuit (a) immediately the connection is made, (b) after 0.05 s, (c) after 0.1 s?

18 A 1 μF capacitor is connected in series with a 1 MΩ resistor. What will be (a) the time constant of the circuit, (b) the voltages across each component after 0.7 time constants has elapsed after a d.c. supply of 10 V has been connected to the circuit?

19 A 0.02 μF capacitor is connected in series with a 15 kΩ resistor. What will be (a) the time constant of the circuit, (b) the circuit current and (c) the voltages across each component 1.5 times constants after a 20 V d.c. supply is connected to the circuit?

20 What is the initial rate at which the voltage across a 10 μF capacitor will change with time when it is charged through a resistance of 1 kΩ by a voltage being switched from 0 to 10 V?

7.4.2 *RC* circuit: discharging

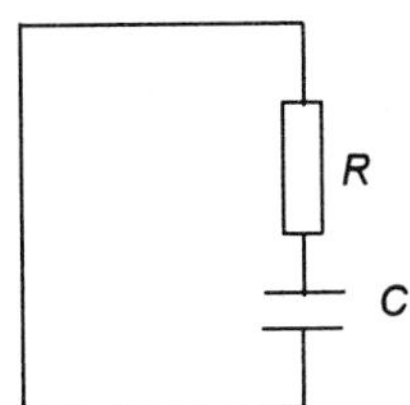

Figure 7.39 *Discharge of a capacitor*

When a voltage is applied to a capacitor and current flows to one of its plates and from the other, it becomes charged. This is what has been considered earlier in this chapter as a consequence of the application of a step voltage input to a *RC* circuit. Now consider what happens if a capacitor that has been charged by a voltage *V* being applied to the *RC* circuit now has the voltage removed and a current path from one terminal to the other provided through the resistor (Figure 7.39). A charged capacitor has a voltage between its terminals and this will result in a current flowing through the resistor and a voltage developing across it. Because there are no other sources of voltage in the circuit, if v_C is the voltage across the capacitor at some instant of time and v_R the voltage across the resistor, we must have $v_C + v_R = 0$ and so $v_R = -v_C$. The circuit current i is thus, at any instant:

$$i = \frac{v_R}{R} = -\frac{v_C}{R} \quad [19]$$

As the charge flows from one plate of the capacitor through the circuit to the other plate, so the charge on the capacitor decreases with time and hence the voltage across the capacitor decreases. Consequently the voltage across the resistor and the circuit current decreases with time.

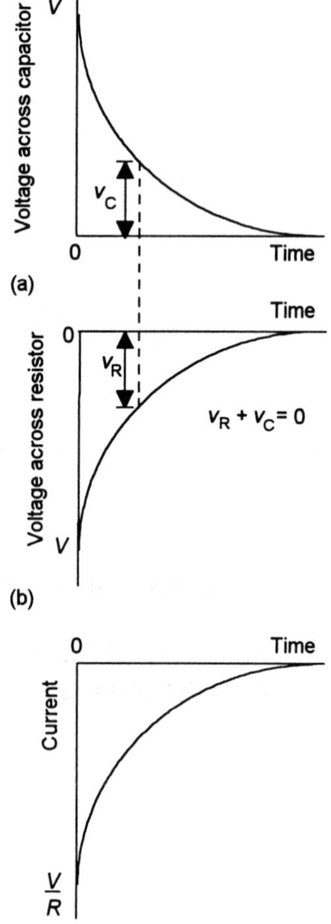

Figure 7.40 *Discharge of a charged capacitor through a resistor*

Figure 7.40 shows how they vary with time. Note that the current flows in the opposite direction to that occurring during charging.

The circuit current i is the rate of movement of charge through the circuit. But each bit of charge moved through the circuit changes the voltage across the capacitor; $q = Cv_C$ and so:

$$i = \text{rate of movement of charge} = \text{rate of change of } Cv_C$$

$$= C \times \text{rate of change of } v_C$$

We can write this, in calculus notation, as:

$$i = \frac{dq}{dt} = \frac{d(Cv_C)}{dt} = C\frac{dv_C}{dt}$$

The current is thus proportional to the rate of change of the voltage across the capacitor. At any instant of time $V = v_C + v_R = 0$. But $v_R = iR$ and so we can write:

$$0 = v_C + v_R = v_C + iR = v_C + RC\frac{dv_C}{dt}$$

and so:

$$-v_C = RC\frac{dv_C}{dt} \qquad [20]$$

Initially, $v_C = V$ and so the initial rate of change with time of the voltage across the capacitor is

initial rate of change of voltage across capacitor $-V/RC$	[21]

The product RC has the unit of time and we define *time constant* $\tau = RC$. The significance of the time constant is that, the bigger the time constant the smaller the initial rate of discharging of the capacitor and the longer it will take to completely discharge.

We can obtain the equation of the graph of v_C with time by integrating equation [20]. Separation of variables gives:

$$\frac{dv_C}{-v_C} = \frac{dt}{RC}$$

Integration then gives:

$$-\int_V^{v_C} \frac{dv_C}{v_C} = \int_0^t \frac{dt}{RC}$$

$$-\ln v_C + \ln V = \frac{t}{RC}$$

$$\frac{V}{v_C} = e^{t/RC}$$

$$\frac{v_C}{V} = e^{-t/RC}$$

and so:

$$v_C = V\,e^{-t/RC} \qquad [22]$$

Since $v_R = -v_C$ then:

$$v_R = -V\,e^{-t/RC} \qquad [23]$$

Since $i = v_R/R$ then:

$$i = -\frac{V}{R}\,e^{-t/RC} \qquad [24]$$

The above three equations are those describing the graphs in Figure 7.40.

The time constant $\tau = RC$. Thus, we can write equation [24] as $v_C = V\,e^{-t/\tau}$. The time taken for v_C to drop from V to $0.5V$ is thus given by:

$$0.5V = V\,e^{-t/\tau}$$

$$e^{-t/\tau} = 0.5$$

$$-\frac{t}{\tau} = \ln 0.5 = -0.693$$

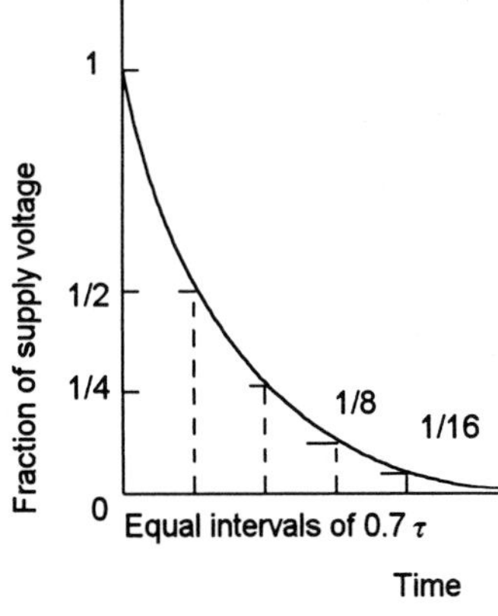

Figure 7.41 *Voltage across the capacitor*

Thus in a time of 0.693τ the voltage will drop to half its initial voltage. The time taken to drop to $0.25V$ is given by:

$$0.25V = V\,e^{-t/\tau}$$

$$e^{-t/\tau} = 0.25$$

$$-\frac{t}{\tau} = \ln 0.25 = -1.386$$

Thus in a time of 1.386τ the voltage will drop to one-quarter of its initial voltage. This is twice the time taken to drop to half the voltage. This is a characteristic of a decaying exponential graph: if t is the time taken to reach half the steady state value, then in $2t$ it will reach one-quarter, in $3t$ it will reach one-eighth, etc. In each of these time intervals it reduces its value by a half (Figure 7.41) (Table 7.3).

When $t = 1\tau$ then $v_C = V\,e^{-1} = 0.632V$. Thus in a time equal to the time constant the voltage across the capacitor drops to 63.2% of the initial voltage. When $t = 2\tau$ then $v_C = V\,e^{-2} = 0.135V$. Thus the voltage across the capacitor drops to 13.5% of the initial voltage. When $t = 3\tau$ then $v_C =$

$V e^{-3} = 0.050V$. Thus the voltage across the capacitor drops to 5.0% of the initial voltage. Table 7.4 summarises the above data.

Table 7.3 *Discharge of a capacitor*

Time	v_C
0	V
$0.7T$	$0.5V$
$1.4T$	$0.25V$
$2.1T$	$0.125V$
$2.8T$	$0.0625V$
$3.5T$	$0.03125V$

Table 7.4 *Discharge of a capacitor*

Time	v_C
0	V
$1T$	$0.632V$
$2T$	$0.135V$
$3T$	$0.050V$

Example

A capacitor is charged to a voltage of 12 V. It is then allowed to discharge through a resistance of 2 kΩ. What will be (a) the initial circuit current, (b) the circuit current when the voltage across the capacitor has dropped to 2 V?

(a) Initially the voltage across the capacitor is 12 V and so the voltage across the resistor is –12 V. The initial current is thus –12/1000 = –12 mA.
(b) When the voltage across the capacitor has dropped to 2 V the voltage across the resistor will be –2 V and so the circuit current is –2/1000 = –2 mA.

Example

Determine after 25 ms the circuit current and voltage across the capacitor of a circuit in which the capacitor of capacitance 0.01 μF, which has been charged to a voltage of 10 V, discharges through a resistance of 5 MΩ.

$RC = 0.01 \times 10^{-6} \times 5 \times 10^{6} = 0.05$ s. Thus:

$$i = -\frac{V}{R} e^{-t/RC} = -\frac{10}{5 \times 10^{6}} e^{-0.025/0.05} = -1.2\ \mu\text{A}$$

$$v_C = V e^{-t/RC} = 10\ e^{-0.025/0.05} = 6.1\ \text{V}$$

Example

What is the time constant for a circuit if it has a capacitor of 10 μF discharging through a resistance of 1 MΩ?

The time constant = $RC = 1 \times 10^6 \times 10 \times 10^{-6} = 10$ s.

Example

A 1000 μF capacitor is fully charged by connection to a 20 V d.c. supply. It is then allowed to discharge through a 200 Ω resistor. What is (a) the time constant of the discharge circuit, (b) the equation describing how the discharge current will vary with time, (c) the discharge current after 0.1 s?

(a) The time constant = $RC = 1000 \times 10^{-6} \times 200 = 0.2$ s.
(b) Equation [24] gives:

$$i = -\frac{V}{R}\,e^{-t/RC} = -\frac{20}{200}\,e^{-t/0.2} = -0.1\,e^{-t/0.2}\text{ A}$$

(c) Using the above equation, $i = -0.1\,e^{-0.1/0.2} = 60.7$ mA.

Revision

21 What is the time constant for a circuit if it has a capacitor of 100 μF discharging through a resistance of 2 MΩ?

22 A 10 μF capacitor is fully charged by connection to a 20 V d.c. supply. It is then allowed to discharge through a 20 kΩ resistor. What is (a) the time constant of the discharge circuit, (b) the initial current, (c) the discharge current after 0.1 s?

23 Determine circuit current and voltage across the capacitor of a circuit in which the capacitor of capacitance 0.1 μF, which has been charged to a voltage of 10 V, after 0.2 s from when it starts to discharges through a resistance of 10 MΩ.

24 A 1 μF capacitor is in series with a 2 MΩ resistor. It is then charged for 2 s by a 10 V d.c. supply. After that time, the supply is disconnected and the capacitor allowed to discharge through the resistor. What will be the voltage across the capacitor (a) at the end of the charging time, (b) 1 s after it starts to discharge?

Activity

Determine the graph for the variation of current with time for the discharge of a charged capacitor. Figure 7.42 shows a possible circuit; possible values to give a discharge graph which will take a reasonable amount of time are C = 500 μF, R = 100 kΩ, 100 μA meter, V = 10 V or C = 10 000 μF, R = 5 kΩ, 1 mA meter, V = 10 V.

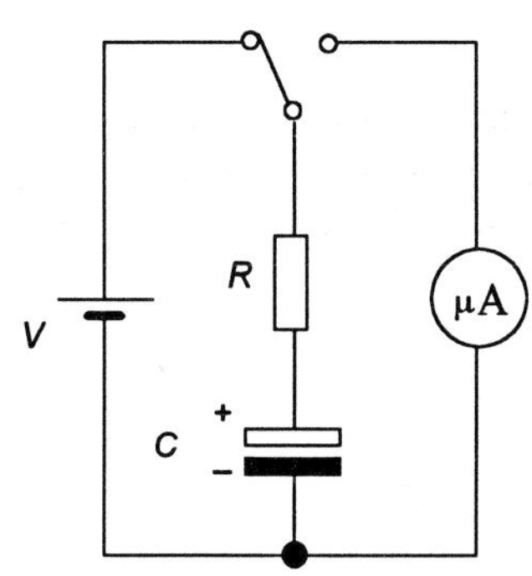

Figure 7.42 *Activity*

7.5 Inductors

A current through a coil produces magnetic flux which links the turns of the coil. Thus when the current though the coil changes, the flux linked by that coil changes. Hence an e.m.f. is induced. This phenomenon is known as *self inductance* or just *inductance*.

The induced e.m.f. is proportional to the rate of change of linked flux (Faraday's law). However, the flux produced by a current is proportional to the size of the current. Thus the rate of change of flux will be proportional to the rate of change of the current responsible for it. Hence the induced e.m.f. e is proportional to the rate of change of current, i.e. $e \propto$ rate of change of current. Thus we can write $e = L \times$ (rate of change of current), where L is the constant of proportionality, and so, writing $\mathrm{d}I/\mathrm{d}t$ for the rate of change of current:

$$e = L\frac{\mathrm{d}I}{\mathrm{d}t} \qquad [25]$$

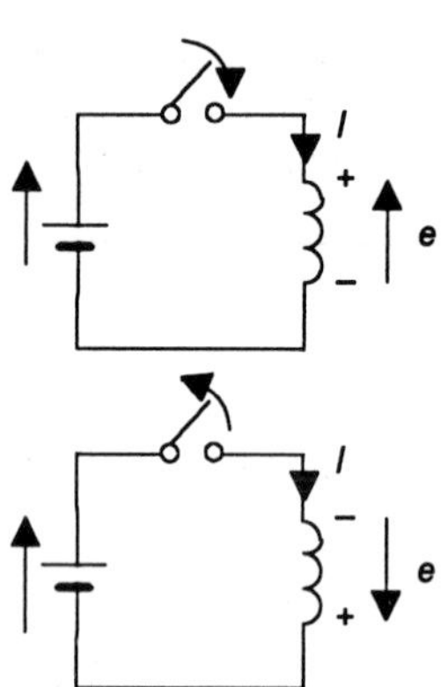

Figure 7.43 *Back e.m.f.*

where L is called the *inductance* of the circuit. The inductance is said to be 1 henry (H) when the e.m.f. induced is 1 V as a result of the current changing at the rate of 1 A/s.

The effect of inductance on the current in a circuit is that, when the applied voltage is switched on or off, the current does not immediately rise to its maximum value or fall to zero but takes some time. When the voltage is switched on and the current starts to increase from zero, then the changing current results in an induced e.m.f. Thus is in such a direction as to oppose the growing current (Lenz's law) and slow its growth. For this reason, the induced e.m.f. is often referred to as a *back e.m.f.* (Figure 7.43). When the voltage is switched off, then the current starts to fall and so produces an induced e.m.f. This is in such a direction as to oppose the current falling (Lenz's law) and so the current takes longer to fall to zero.

Application
The basic principle involved in a car ignition system is that opening and closing switch contacts causes large rates of changes of current through a coil and hence large voltages are induced in the circuit and routed to the spark plugs.

Example

What is the average back e.m.f. induced in a coil of inductance 500 mH when the current through it is increased from 1.0 A to 3.0 A in 0.05 s?

$$e = L\frac{\mathrm{d}I}{\mathrm{d}t} = \frac{0.500 \times (3.0 - 1.0)}{0.05} = 20 \text{ V}$$

Application
While it is generally possible in circuit anlysis to treat resistors and capacitors as ideal and only have the property of resistance or capacitance, this is not true of inductors which always have both inductance and resistance and can be considered to behave as an ideal inductance in series with a pure resistance.

Large inductors, often called *chokes*, are normally constructed as required from a ferromagnetic core and insulated copper wire. Smaller, commercially available, inductors generally have ferrite cores. Ferrite is used because it has a high resistivity and so reduces losses due to eddy currents

Revision

25 What is the back e.m.f. induced in a motor coil of 200 mH when the current through it is switched off and changes at the rate of 100 A/s?

7.5.1 *RL* circuit: current growth

Consider a circuit of an inductance in series with resistance and to which a step voltage is applied (Figure 7.44). When the switch is closed the current in the circuit starts to grow. The changing current in the

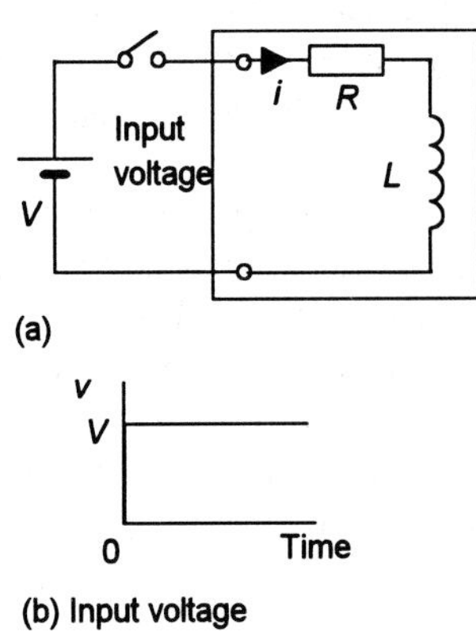

Figure 7.44 *Step voltage applied to RL circuit*

inductance generates a changing magnetic field in the inductance coil and generates a back e.m.f. which slows down the changing current. The back e.m.f. e depends on the rate of increase of the current di/dt through the inductor:

$$e = -L\frac{di}{dt}$$

where L is the inductance. To maintain a current through the inductor, and hence the circuit, the source must supply a voltage across the inductor v to cancel out the induced e.m.f. Thus the voltage drop across the inductor when there is a current i is

$$\text{voltage across inductor} = L\frac{di}{dt}$$

Thus for the circuit we have V = voltage across resistor + voltage across inductor and so:

$$V = iR + L\frac{di}{dt} \qquad [26]$$

At the instant the switch is closed and the voltage is applied to the circuit, there is zero current in the circuit and, since the voltage across the resistor will then be zero, the rate of change of current must be such that the induced e.m.f. equals V. Equation [26] can be written as:

$$\frac{di}{di} = \frac{V}{L} - \frac{R}{L}i$$

Thus the initial rate of change of current is:

$$\text{initial rate of change of current with time} = \frac{V}{L} \qquad [27]$$

As the current in the circuit increases, so the voltage across the resistor increases and hence the voltage across the inductor decreases. This can only mean that the rate of change of current with time is decreasing. Eventually the entire voltage V is across the resistor, there then being no voltage across the inductor and so the current ceases to change. When this occurs the current has reached its steady state value of V/R (Figure 4.45).

We can obtain the equation of the graph of the current i with time by integrating equation [26]. Separation of variables gives:

$$\frac{di}{(V/R) - i} = \frac{dt}{R/L}$$

Integration then gives:

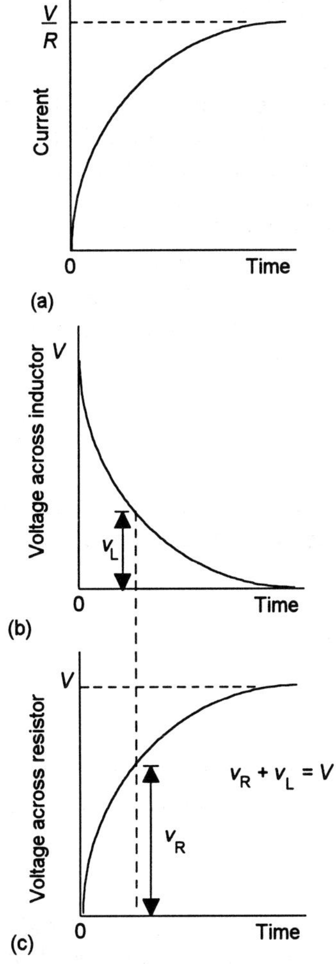

Figure 7.45 *Series RL*

$$\int_0^i \frac{di}{(V/R) - i} = \int_0^t \frac{dt}{R/L}$$

$$-\ln[(V/R) - i] + \ln(V/R) = \frac{Rt}{L}$$

$$\ln\left[\frac{(V/R) - i}{(V/R)}\right] = -\frac{Rt}{L}$$

$$\frac{(V/R) - i}{(V/R)} = e^{-Rt/L}$$

$$i = \frac{V}{R}(1 - e^{-Rt/L}) \qquad [28]$$

The voltage across the resistor $v_R = iR$ and so:

$$v_R = V(1 - e^{-Rt/L}) \qquad [29]$$

The voltage across the inductor $v_L = V - v_R$ and so is:

$$v_L = V e^{-Rt/L} \qquad [30]$$

The above three equations describe the graphs in Figure 7.45.

For a series *LR* circuit we can write the time constant τ as *L/R*. Thus equation [28] can be written as:

$$i = \frac{V}{R}(1 - e^{-t/\tau})$$

When τ is very large then the exponential term becomes 0 and so the current becomes the steady state current $I = V/R$. What time will be required for i to reach 0.5I ?

$$0.5I = I(1 - e^{-t/\tau})$$

$$e^{-t/\tau} = 0.5$$

$$-\frac{t}{\tau} = \ln 0.5 = -0.693$$

Thus in a time of 0.693τ the current will reach half its steady state current. The time taken to reach $0.75I$ is given by:

$$0.75I = I(1 - e^{-t/\tau})$$

$$e^{-t/\tau} = 0.25$$

$$-\frac{t}{\tau} = \ln 0.25 = -1.386$$

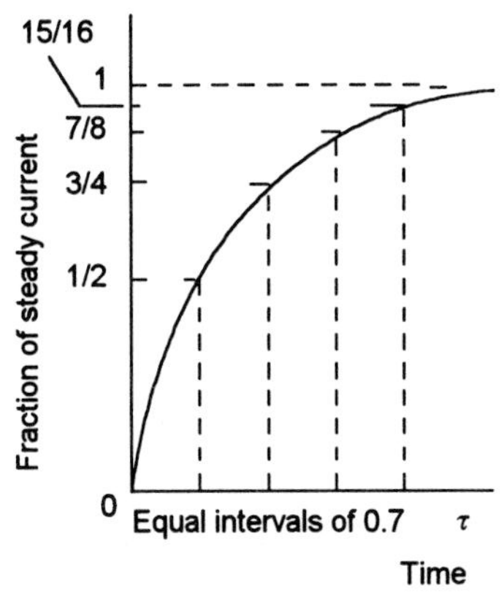

Figure 7.46 *Current in series LR circuit*

Thus in a time of 1.386τ the current will reach three-quarters of its steady state value. This is twice the time taken to reach half the steady state current. This is a characteristic of exponential graphs: if t is the time taken to reach half the steady state value, then in $2t$ it will reach three-quarters, in $3t$ it will reach seven-eighths, etc. In each of these time intervals it reduces its value by a half (Figure 7.46) (Table 7.5).

Table 7.5 *Growth of current in LR circuit*

Time	i_L
0	0
$0.7T$	$0.5I$
$1.4T$	$0.75I$
$2.1T$	$0.875I$
$2.8T$	$0.938I$
$3.5T$	$0.969I$

When $t = 1\tau$ then $i = I(1 - e^{-1}) = 0.632I$. Thus in a time equal to the time constant the current rises to 63.2% of the steady state current. When $t = 2\tau$ then $i = I(1 - e^{-2}) = 0.865I$. Thus the current rises to 86.5% of the steady state current. When $t = 3\tau$ then $i = I(1 - e^{-3}) = 0.950I$. Thus the current rises to 95.0% of the steady state current. Table 7.6 summarises this data.

Table 7.6 *Growth of current in LR circuit*

Time	i_L
0	0
$1T$	$0.632I$
$2T$	$0.865I$
$3T$	$0.950I$

Example

A coil has an inductance of 0.5 H. What will be the maximum rate of change of current with time when a d.c. voltage of 100 V is connected across the coil?

The maximum rate of change of current with time occurs at the instant the voltage is applied and is given by equation [27] as V/L = 100/0.5 = 200 A/s.

Example

A coil has a resistance of 20 Ω and an inductance of 0.5 H. What is its time constant?

Time constant = L/R = 0.5/20 = 0.025 s

Example

A series *RL* circuit with a d.c. voltage supply of 24 V has a resistance of 100 Ω and an inductance of 50 mH. Determine (a) the time constant, (b) the voltage across the inductor and (c) the circuit current 0.4 ms after the supply is switched on, (d) the steady state current.

(a) The time constant = $L/R = 50 \times 10^{-3}/100 = 0.5$ ms.
(b) The voltage across the inductor after 0.4 ms is:

$$v_L = V\,e^{-Rt/L} = V\,e^{-t/\tau} = 24\,e^{-0.4/0.5} = 10.8 \text{ V}$$

(c) The current after 0.4 ms is:

$$i = \frac{V}{R}(1 - e^{-Rt/L}) = \frac{24}{100}(1 - e^{-0.4/0.5}) = 0.13 \text{ A}$$

(d) The steady state current is $I = V/R = 24/100 = 0.24$ A.

Revision

26 A coil has an inductance of 200 mH. What will be the maximum rate of change of current with time when a d.c. voltage of 10 V is connected across the coil?

27 An inductor of 50 mH is in series with a resistor of 100 Ω. What is (a) the time constant of the circuit, (b) the initial rate of change of current when a 24 V d.c. supply is connected, (c) the voltage across the inductor after 0.4 ms, (d) the voltage across the resistor after 0.4 ms, (e) the steady state current?

28 An inductor of 50 mH is in series with a resistor of 200 Ω and connected to a 12 V d.c. supply. What is the voltage across the inductor (a) immediately after connecting the voltage supply, (b) after a time equal to three time constants?

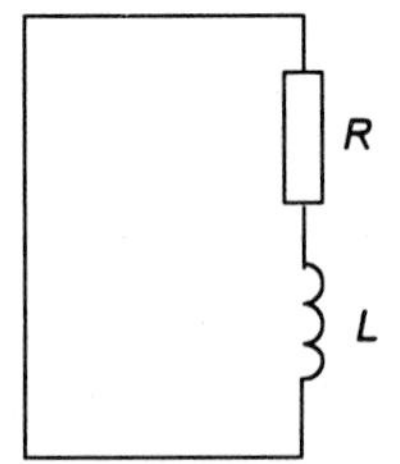

Figure 7.47 *RL circuit*

7.5.2 *RL* circuit: current decay

When the current through a series *RL* circuit is switched off (Figure 7.47), the magnetic field of the inductance changes and induces an e.m.f. in the inductance which opposes the decreasing current. As a result the current does not abruptly drop to a zero value when the current is switched off but decays exponentially to a zero value. Figure 7.48 shows how the circuit current and the voltages across the resistor and inductor vary with time.

The voltage across resistor + voltage across inductor = 0, and so:

$$Ri + L\frac{di}{dt} = 0 \qquad [31]$$

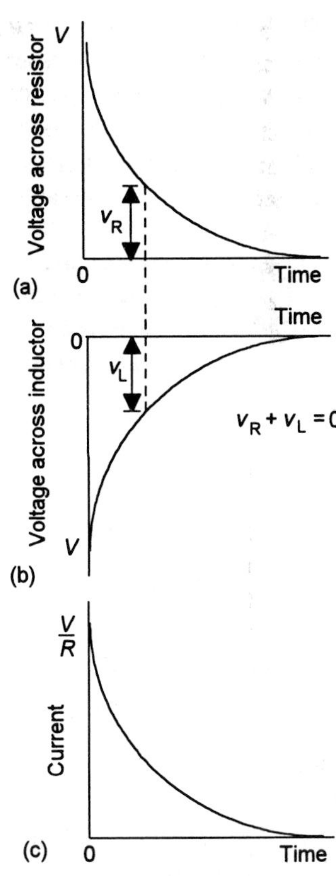

Figure 7.48 *(a) Voltage across R, (b) voltage across L, (c) current*

We can write the above equation as:

$$\frac{di}{dt} = -\frac{R}{L}i$$

Initially the current in the circuit is $I = V/R$ and so immediately the current is switched off the rate of change of current is:

$$\text{initial rate of change of current} = -\frac{R}{L}I \qquad [32]$$

We can obtain the equation of the graph of the current i with time by integrating equation [31]. Separation of variables gives:

$$-\frac{di}{i} = \frac{dt}{R/L}$$

Integration then gives:

$$-\int_I^i \frac{di}{i} = \int_0^t \frac{dt}{R/L}$$

$$-\ln i + \ln I = \frac{Rt}{L}$$

$$\ln\left(\frac{i}{I}\right) = -\frac{Rt}{L}$$

$$i = I\,e^{-Rt/L} \qquad [33]$$

The voltage across the resistor $v_R = iR$ and so:

$$v_R = IR\,e^{-Rt/L} \qquad [34]$$

The voltage across the inductor $v_L = -v_R$ and so is:

$$v_L = -IR\,e^{-Rt/L} \qquad [35]$$

The above three equations describe the graphs in Figure 7.48.

For such a LR circuit, we can write the time constant $\tau = L/R$. Thus:

$$i = I\,e^{-t/\tau}$$

The time taken for the current to drop from I to $0.5I$ is given by $0.5I = I\,e^{-t/\tau}$ and so $e^{-t/\tau} = 0.5$ and we have:

$$-\frac{t}{\tau} = \ln 0.5 = -0.693$$

Thus in a time of 0.693τ the current will drop to half its initial voltage. The time taken to drop to $0.25I$ is given by $0.25I = I\,e^{-t/\tau}$ and so $e^{-t/\tau} = 0.25$ and we have:

$$-\frac{t}{\tau} = \ln 0.25 = -1.386$$

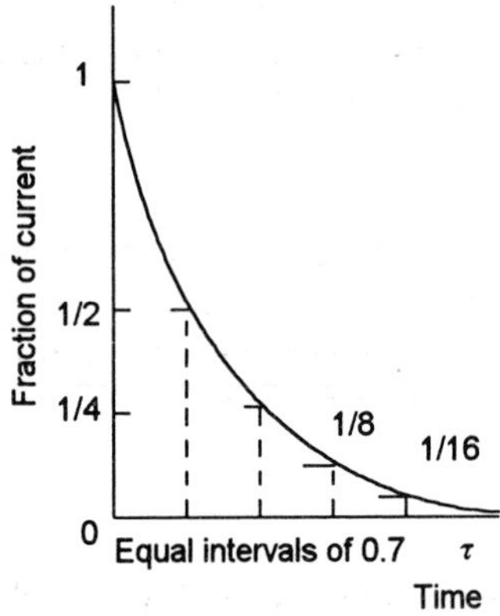

Figure 7.49 *Current decay*

Thus in a time of 1.386τ the current will drop to one-quarter of its initial voltage. This is twice the time taken to drop to half the voltage. This is a characteristic of a decaying exponential graph: if t is the time taken to reach half the steady state value, then in $2t$ it will reach one-quarter, in $3t$ it will reach one-eighth, etc. In each of these time intervals it reduces its value by a half (Figure 7.49) (Table 7.7).

Table 7.7 *Decay of current in series LR circuit*

Time	i_C
0	I
$0.7T$	$0.5I$
$1.4T$	$0.25I$
$2.1T$	$0.125I$
$2.8T$	$0.0625I$
$3.5T$	$0.03125I$

When $t = 1\tau$ then $i = I\,e^{-1} = 0.632I$. Thus in a time equal to the time constant the current drops to 63.2% of the initial voltage. When $t = 2\tau$ then $i = I\,e^{-2} = 0.135I$. Thus the current drops to 13.5% of the initial current. When $t = 3\tau$ then $i = I\,e^{-3} = 0.050I$. Thus the current drops to 5.0% of the initial current. Table 7.8 summarises this data.

Table 7.8 *Decay of current in series LR circuit*

Time	i_C
0	I
$1T$	$0.632I$
$2T$	$0.135I$
$3T$	$0.050I$

Example

A coil with a resistance of 15 Ω and an inductance of 2 H has been connected to a 24 V supply for some time. Determine the current through the coil and the initial rate of change of the current with time when the supply is removed and replaced by a shorting link.

The current through the coil is $V/R = 24/15 = 1.6$ A. When the coil is shorted the initial rate of change of current with time is $-RI/L = -15 \times 1.6/2 = 12$ A/s.

Example

A coil of inductance 200 mH and resistance 8 kΩ is connected to a d.c. voltage source of 16 V. Some time after a steady current exists, the voltage source is short-circuited. What will be (a) the current at the time the short-circuit occurs, (b) the current 10 μs later, (c) the time taken for the current to fall to 10% of its initial value?

(a) The initial current is $I = V/R = 16/8000 = 2$ mA.
(b) The time constant of the circuit $= L/R = 0.200/8000 = 25$ μs. Thus the current after 10 μs is $i = I\,e^{-t/\tau} = 2\,e^{-10/25} = 1.3$ mA.
(c) The time taken for the i to equal $0.1I$, is given by $i = I\,e^{-t/\tau}$ and so $0.1I = I\,e^{-t/25}$. Thus $\ln 0.1 = -t/25$ and hence $t = 57.6$ μs.

Revision

29 A coil has an inductance of 10 H and is carrying a current of 3 A. What will be the current 1 s after the coil is shorted by a resistance of 20 Ω?

30 A coil of inductance 0.4 H and resistance 100 Ω is connected to a d.c. voltage of 10 V. What will be (a) the steady state current, (b) the current 1 ms after the voltage supply is short-circuited?

31 A coil with a resistance of 10 Ω and an inductance of 200 mH has been connected to a 20 V supply for some time. Determine the current through the coil and the initial rate of change of the current with time when the supply is shorted.

Activity

Determine the inductance L of an inductor by measuring the initial rate of change of current through it when a voltage V is first applied to a series circuit of it and a resistor; the initial rate of change of current is V/L. An oscilloscope can be used to monitor the way the voltage across the resistor varies with time and hence, since $i = v_R/R$ give the circuit current. To give a large inductance, the inductor can be a coil mounted on an iron magnetic circuit with the resistance being perhaps 100 Ω and the voltage $V = 1.5$ V.

7.6 D.c. motor

The basic principle of a d.c. motor is a loop of wire which is free to rotate in the field of a magnet (Figure 7.50). When a current is passed through the loop, the resulting forces acting on its sides at right angles to the field cause forces to act on those sides to give rotation.

Coils of wire are mounted in slots on a cylinder of magnetic material called the *armature* which is mounted on bearings and free to rotate. It is mounted in the magnetic field produced by *field poles*. This magnetic field might be produced by, for small motors, permanent magnets or a current in, so-termed, *field coils*. Whether permanent magnet or field coils, these generally are part of the outer casing of the motor and are termed the *stator*. Figure 7.51 shows the basic elements of a d.c. motor with the magnetic field produced by field coils. In practice there will be more than one armature coil and more than one set of stator poles. The ends of the armature coil are connected to adjacent segments of a segmented ring called the *commutator* with electrical contacts made to the segments through fixed carbon contacts called *brushes*. They carry direct current to the armature coil. As the armature rotates, the commutator reverses the current in each coil as it moves between the field poles. This is necessary if the forces acting on the coil are to remain acting in the same direction and so the rotation continue.

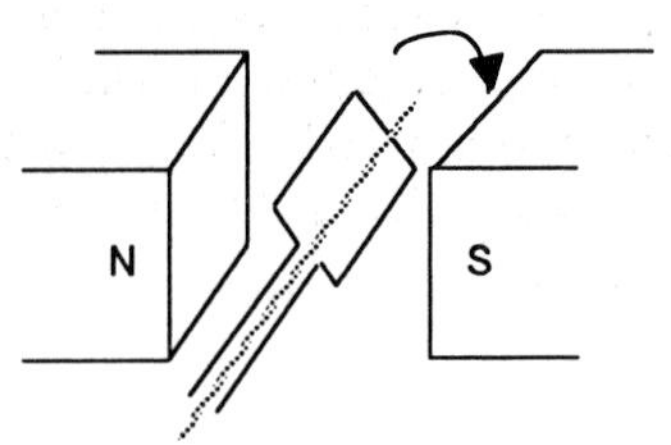

Figure 7.50 *D.c. motor principle*

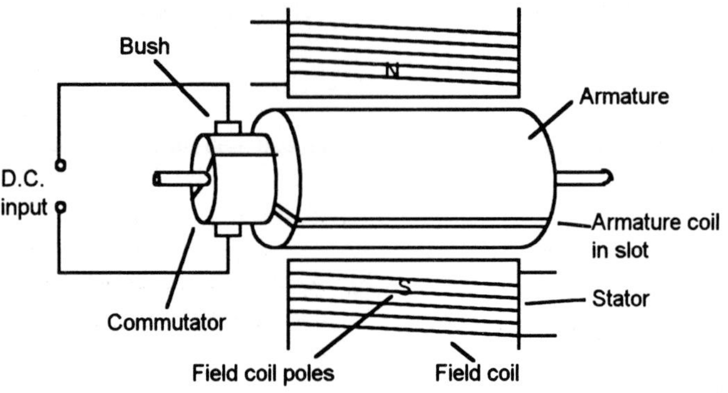

Figure 7.51 *D.c. motor*

7.6.1 Basic theory of the d.c. motor

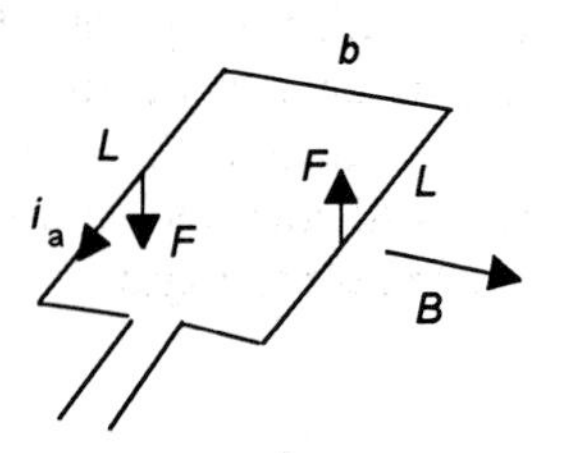

Figure 7.52 *Force on a single armature turn*

Consider a d.c. motor with a flux density B at right angles to an armature loop of length L and carrying a current i_a (Figure 7.52). The force acting on the conductor is Bi_aL. The forces result in a torque about the coil axis of Fb, with b being the breadth of the coil. Thus:

$$\text{torque per turn} = BLi_ab = \Phi i_a \quad [36]$$

where Φ is the flux linked per armature turn. In practice there will be more than one armature loop and more than one set of poles, the torque will, however, be proportional to Φi_a and so we can write:

$$\text{torque acting on armature} = k_t \Phi i_a \quad [37]$$

where k_t is the torque constant.

Since an armature coil is rotating in a magnetic field, a back e.m.f. v_b will be induced which is equal to the rate at which the flux linked by the coil changes and hence, for an armature rotating with an angular velocity ω, is proportional to $\Phi\omega$. Thus we can write:

$$\text{back e.m.f. } v_b = k_v\Phi\omega \quad [38]$$

where k_v is the back e.m.f. constant.

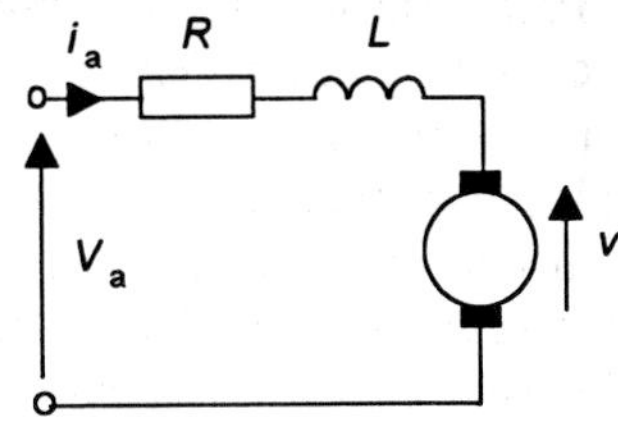

Figure 7.53 *Armature circuit*

We can consider a d.c. motor to be the armature coil, represented by a resistor R in series with an inductance L, in series with a source of back e.m.f. (Figure 7.53). If we are just concerned with steady-state conditions we can neglect the inductance of the armature coil. The voltage providing the current i through the resistance is the applied armature voltage V_a minus the back e.m.f., i.e. $V - v_b$. Hence:

$$V_a = v_b + Ri_a$$

$$i_a = \frac{V_a - v_b}{R} = \frac{V_a - k_v\Phi\omega}{R}$$

and so the torque T is:

$$T = k_t\Phi i_a = \frac{k_t\Phi}{R}(V_a - k_v\Phi\omega) \qquad [39]$$

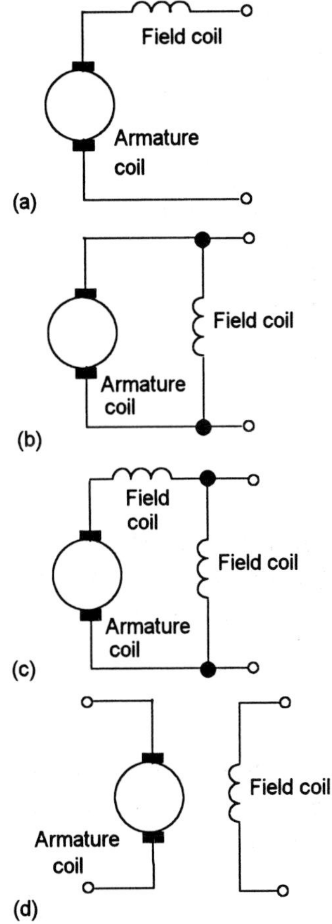

Figure 7.54 *D.c. motors: (a) series (b) shunt, (c) compound, (d) separately wound*

There are a number of forms of d.c. motor, one where a permanent magnet is used to provide the field and ones with field coils which are classified as series, shunt, compound and separately excited according to how the field windings and armature windings are connected (Figure 7.54):

1 *Permanent magnet motor*
For a constant value of flux, e.g. a permanent magnet motor, the starting torque, i.e. the torque when $\omega = 0$, is proportional to the applied voltage and the torque then, as equation [39] indicates, decreases with increasing speed. If the load driven by a d.c. motor is increased, the speed will drop and hence the current through the armature. The relationship between the speed and the armature current determines how the motor will react to load changes.

2 *Series wound motor*
With the series wound motor the armature and field coils are in series and thus carry the same current. Thus the flux Φ depends on the armature current and so the

torque acting on the armature $= k_t\Phi i_a = ki_a^2$

When $\omega = 0$ then $i_a = V/R$ and so:

starting torque $= k(V/R)^2$

Thus, such a motor exerts a high starting torque and has a high no-load speed. Such motors are used where high starting torques are required, e.g. hoists and car engine starters. As the speed is increased, so the torque decreases. Since Ri is small, $V_a = v_b + Ri \simeq v_b$ and so, since $v_b = k_v\Phi\omega$ and Φ is proportional to i, we have V_a proportional to $i\omega$. To a reasonable approximation V_a is constant and so the speed is inversely proportional to the current. The speed thus drops quite markedly when the load is increased. Reversing the polarity of the supply to the coils has no effect on the direction of rotation of the motor; it will continue rotating in the same direction since both the field and armature currents have been reversed. If the core is laminated, to reduce eddy currents, the series motor can be used with a single-phase a.c. supply and is then known as a *universal motor*. Such motors are used in portable power tools and domestic appliances such as food mixers and vacuum cleaners.

3 *Shunt wound motor*
With the shunt wound motor the armature and field coils are in parallel. The field coil is wound with many turns of fine wire and so has a more larger resistance than the armature coil; thus, with a constant supply voltage, the field current is virtually constant. Thus, to a reasonable approximation, we have:

$$T = k_1V - k_2\omega$$

It provides the lowest starting torque and a much lower no-load speed. Since Ri is small, $V_a = v_b + Ri \simeq v_b$ and so, since $v_b = k_v\Phi\omega$ and Φ is virtually constant, we have V_a proportional to ω. With V_a virtually constant, the motor gives almost constant speed regardless of load and such motors are very widely used because of this characteristic. To reverse the direction of rotation, either the armature or field supplied must be reversed.

4 *Compound motor*
The compound motor has two field windings, one in series with the armature and one in parallel. Compound wound motors aim to get the best features of the series and shunt wound motors, namely a high starting torque and constant speed regardless of load.

5 *Separately excited motor*
The separately excited motor has separate control of the armature and field currents.

Example

A permanent d.c. motor develops a torque of 4 N m when the armature current is 2 A. What will be the torque when the armature current is 0.5 A?

Torque = $k_t\,\Phi i_a = ki_a$ and thus $k = 4/2 = 2$ N m/A. When the current is 0.5 A, then torque = $2 \times 0.5 = 1$ N m.

Problems

1 A 100 Ω resistor has a power rating of 250 mW. What should be the maximum current used with the resistor?

2 A 270 Ω resistor has a power rating of 0.5 W. What should be the maximum potential difference applied across the resistor?

3 A 10 Ω resistor is required for use in a circuit where the potential difference across it will be 4 V. What should its power rating be?

4 Resistors of resistances 20 Ω, 30 Ω and 50 Ω are connected in series across a d.c. supply of 20 V. What is the current in the circuit and the potential difference across each resistor?

5 The current to a light-emitting diode (LED) of resistance 250 Ω has to be limited to 5 mA. What resistor should be connected in series with the LED when it is connected to a 5 V d.c. supply?

6 A potential divider circuit is to be used to reduce a 24 V d.c. supply to 5 V. What resistors should be used if the circuit current should not exceed 10 mA?

7 Determine the power dissipated in each resistor and the total power dissipated when a voltage of 10 V is applied across (a) a series-connected (b) a parallel-connected pair of resistors with resistances of 20 Ω and 50 Ω.

8 Calculate the equivalent resistance of three resistors of 5 Ω, 10 Ω and 15 Ω which are connected in (a) series, (b) parallel.

9 Three resistors of 15 kΩ, 20 kΩ and 24 kΩ are connected in parallel. What is (a) the total resistance and (b) the power dissipated if a voltage of 24 V is applied across the circuit?

10 A circuit consists of two resistors in parallel. The total resistance of the parallel arrangement is 4 Ω. If one of the resistors has a resistance of 12 Ω, what will be the resistance of the other one?

11 A circuit consists of two resistors in series. If they have resistances of 4 Ω and 12 Ω, what will be (a) the circuit current and (b) the potential difference across each resistor when a potential difference of 4 V is applied across the circuit?

12 A circuit consists of two resistors in parallel. If they have resistances of 4 Ω and 12 Ω, what will be (a) the current through each resistor and (b) the total power dissipated when a potential difference of 4 V is applied across the circuit?

13 A circuit consists of two parallel-connected resistors of 2.2 kΩ and 3.9 kΩ in series with a 1.5 kΩ resistor. With a supply voltage of 20 V, what is the current drawn from the supply?

14 A circuit consists of three parallel-connected resistors of 330 Ω, 560 Ω and 750 Ω in series with a 800 Ω resistor. If the supply

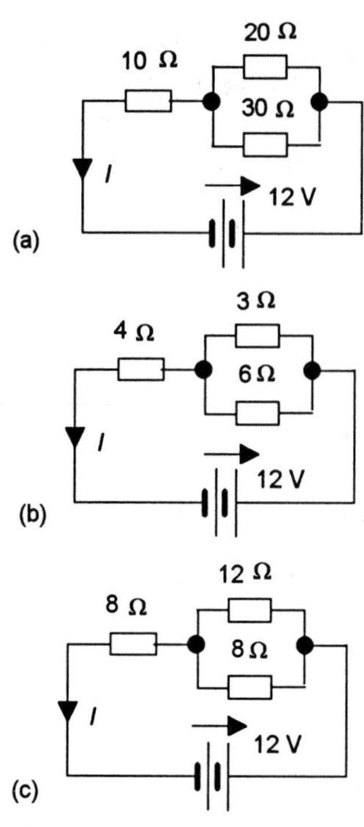

Figure 7.55 *Problem 16*

voltage connected to the circuit is 12 V, what are the voltages across the parallel resistors and the series resistor?

15 Three resistors of 6 Ω, 12 Ω and 24 Ω are connected in parallel across a voltage supply. What fraction of the supply current will flow through each resistor?

16 Determine the total resistances of the circuits shown in Figure 7.55 and the circuit currents *I*.

17 Determine the total resistance and the total circuit drawn from the voltage source in each of the series–parallel arrangements of resistors shown in Figure 7.56.

18 Determine the current through and the voltage drop across each of the resistors in the series–parallel circuits shown in Figure 7.57.

19 Determine the potential difference across the 10 Ω resistor in the circuit shown in Figure 7.58.

20 Determine, using (a) node analysis and (b) mesh analysis, the current *I* in each of the circuits shown in Figure 7.59.

21 What is the potential difference between the plates of a 4 μF capacitor when there is charge of 16 μC on its plates?

22 A 2 μF capacitor and a 4 μF capacitor are connected in series across a 12 V d.c. supply. What will be the resulting (a) charge on the plates of each capacitor and (b) the potential difference across each?

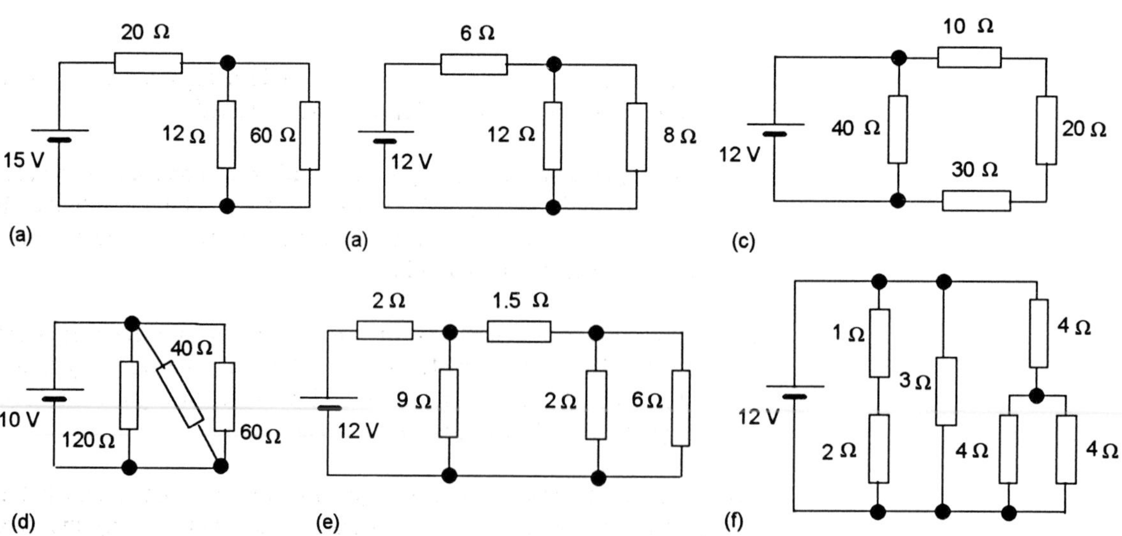

Figure 7.56 *Problem 17*

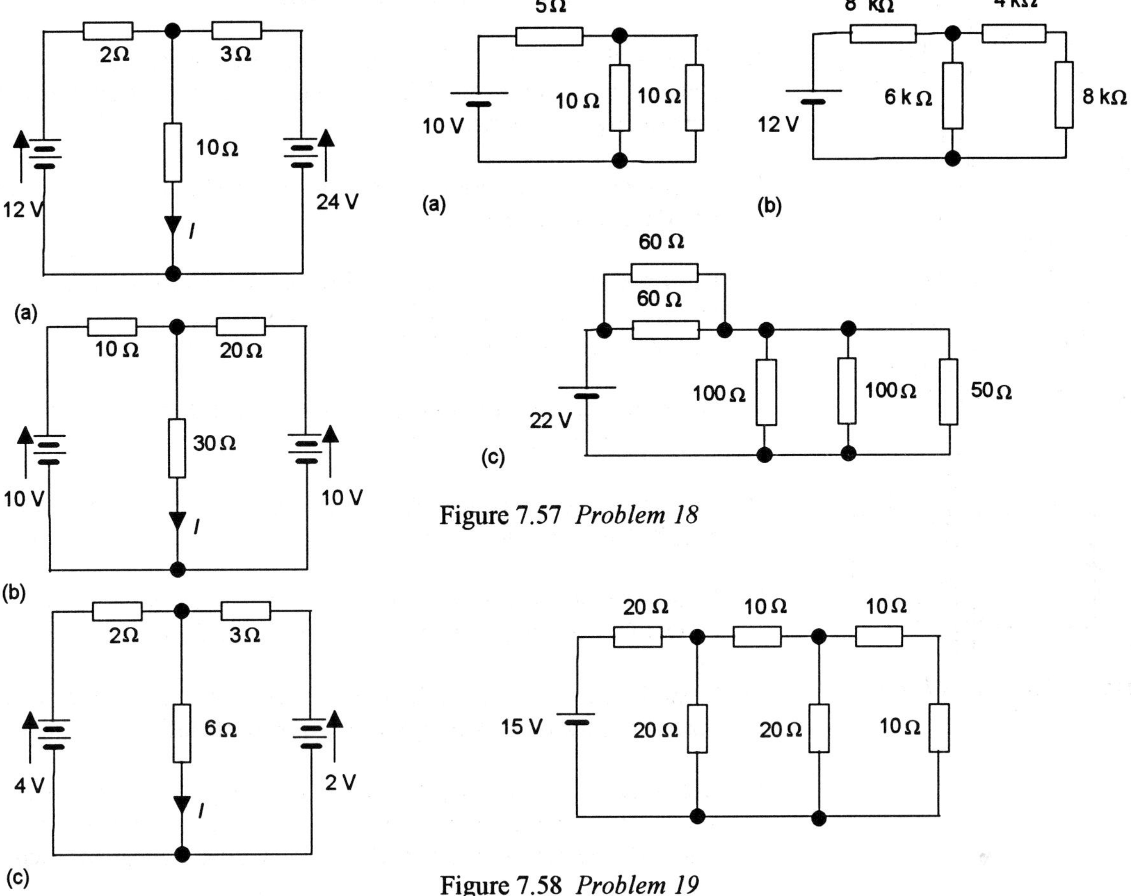

Figure 7.57 *Problem 18*

Figure 7.58 *Problem 19*

Figure 7.59 *Problem 20*

23 What will be the total capacitances when three capacitors, 2 μF, 4 μF and 8 μF, are connected in (a) series, (b) parallel?

24 Three capacitors, 1 μF, 2 μF and 3 μF, are connected in parallel across a 10 V d.c. supply. What will be (a) the total capacitance, (b) the resulting charge on each capacitor, (c) the resulting potential difference across each capacitor?

25 Three capacitors, 20 μF, 30 μF and 60 μF, are connected in series across a 36 V d.c. supply. What will be (a) the total equivalent capacitance, (b) the charge on the plates of each capacitor and (c) the potential differences across each capacitor?

26 Three capacitors, 50 pF, 100 pF and 220 pF, are connected in parallel across a 10 V d.c. supply. What will be (a) the total capacitance, (b) the resulting charge on each capacitor, (c) the resulting potential difference across each capacitor?

27 What is the initial rate at which the voltage across a 0.1 μF capacitor will change with time when it is charged through a resistance of 5 kΩ by a voltage being switched from 0 to 10 V?

28 What is the initial rate at which the voltage across a 10 μF capacitor will change with time when it is charged through a resistance of 20 kΩ by a voltage being switched from 0 to 6 V?

29 A series *RC* circuit has a resistance of 50 kΩ and a capacitance of 20 μF. Determine the current (a) initially and (b) 1 s, after a steady voltage of 20 V is connected to the circuit.

30 A series *RC* circuit has a resistance of 15 kΩ and a capacitance of 0.02 μF and a steady voltage of 30 V is connected to the circuit. Determine (a) the time constant τ of the circuit, (b) the current after 1τ, (c) the current after 1.5τ, (d) the steady state current.

31 A series *RC* circuit has a resistance of 15 kΩ and a capacitance of 0.02 μF and a steady voltage of 30 V is connected to the circuit. Determine the voltage across the resistor (a) initially, (b) after 0.3 ms, (c) after 1 ms.

32 A series *RC* circuit has a resistance of 0.5 kΩ and a capacitance of 500 μF and a steady voltage of 10 V is connected to the circuit. Determine (a) the time constant τ of the circuit, (b) the voltage across the capacitor after 0.25 s, (c) the voltage across the capacitor after 0.5 s.

33 A 16 μF capacitor is charged to 10 V and then discharged through a resistance of 50 kΩ. Determine (a) the time taken for the voltage across the capacitor to drop to 2 V, (b) the circuit current after 0.5 s.

34 A 5 μF capacitor is charged to 10 V and then discharged through a resistance of 2 MΩ. Determine (a) the initial current, (b) the time taken for the voltage across the capacitor to drop to 6.3 V.

35 A 0.2 μF capacitor is charged to 5 V and then discharged through a resistance of 40 kΩ. Determine (a) the voltage across the capacitor after 15 ms, (b) the voltage across the resistor after 15 ms.

36 What is the inductance of a coil if an e.m.f. of 20 V is induced in it when the current changes at the rate of 10 A/s?

37 What is the e.m.f. induced in a coil of inductance 100 mH when the current through it is changing at the rate of 4 A/s?

38 A coil with a resistance of 50 Ω and an inductance of 2.5 H has a d.c. voltage of 100 V connected across it. Determine (a) the time constant of the circuit, (b) the initial rate of change of current, (c)

the circuit current 0.15 s after the voltage is applied to the coil, (d) the final steady state current.

39 A coil with a resistance of 25 Ω and an inductance of 2.5 H has a d.c. voltage of 100 V connected across it. Determine (a) the time constant of the circuit, (b) the initial rate of change of current, (c) the circuit current 0.15 s after the voltage is applied to the coil, (d) the final steady state current.

40 A series *RL* circuit has a d.c. voltage of 10 V applied to a resistance of 125 Ω and an inductance of 0.25 H. Determine the circuit current after 5 ms.

41 A coil of inductance 5 H and resistance 20 Ω is connected to a d.c. voltage source of 100 V. Some time after a steady current exists, the voltage source is short-circuited. What will be the current (a) at the time the short-circuit occurs, (b) 1 s later?

42 A relay coil has a resistance of 100 Ω and an inductance of 100 mH. A rectangular voltage pulse of amplitude 5 V and duration 5 ms is applied to the coil. If the relay contacts close when the current in the coil reaches 40 mA and open when it drops to 15 mA, determine the length of time for which the contacts remain open.

43 A coil of inductance 20 H and resistance 5 Ω is connected to a d.c. voltage source. When the voltage is applied, the initial rate of increase of current is 4 A/s. What is (a) the value of the applied voltage, (b) the rate of growth of the current when the circuit current is 5 A?

44 A coil of inductance 10 H and resistance 10 Ω is connected to a d.c. voltage source of 100 V. What is (a) the current 0.1 s after switching the voltage on, (b) the time taken for the current to decrease to half its initial value?

45 If the time constant of a coil is measured and found to be 50 ms, what will be its inductance if it has a resistance of 2 Ω?

46 When a d.c. voltage of 100 V is applied to the field coil of a machine, after 2 s the current has risen to 6.32 A and eventually it reaches 10 A. What is the inductance of the coil?

8 A.c. theory

8.1 Introduction

This chapter considers the terms used to describe sinusoidal alternating voltages and currents and their representation by phasors. Phasors are then used in a discussion of the basic principles of a.c. theory and the solution of problems involving circuits containing resistance, inductance and capacitance in series and in parallel. The approach is in a general manner involving the consideration of phasors and their addition by means of 'phasor diagrams'.

The term *direct* voltage or current is used when the voltage or current is always in the same direction. The term *alternating* voltage or current is used when the polarity or direction of flow of the voltage or current alternates, continually changing with time. Alternating waveforms oscillate from positive to negative values in a regular, periodic manner. One complete sequence of such an oscillation is called a *cycle* (Figure 17.2). The time T taken for one complete cycle is called the *periodic time* and the number of cycles occurring per second is called the *frequency* f. Thus $f = 1/T$. The unit of frequency is the hertz (Hz), 1 Hz being 1 cycle per second.

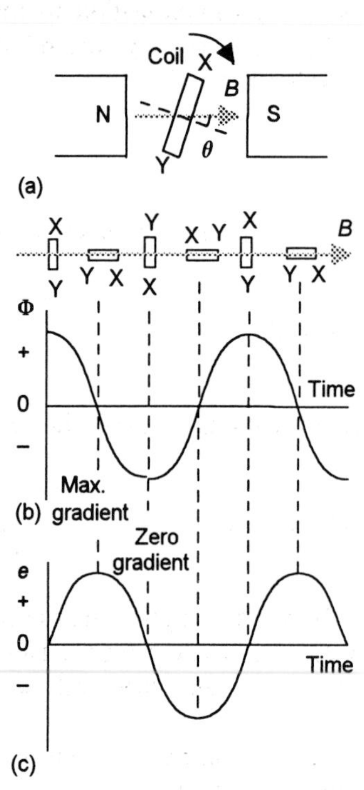

Figure 8.1 *Basic ac generator*

8.1.1 A.c. generators

Faraday's law of electromagnetic induction states that the size of the induced e.m.f is proportional to the rate of change of flux linked by a coil. One way of changing the flux linked by a coil when in a magnetic field is to rotate the coil so that the angle between the field direction and the plane of the coil changes (Figure 8.1(a)). When the plane of the coil is at right angles to a field of flux density B then the flux linked per turn of wire is a maximum and given by BA, where A is the area of the coil. When the plane of the coil is parallel to the field then there is no flux linked by the coil. When the axis of the coil is at an angle θ to the field direction then the flux linked per turn is $\Phi = BA \cos \theta$. The flux linked thus changes as the coil rotates. If the coil rotates with an angular velocity ω then its angle θ at a time t is ωt and so the flux linked per turn varies with time according to:

$$\Phi = BA \cos \omega t$$

Figure 8.1(b) shows how the flux linked varies with time. The induced e.m.f. produced per turn is the rate at which the flux changes with time. It is thus the gradient of the graph of flux against time. Thus when the gradient is zero there is no induced e.m.f., when the gradient is a maximum the induced e.m.f. is a maximum. The direction of the induced e.m.f is such as to oppose the change producing it, this is known as *Lenz's law*, and so the induced e.m.f. e is $-\mathrm{d}\Phi/\mathrm{d}t$. Thus, differentiating $\Phi = BA \cos \omega t$ gives for the rate of change of flux with time $\mathrm{d}\Phi/\mathrm{d}t = -BA \sin \omega t$ and thus, since $e = -\mathrm{d}\Phi/\mathrm{d}t$, obtain

$$e = BA\omega \sin \omega t = E_{max} \sin \omega t \qquad [1]$$

where the maximum e.m.f. E_{max} is $BA\omega$. The induced e.m.f. is given by a sine graph (Figure 8.1(c)). Thus rotating the coil in a magnetic field has led to a basic alternating voltage *generator*.

Since the angular velocity $\omega = 2\pi f$, where f is the frequency of rotation, then equation [1] can be written as:

$$e = E_{max} \sin 2\pi ft \qquad [2]$$

8.1.2 Sinusoidal voltages and currents

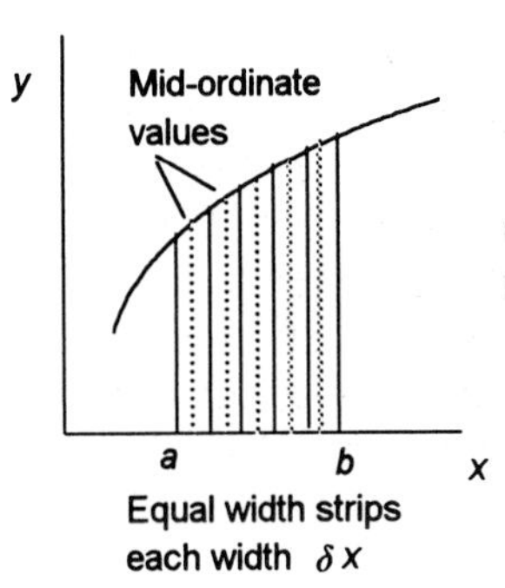

Figure 8.2 *Mean value*

Two terms used to describe the value of an alternating current or voltage are the mean value and root-mean-square value.

The *mean* value of a set of numbers is their sum divided by the number of numbers summed. The mean value of some function between specified limits that is described by a graph can be considered to be the mean value of all the ordinates representing the values between these limits. We can give an approximation of this if we divide the area into a number of equal width strips (Figure 8.2); an approximation to the average value is then the sum of all the mid ordinates y_1, y_2, y_3, etc. of the strips divided by the number n of strips considered.

$$\text{mean value} = \frac{\text{sum of mid-ordinate values}}{\text{number of mid-ordinates}}$$

If each strip has a width δx then between $x = a$ and $x = b$ we have $(b - a)/\delta x$ strips and so:

$$\text{mean value} = \frac{(\text{sum of mid-ordinate values})\delta x}{b - a}$$

But each mid-ordinate value multiplied by its width is its area. Thus the mean value between a and b is:

$$\text{mean value} = \frac{\text{area under graph between } a \text{ and } b}{b - a} \qquad [3]$$

For a sinusoidal signal, because the waveform has a negative half cycle which is just the mirror image of the positive half cycle, then the mean over a full cycle must be zero. For every positive value there will be a corresponding negative value.

Over one half cycle the average value of the sinusoidal waveform is the area, which can be obtained by integration, divided by π.

$$\text{mean value} = \frac{1}{b - a}\int_a^b y \, \mathrm{d}x = \frac{1}{\pi - 0}\int_0^\pi \sin x \, \mathrm{d}x$$

$$= \frac{1}{\pi}[-\cos x]_0^\pi = \frac{2}{\pi} = 0.637$$

The average value for a sinusoidal current waveform of maximum value I_m, or a voltage waveform of maximum value V_m, over half a cycle is thus 0.637 times its maximum value.

Since we are frequently concerned with the power developed by a current passing through a circuit component, a useful measure of an alternating current is in terms of the direct current that would give the same power dissipation in a resistor. For an alternating current, the power at an instant of time is i^2R, where i is the current at that instant and R is the resistance. Thus to obtain the power developed by an alternating current over a cycle, we need to find the average power developed over that time. In terms of mid-ordinates we add together all the values of power given at each mid-ordinate of time in the cycle and divide by the number of mid-ordinates considered. Because we are squaring the current values, negative currents give positive values of power. Thus the powers developed in each half cycle add together.

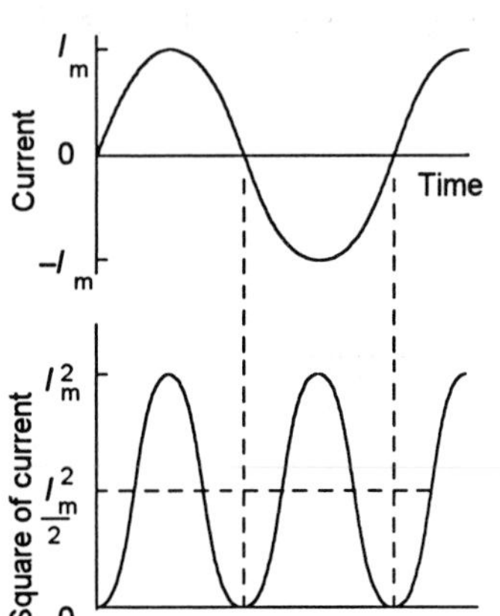

Figure 8.3 *Root-mean-square current*

We can determine the root-mean-square value for a sinusoidal waveform by considering the form of the graph produced by plotting the square of the current values, or the voltage values. Figure 8.3 shows the graph. The squares of the positive and negative currents are all positive quantities and so the resulting graph oscillates between a maximum value of I_m^2 and 0. The mean value of the i^2 graph is $I_m^2/2$, the i^2 graph being symmetrical about this value. The mean power is thus $R\,I_m^2/2$. For a direct current I to give the same power as the alternating current, we must have $I^2R = P_{av}$ and so the root-mean-square current I_{rms} is given by:

$$I_{rms}^2 R = R\frac{I_m^2}{2}$$

$$I_{rms} = \frac{I_m}{\sqrt{2}} \qquad [4]$$

The root-mean-square value for a sinusoidal current waveform of maximum value I_m, or a voltage waveform of maximum value V_m, over half a cycle is thus its maximum value divided by √2.

Example

For a sinusoidal current of maximum value 3 A, determine (a) the mean value over half-a-cycle, (b) its root-mean-square value of a current.

(a) The mean value over half-a-cycle is $0.637I_m = 1.9$ A.
(b) The root-mean-square value = $I_m/\sqrt{2} = 3/\sqrt{2} = 2.1$ A.

Revision

1 Determine the mean value over half-a-cycle and the root-mean-square value of (a) a sinusoidal voltage of maximum value 4 V, (b) a sinusoidal current of maximum value 1.2 A.

8.2 Phasors

We can represent a voltage v which varies sinusoidally with time t by the equation $v = V \sin \omega t$, where V is the maximum value of the voltage and ω the angular frequency and equal to $2\pi f$. We can imagine such a signal being produced by the vertical projection of a radial line of length V rotating with a constant angular velocity ω from some initial start position (Figure 8.4). Thus instead of specifying the variation of the voltage with time by the above equation, we can specify it by the length of the line V and whether it starts at $t = 0$ at some angle, termed the *phase angle* ϕ, to the reference axis which is usually taken as the horizontal axis. Such lines are termed *phasors*.

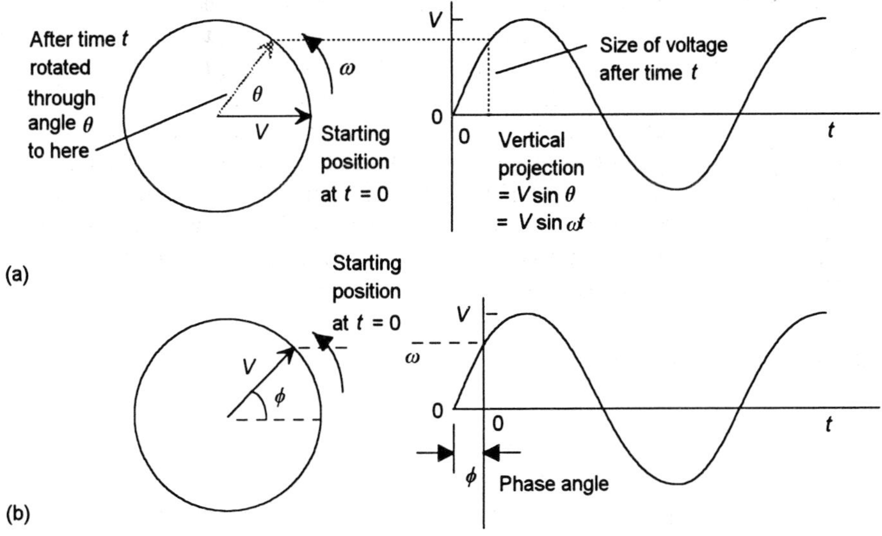

Figure 8.4 *(a) $v = V \sin \omega t$, (b) $v = V \sin(\omega t + \phi)$*

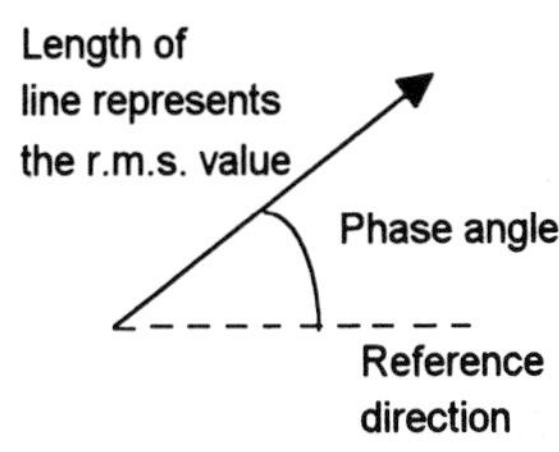

Figure 8.5 *A phasor*

A *phasor* can be described by drawing an arrow-headed line, the length of the line representing the amplitude and its direction, relative to a reference direction, as the phase angle (Figure 8.5). Because with alternating currents and voltages we are normally concerned with root-mean-square (r.m.s.) values rather than maximum value, for sinusoidal waves the maximum value is just the r.m.s. value divided by $\sqrt{2}$, generally when the term phasor is used for an arrow-headed line describing alternating currents and voltages the length of the line represents the r.m.s. value.

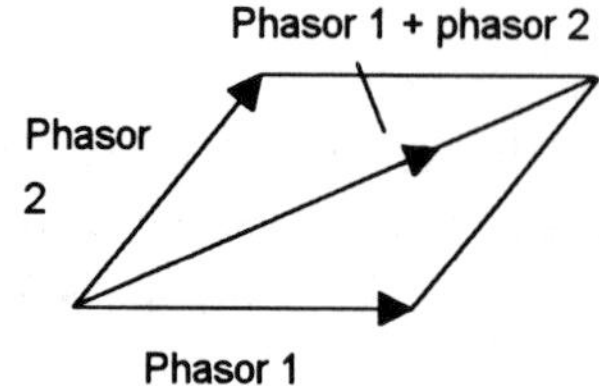

Figure 8.6 *Addition of phasors by the parallelogram law*

The phase angle is the angle between a phasor and some reference direction. In the case of a series circuit it is customary to use the direction of the current phasor for the circuit as the reference direction, the current being the same for all the series components. For a parallel circuit it is customary to use the direction of the voltage phasor for the parallel circuit as the reference phasor, the voltage being the same for all parallel components.

In textbooks the common practice to indicate that a symbol represents a phasor is to use bold print, e.g. **V** represents a voltage phasor. The voltage representing the length of the phasor would be given by the italic, non-bold, symbol *V*. The instantaneous value of the voltage is represented by a lower case *v*.

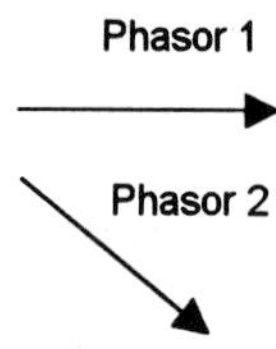

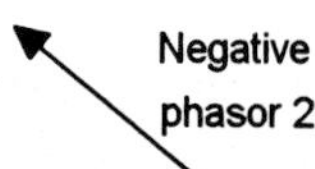

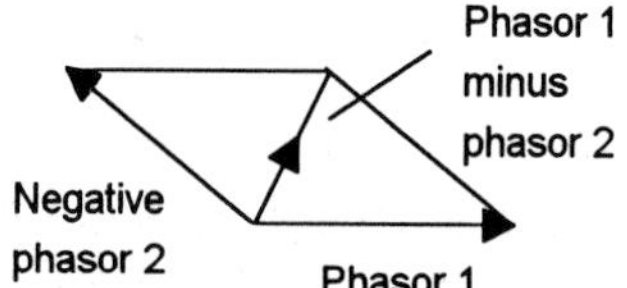

Figure 8.7 *Subtraction of one phasor from another*

8.2.1 Addition and subtraction of phasors

Phasors can be added, or subtracted, by the methods used to add, or subtract, vector quantities. One method to add two phasors is the *parallelogram law* (the same parallelogram law as used for the addition of vector quantities). If the phasors are drawn to scale as arrow-headed lines then we draw one phasor, then draw the next phasor so that its tail is attached to the tail of the preceding one, and then complete a parallelogram with lines parallel to these two phasors; the line from junctions of the tails of the two phasors to give the diagonal of the parallelogram represents the result of the addition (Figure 8.6). Since a negative phasor –**V** is just **V** with the direction reversed, subtraction is accomplished by the addition of the negative of a phasor (Figure 8.7).

Example

Determine the phasor to represent the voltage across a circuit if it is the sum of the phasors 10 V with phase angle 0° and 5 V with phase angle 90°.

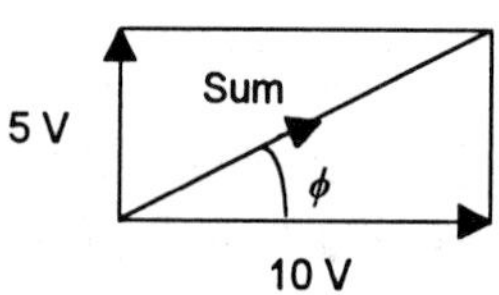

Figure 8.8 *Example*

Figure 8.8 shows the phasor parallelogram. The phasor representing the sum is the diagonal of the parallelogram from the junction of the two tails of the phasor. Using the Pythagoras theorem,

$$(\text{diagonal})^2 = 5^2 + 10^2$$

Hence the sum phasor has a magnitude of 11.2 V. This is at a phase angle ϕ where $\tan \phi = 5/10$ and so ϕ is 26.6°. The voltage leads the 10 V voltage by 26.6°. The term 'lead' is used when the angle is an anticlockwise rotation from the reference phasor, in this case the 10 V phasor. The phasor is thus 11.2 V leading by 26.6°.

Example

The phasor representing the current entering an arrangement of two parallel components is 4 A at zero phase angle. If the phasor representing the current through one component is 2 A at a phase

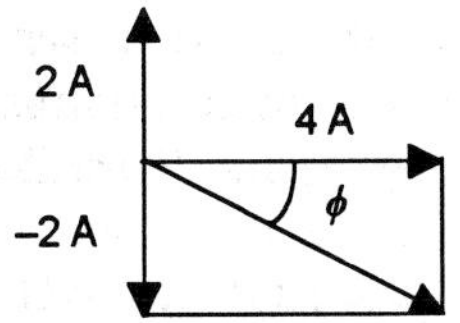

Figure 8.9 *Example*

angle of 90°, determine the phasor to represent the current through the other component .

We need to subtract the phasor for the current through one component from the phasor for the total current. We can do this by adding to the phasor for the total current the negative phasor for the current through the component. Figure 8.9 shows the phasors and the 'sum' phasor. The current through the second component is thus given by the Pythagoras theorem as $\sqrt{(4^2 + 2^2)} = 4.5$ A at a phase angle given by $\tan \phi = -2/4$ and thus $\phi = -25.6°$. The current lags the overall current by 25.6°. The term 'lag' is used when the current phasor is at a phase which involves a clockwise rotation from the reference phasor.

Revision

2 Determine the phasor to represent the voltage across a circuit if it is the sum of the phasors 5 V with phase angle 0° and 2 V with phase angle 90°.

3 Determine the phasor to represent the current leaving a parallel circuit if it is the sum of the phasors 2 A with phase angle 0° and 3 A with phase angle 90°.

4 If the voltage across two series components is represented by a 10 V phasor with zero phase angle and the voltage phasor across one of them is 4 V at 90° phase angle, determine the voltage phasor for the other component.

8.3 Reactance and susceptance

The *reactance* X of a component is the maximum voltage V_m across the component divided by the maximum current I_m through it and has the unit of ohms. Note that these maximum values may not occur at the same time.

$$X = \frac{V_m}{I_m} \quad [5]$$

Since, with sinusoidal signals, the root-mean-square values $V_{r.m.s} = V_m/\sqrt{2}$ and $I_{r.m.s} = I_m/\sqrt{2}$, we can also write:

$$X = \frac{V_{r.m.s}}{I_{r.m.s}} \quad [6]$$

With d.c. we have the term conductance for the reciprocal of resistance; with a.c. the term *susceptance* B is used for the reciprocal of reactance X and has the unit of /Ω which is given the name siemen (S).

$$B = \frac{1}{X} \qquad [7]$$

Example

What are the reactance and susceptance of a component if the alternating voltage drop across it has a root-mean-square value of 20 V when the root-mean-square current through it is 50 mA.

The reactance X is $V_{\text{r.m.s.}}/I_{\text{r.m.s.}}$ = 20/0.050 = 400 Ω. The susceptance is the reciprocal of reactance and so is 1/400 = 0.0025 S.

Revision

5 What are the reactance and susceptance of a component if the alternating voltage drop across it has a root-mean-square value of 5 V when the root-mean-square current through it is 2 mA?

8.4 Phasor relationships for pure components

The term 'pure' is used since it is assumed that the components only have the single property concerned. Thus a pure inductor is assumed to have only inductance and no resistance or capacitance.

For a pure resistor R the voltage drop V across it is $v = Ri$ and thus for $i = I \sin \omega t$ we have $v = Ri \sin \omega t$ and so the voltage is in phase with the current (Figure 8.10(a)). The maximum value of the current occurs at the same time as the maximum value of the voltage and the ratio of the maximum voltage to the maximum current or the r.m.s. voltage to the r.m.s. current is the resistance.

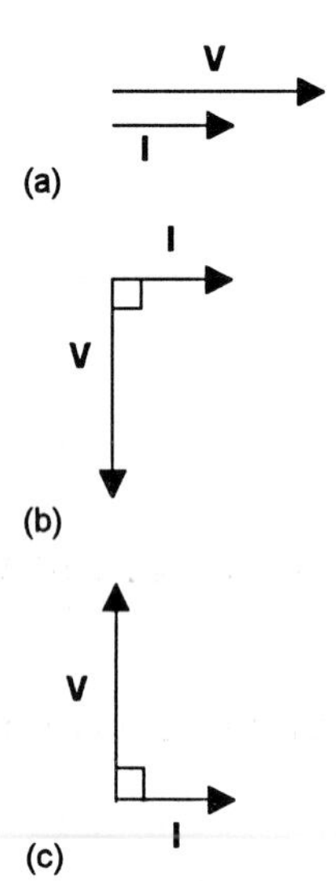

Figure 8.10 *Phasors for: (a) resistors, (b) capacitors, (c) inductors*

For a pure capacitor of capacitance C the charge $q = Cv$ and thus, for $v = V \sin \omega t$, we have $q = CV \sin \omega t$ and so the current $i = \mathrm{d}q/\mathrm{d}t = \omega CV \cos \omega t = \omega CV \sin(\omega t + 90^\circ)$. The current leads the voltage across the capacitor by 90° (Figure 8.10(b)). Alternatively we can say that the voltage lags the current by 90°. Thus the maximum voltage and maximum current do not occur at the same time. The ratio of the maximum voltage to maximum current or the r.m.s. voltage to r.m.s. current, i.e. the capacitive reactance X_C, is:

$$X_C = \frac{1}{2\pi fC} = \frac{1}{\omega C} \qquad [8]$$

For a pure inductor of inductance L the voltage drop across it $v = L\,\mathrm{d}i/\mathrm{d}t$ and thus for a current of $i = I \sin \omega t$ we have $v = \omega LI \cos \omega t = \omega LI \sin(\omega t + 90^\circ)$. The voltage leads the current by 90° or alternatively we can say the current lags the voltage by 90° (Figure 8.10(c)). Thus the maximum voltage and the maximum current do not occur at the same time. The ratio of the maximum voltage to maximum current or the r.m.s. voltage to r.m.s. current, i.e. the inductive reactance X_L, is:

$$X_L = 2\pi fL = \omega L \qquad [9]$$

A useful way of remembering the phase relationships with capacitors and inductors is the mnemonic:

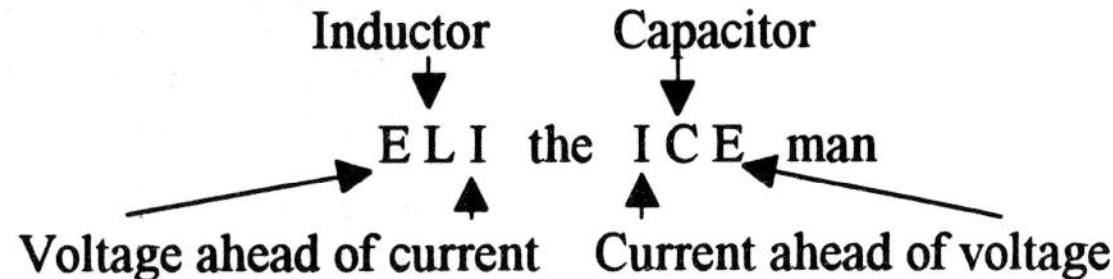

Another mnemonic that can be used is:

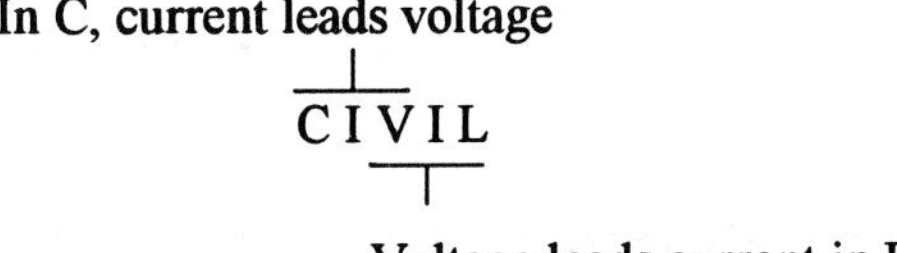

Example

Determine the reactance of a pure inductor if it has an inductance of 200 mH and is used in a circuit where the frequency of the alternating current is 50 Hz. What will be the r.m.s. current through the inductor when the r.m.s. voltage drop across it is 20 V?

The reactance is given by equation [9] as:

$$X_L = 2\pi fL = 2\pi \times 50 \times 0.200 = 62.8\ \Omega$$

The r.m.s. current $I_{r.m.s}$ is given by:

$$I_{r.m.s} = \frac{V_{r.m.s}}{X_L} = \frac{20}{62.8} = 0.32\text{ A}$$

Example

If the alternating current through a 20 μF capacitor is 3 sin 800*t* A, what is the voltage across it?

With ω = 800 rad/s, the reactance of the capacitor is given by equation [8] as:

$$X_C = \frac{1}{\omega C} = \frac{1}{800 \times 20 \times 10^{-6}} = 62.5\ \Omega$$

The maximum value of the current is 3 A. Hence the maximum value of the voltage is $V = IX_C = 3 \times 62.5 = 187.5$ V. The voltage is thus 187.5 sin 800*t* V.

Revision

6 Determine the reactance of a pure capacitor if it has a capacitance of 8 μF and is used in a circuit where the frequency of the alternating current is 1 kHz. What will be the r.m.s. current through the capacitor when the r.m.s. voltage drop across it is 10 V?

7 Determine the capacitive reactance of a 0.5 μF capacitor, and the current through it, when the voltage across it is 16 sin 2000*t* V.

8 Determine the inductive reactance of a pure inductor of inductance 80 mH when the current through it is 100 sin 400*t* mA and the voltage drop across it.

8.5 Impedance and admittance

In discussing series and parallel a.c. circuits involving resistors, capacitors and inductors, the term impedance will be used.

> The *impedance Z* of a circuit is the phasor value of the voltage across the circuit divided by the phasor value of the current through it., i.e. the values taking into account not only their magnitudes but also their phase angles.

Impedance is not a sinusoidally varying quantity and thus is not a phasor, so bold print is not used for it in this book (though some textbooks use bold print because it is a complex quantity). Impedance has the unit of ohms.

$$Z = \frac{\mathbf{V}}{\mathbf{I}} \qquad [10]$$

Admittance Y is the reciprocal of impedance *Z*. Impedance is a measure of how well a component impedes the current, admittance is a measure of how it admits, i.e. allows, the current.

$$Y = \frac{1}{Z} \qquad [11]$$

The unit of admittance is /Ω or siemen (S). Thus an impedance of 10 Ω is an admittance of 1/10 = 0.1 S.

8.6 Series a.c. circuits

For a series d.c. circuit there is the same current through each component and the voltage drop across all the components is the sum of the voltage drops across each. For a series a.c. circuit there is the same current phasor for the current through each component and the phasor

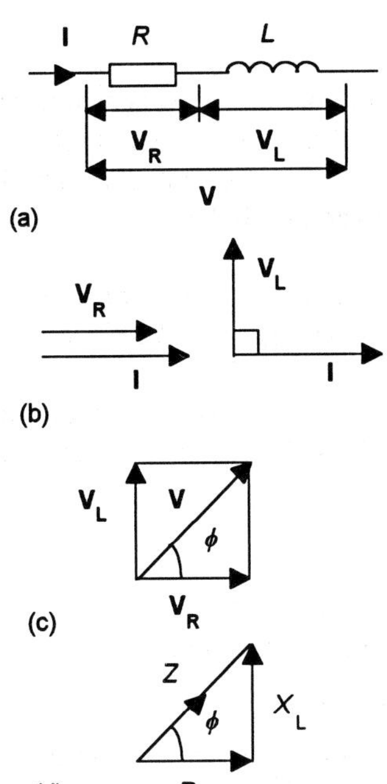

Figure 8.11 *RL series circuit*

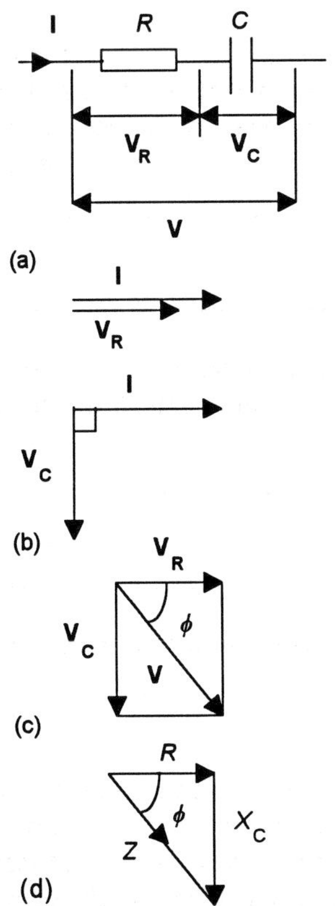

Figure 8.12 *Series RC circuit*

for the voltage drop across all the components is the sum of the phasors for the voltage drops across each component.

For a *series circuit containing resistance and inductance* (Figure 8.11(a)), the voltage for the resistance is in phase with the current and the voltage for the inductor leads the current by 90° (Figure 8.11(b)). Thus the phasor for the sum of the voltage drops across the two series components is given by Figure 18.11(c) as a voltage phasor **V** with a phase angle ϕ. We can use the Pythagoras theorem to give the magnitude V of the voltage:

$$V^2 = V_R^2 + V_L^2 \qquad [12]$$

and trigonometry to give the phase angle ϕ, i.e. the angle by which the voltage leads the current, which is in the direction of $\mathbf{V}_R$:

$$\tan\phi = \frac{V_L}{V_R} \qquad [13]$$

or

$$\cos\phi = \frac{V_R}{V} \qquad [14]$$

Since $V_R = IR$ and $V_L = IX_L$ and the magnitude V of **V** is given by $V = IZ$, then by substitution in equations [12], [13] and [14]:

$$Z^2 = R^2 + X_L^2 \qquad [15]$$

$$\tan\phi = \frac{X_L}{R} \qquad [16]$$

$$\cos\phi = \frac{R}{Z} \qquad [17]$$

Alternatively, we can arrive at the above equations taking the voltage triangle formed by half the parallelogram in Figure 8.11(c) and, since the voltages are IR, IX_L and IZ, if we divide each side by I we end up with an impedance triangle shown in Figure 8.11(d). The values of the impedance and the phase angle can then be determined from the impedance triangle by the use of the Pythagoras theorem and trigonometry:

For a *series circuit containing resistance and capacitance* (Figure 8.12(a)), the voltage for the resistance is in phase with the current and the voltage for the capacitor lags the current by 90° (Figure 8.12(b)). Thus the phasor for the sum of the voltage drops across the two series components is given by Figure 8.12(c) as a voltage phasor with a phase angle ϕ. We can use the Pythagoras theorem to give the magnitude V of the voltage:

Application

A circuit with an inductor in series with a resistor has a voltage across the resistor which depends on the frequency of the voltage supply. The lower the frequency the lower the reactance of the inductor and so the higher the voltage across the resistor. Such a circuit thus acts as a *low-pass filter* in that the high frequency voltage inputs are effectively blocked and only the low frequency voltages appear across the resistor.

A circuit with a capacitor in series with a resistor has a voltage across the resistor which depends on the frequency of the voltage supply. The higher the frequency the lower the reactance of the capacitor and so the higher the voltage across the resistor. Such a circuit thus acts as a *high-pass filter* in that low frequency voltage inputs are effectively blocked and only the high frequency voltages appear as output voltage across the resistor.

The term *cut-off frequency* is used for the frequency at which the power delivered to the resistor is half its maximum value. Thus, with the high-pass filter of series capacitor and resistor, at high frequencies the capacitive reactance is negligible and virtually all the supply voltage V is across the resistance and so the current is at its highest and V/R; the power delivered to the resistor is thus V^2/R and the cut-off frequency is when the power is $\frac{1}{2}V^2/R$.

$$V^2 = V_R^2 + V_C^2 \qquad [18]$$

and trigonometry to give the phase angle ϕ, i.e. the angle by which the current leads the voltage:

$$\tan\phi = \frac{V_C}{V_R} \qquad [19]$$

$$\cos\phi = \frac{V_R}{V} \qquad [20]$$

Since $V_R = IR$ and $V_C = IX_C$ and the magnitude of **V** is I multiplied by the magnitude of the impedance, i.e. $V = IZ$, by substitution in equations [18], [19] and [20] we obtain;

$$Z^2 = R^2 + X_C^2 \qquad [21]$$

$$\tan\phi = \frac{X_C}{R} \qquad [22]$$

$$\cos\phi = \frac{R}{Z} \qquad [23]$$

Alternatively as the voltage parallelogram in Figure 8.12(c) has sides of lengths IR, IX_C and IZ, if we divide each side by I we end up with an impedance triangle shown in Figure 8.12(d). The above equations for the impedance and the phase angle then can be determined by the use of the Pythagoras theorem and trigonometry:

For a *series circuit containing resistance, capacitance and inductance* (Figure 8.13), the voltage across the resistance is in phase with the current, the voltage across the capacitor lags the current by 90° and the voltage across the inductor leads the current by 90°.

For $V_L > V_C$, i.e. $X_L < X_C$ (Figure 8.14): because the voltage phasors for the inductor and capacitor are in opposite directions we can subtract them to give a phasor for the voltage drop across the inductor and capacitor of $V_L - V_C$. Then, for the voltage triangle:

$$V^2 = V_R^2 + (V_L - V_C)^2 \qquad [24]$$

$$\tan\phi = \frac{V_L - V_C}{V_R} \qquad [25]$$

$$\cos\phi = \frac{V_R}{V} \qquad [26]$$

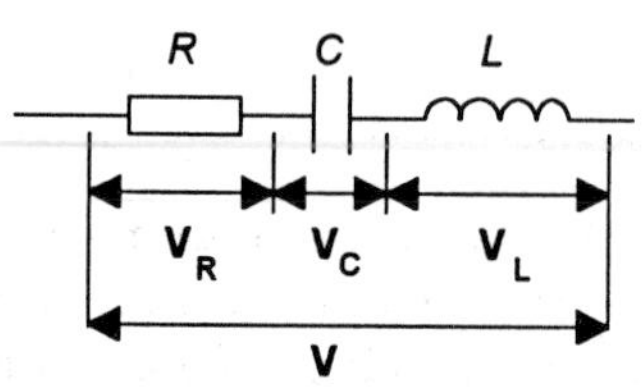

Figure 8.13 *Series RCL circuit*

As a result the circuit behaves as though it was inductance in series with resistance, the voltage across the series arrangement leading the current by ϕ. Since $V_R = IR$, $V_L = IX_L$ and $V_C = IX_C$ and the magnitude of **V** is I

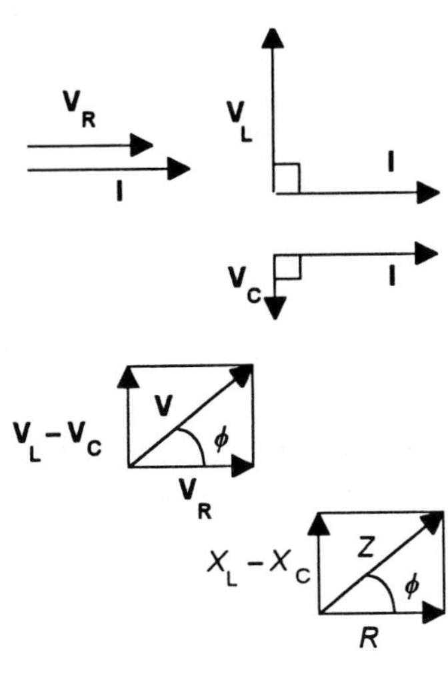

Figure 8.14 $V_L > V_C$

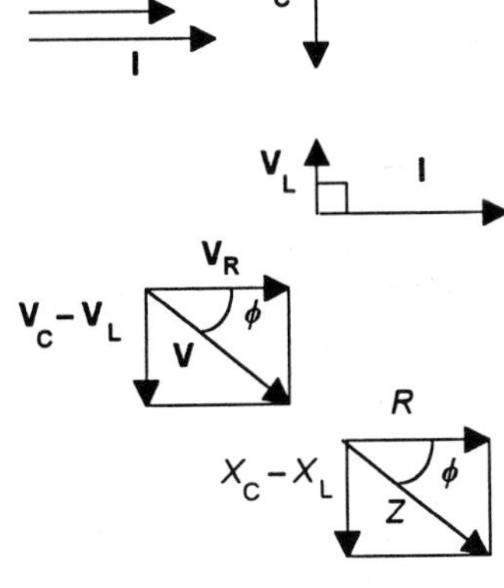

Figure 8.15 $V_L < V_C$

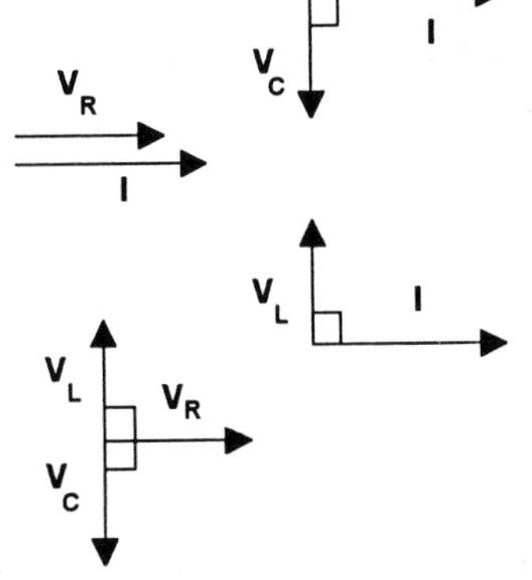

Figure 8.16 $V_L = V_C$

multiplied by the magnitude of the impedance, i.e. $V = IZ$, by substitution for the voltages in equations [24], [25] and [26]; or alternatively by constructing the impedance triangle:

$$Z^2 = R^2 + (X_L - X_C)^2 \quad [27]$$

$$\tan\phi = \frac{X_L - X_C}{R} \quad [28]$$

$$\cos\phi = \frac{R}{Z} \quad [29]$$

For $V_L < V_C$, i.e. $X_L < X_C$ (Figure 8.15): because the voltage phasors for the inductor and capacitor are in opposite directions we can subtract them to give a phasor for the voltage drop across the inductor and capacitor of $\mathbf{V_C} - \mathbf{V_L}$. For the voltage triangle:

$$V^2 = V_R^2 + (V_C - V_L)^2 \quad [30]$$

$$\tan\phi = \frac{V_C - V_L}{V_R} \quad [31]$$

$$\cos\phi = \frac{V_R}{V} \quad [32]$$

As a result the circuit behaves as though it was capacitance in series with resistance, the voltage across the series arrangement lagging behind the current by ϕ. Since $V_R = IR$, $V_L = IX_L$ and $V_C = IX_C$ and the magnitude of **V** is I multiplied by the magnitude of the impedance, i.e. $V = IZ$, by substitution for the voltages in equations [30], [31] and [32]; or alternatively by constructing the impedance triangle:

$$Z^2 = R^2 + (X_C - X_L)^2 \quad [33]$$

$$\tan\phi = \frac{X_C - X_L}{R} \quad [34]$$

$$\cos\phi = \frac{R}{Z} \quad [35]$$

For $V_L = V_C$, i.e. $X_L = X_C$ (Figure 8.16): the voltage phasor for the capacitor is equal in magnitude to the voltage phasor for the inductor but in exactly the opposite direction. Thus the two voltage phasors when added cancel each other out. The total voltage is thus just $\mathbf{V_R}$. The result is that the circuit behaves as though it was just the resistance with the impedance $Z = R$ and $\phi = 0°$.

Example

In a series *RL* circuit, the resistance is 10 Ω and the inductance 50 mH. Determine the value of the current and its phase angle with respect to the voltage if a 10 V r.m.s., 50 Hz supply is connected to the circuit.

The inductive reactance is $X_L = 2\pi fL = 2\pi \times 50 \times 0.05 = 15.7\ \Omega$. Thus equations [15] and [16] give:

$$Z = \sqrt{R^2 + X_L^2} = \sqrt{10^2 + 15.7^2} = 18.6$$

$$\phi = \tan^{-1}\frac{X_L}{R} = \tan^{-1}\frac{15.7}{10} = 57.5°$$

Hence the magnitude of the current I is given by:

$$I = \frac{V}{Z} = \frac{10}{18.6} = 0.54\ \text{A}$$

and, since the current is in the direction of the phasor $\mathbf{V_R}$, its phase angle is 57.5° lagging behind the applied voltage **V**.

Example

In a series *RC* circuit the resistance is 20 Ω and the capacitance 50 μF. Determine the value of the current and its phase angle with respect to the voltage if a 240 V r.m.s., 50 Hz supply is connected to the circuit.

The capacitive reactance is $1/2\pi fC = 1/(2\pi \times 50 \times 50 \times 10^{-6}) = 63.7\ \Omega$. Thus equations [21] and [22] give:

$$Z = \sqrt{R^2 + X_C^2} = \sqrt{20^2 + 63.7^2} = 66.8\ \Omega$$

$$\phi = \tan^{-1}\frac{X_C}{R} = \tan^{-1}\frac{63.7}{20} = 72.6°$$

Hence the magnitude of the current I is given by:

$$I = \frac{V}{Z} = \frac{240}{66.8} = 3.59\ \text{A}$$

and, since the current is in the direction of the phasor $\mathbf{V_R}$, its phase angle is 72.6° leading the applied voltage **V**.

Example

In a series *RLC* circuit the resistance is 10 Ω, the inductance 60 mH and the capacitance 300 μF. Determine the value of the current and its phase angle with respect to the voltage if a 24 V r.m.s., 50 Hz supply is connected to the circuit.

The inductive reactance is $X_L = 2\pi fL = 2\pi \times 50 \times 0.060 = 18.8\ \Omega$ and the capacitive reactance is $X_C = 1/(2\pi fC) = 1/(2\pi \times 50 \times 300 \times 10^{-6}) = 10.6\ \Omega$. Since the inductive reactance is greater than the capacitive reactance the situation is like that shown in Figure 8.14. The circuit impedance Z is:

$$Z = \sqrt{R^2 + (X_L - X_C)^2} = \sqrt{10^2 + (18.8 - 10.6)^2} = 12.9$$

$$\phi = \tan^{-1}\frac{X_L - X_C}{R} = \tan^{-1}\frac{18.8 - 10.6}{10} = 39.4°$$

The circuit current has a magnitude I of:

$$I = \frac{V}{Z} = \frac{24}{12.9} = 1.86\ \text{A}$$

and because the circuit behaves as an inductive load the current lags the voltage by 39.6°.

Revision

9 Determine the total impedance, and the phase angle between the overall voltage and the current, of a circuit containing a resistance of 300 Ω in series with an inductance of 200 mH when a voltage of 10 sin 2000t V is applied.

10 A voltage of root-mean-square value 30 V at an angular frequency of 1000 rad/s is applied to a circuit consisting of a resistance of 200 Ω in series with an inductance of 100 mH. Determine the current.

11 Determine the total impedance, and the phase angle between the overall voltage and the current, of a circuit containing a resistance of 3.3 kΩ in series with a capacitance of 2.2 μF when a voltage of 10 sin 240t V is applied.

12 A voltage of root-mean-square value 12 V at an angular frequency of 120 rad/s is applied to a circuit consisting of a resistance of 200 Ω in series with a capacitance of 2 μF. Determine the current.

13 A series a.c. circuit consists of a resistance of 100 Ω, a capacitance of 8 μF and an inductance of 100 mH. What is the total circuit impedance and the phase angle between the overall voltage and the current when the angular frequency is 1000 rad/s?

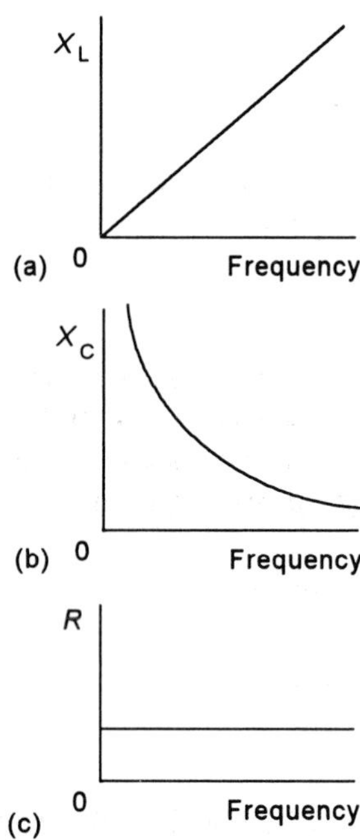

Figure 8.17 *Effect of changes in frequency*

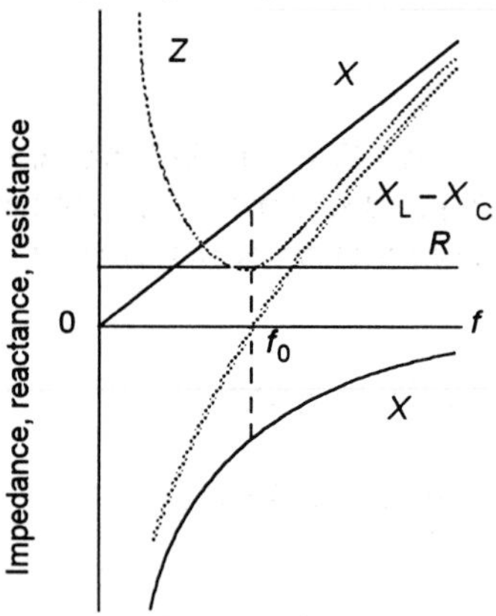

Figure 8.18 *Resonance*

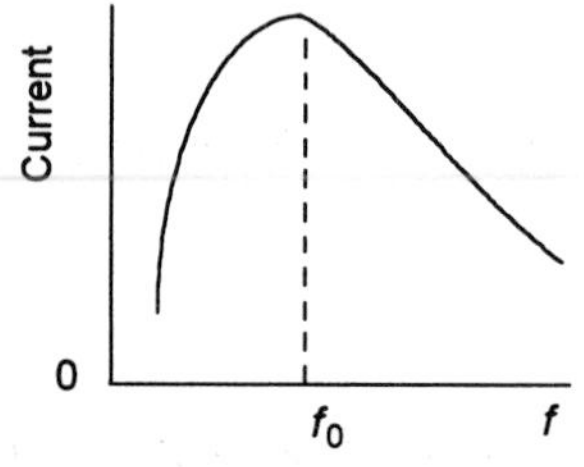

Figure 8.19 *Current*

8.6.1 Series resonance

Inductive reactance is given by $X_L = 2\pi fL$ and so increases as the frequency f increases (Figure 8.17(a)). Capacitive reactance $X_C = 1/2\pi fC$ and so decreases as the frequency f increases (Figure 8.17(b)). Resistance does not vary with frequency (Figure 8.17(c)). Thus for a series *RLC* circuit connected to a variable frequency voltage supply, there will be a frequency f_0 at which we have $X_L = X_C$ and so:

$$2\pi f_0 L = \frac{1}{2\pi f_0 C}$$

$$f_0 = \frac{1}{2\pi\sqrt{LC}} \qquad [36]$$

This frequency is known as the *resonant frequency*.

The impedance Z of a *RCL* circuit is given by:

$$Z = \sqrt{R^2 + (X_L \sim X_C)^2} \qquad [37]$$

The 'curly minus' sign is used to indicate that it is the difference between the two reactances we are concerned with and it is either $X_L - X_C$ or $X_C - X_L$ depending which reactance is the greater. Figure 8.18 shows how the impedance varies with frequency. When $X_L = X_C$, i.e. at the resonant frequency, the impedance is just R. At all other frequencies the impedance is greater than R. The circuit thus has a minimum impedance at the resonant frequency.

Since the circuit current $I = V/Z$, the consequence of the circuit impedance being a minimum at the resonant frequency f_0 is that the circuit current is at its maximum value, being V/R (Figure 8.18). At frequencies much lower than the resonant frequency and much higher, the impedance is high and so the current is very low.

Example

A circuit, of a resistance of 4 Ω, an inductance of 0.5 H and a variable capacitance in series, is connected across a 10 V, 50 Hz supply. Determine (a) the capacitance to give resonance and (b) the voltages across the inductance and the capacitance.

(a) For resonance the inductive reactance equals the capacitive reactance and thus:

$$2\pi fL = \frac{1}{2\pi fC}$$

$$C = \frac{1}{4\pi^2 f^2 L} = \frac{1}{4\pi^2 \times 50^2 \times 0.5^2} = 20.3 \times 10^{-6}\ \text{F} = 20.3\ \mu\text{F}$$

(b) At resonance the impedance equals the resistance of 4 Ω. Thus:

Application
Radio receivers need to be able to be tuned to just select just one radio signal from the vast range of frequencies transmitted. A series capacitor–coil resonant circuit can be used with the capacitor being adjusted so that its value gives the resonance condition, and hence the maximum current, for the required radio frequency signal.

$$I = \frac{V}{R} = \frac{10}{4} = 2.5 \text{ A}$$

$$V_L = IX_L = 2.5 \times 2\pi \times 50 \times 0.5 = 392.5 \text{ V}$$

Since we have $X_C = X_L$, the p.d. across the capacitance equals the p.d. across the inductance and so is also 392.5 V.

Example

A series *RLC* circuit has a resonant frequency of 50 Hz, a resistance of 20 Ω, an inductance of 300 mH and a supply voltage of 24 V. Calculate (a) the value of the capacitance, (b) the circuit current at resonance, and (c) the values of the voltages across each component.

(a) The resonant frequency $f_0 = 1/2\pi\sqrt{(LC)}$ and so:

$$C = \frac{1}{4\pi^2 f_0^2 L} = \frac{1}{4\pi^2 \times 50^2 \times 0.300} = 33.8\ \mu\text{F}$$

(b) At resonance $I = V/R = 24/20 = 1.2$ A.
(c) At resonance the voltage across the resistor will be the entire supply voltage of 24 V. The inductive reactance $X_L = 2\pi fL = 2\pi \times 50 \times 0.300 = 94.2$ Ω. At resonance $X_L = X_C$ and so the capacitive reactance is also 94.2 Ω. Hence the value of the voltage across the inductor and across the capacitor is $IX = 1.2 \times 94.2 = 113.0$ V.

Example

A series *RLC* circuit has a resistance of 50 Ω, an inductance of 200 mH and a capacitance of 0.2 μF. Determine the resonant frequency and the current at resonance when the supply voltage is 20 V.

The resonant frequency is given by equation [36] as:

$$f_0 = \frac{1}{2\pi\sqrt{LC}} = \frac{1}{2\pi\sqrt{0.200 \times 0.2 \times 10^{-6}}} = 795.8 \text{ Hz}$$

At resonance the circuit impedance is the resistance and so 50 Ω. Hence the circuit current is:

$$\mathbf{I} = \frac{\mathbf{V}}{Z} = \frac{20}{50} = 0.4 \text{ A}$$

Because, at resonance, there is only resistance, the current is in phase with the supply voltage.

Revision

14 A series *RLC* circuit has a resistance of 40 Ω, an inductance of 10 mH and a capacitance of 0.1 μF. Determine the resonant

Application

Consider the series *RLC* circuit when used as a bandpass filter. Such a filter is one that is used to pass a band of frequencies out of a spectrum of frequencies. Figure 8.20 shows the circuit, the output being the voltage across the resistor. At the resonant frequency the reactance of the inductor cancels out the reactance of the capacitor and so the entire input voltage appears across the resistor. At higher frequencies the reactance of the inductor is large and that of the capacitor small, thus the circuit impedance increases and so the circuit current decreases and the voltage drop across the resistor is smaller. At lower frequencies the reactance of the capacitor is large and that of the inductor small, thus the circuit impedance increases and so the circuit current decreases and the voltage drop across the resistor is smaller. Figure 8.21 shows how the voltage across the resistor changes with frequency for different *Q*-factors. A high *Q*-factor gives a filter with a smaller bandwidth and so greater selectivity.

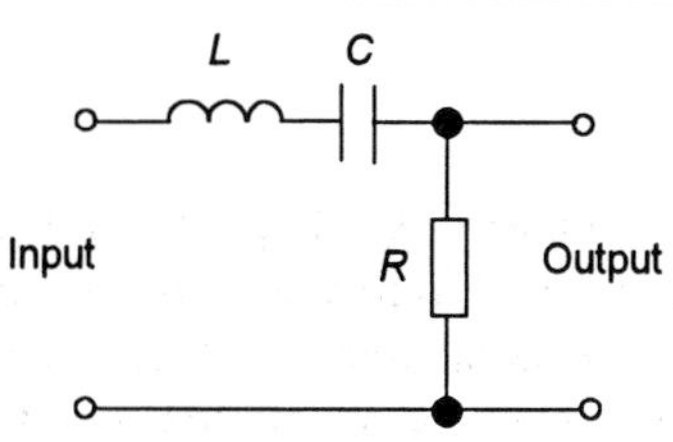

Figure 8.20 *Bandpass filter*

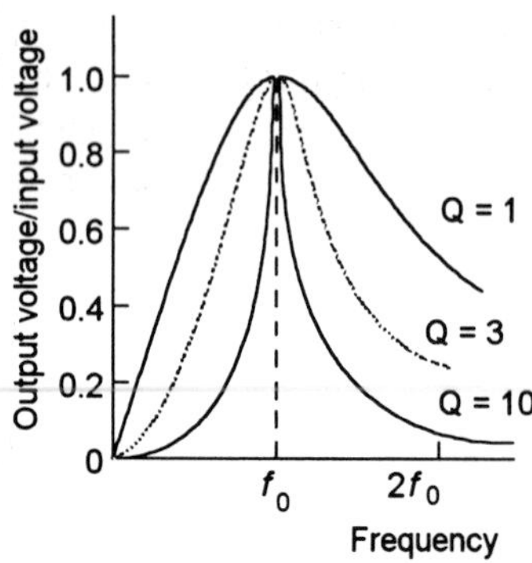

Figure 8.21 *Bandpass filter*

frequency and the current at resonance when the supply voltage is 10 V.

15 A coil with a resistance of 20 Ω and an inductance of 75 mH is connected in series with a capacitor of capacitance 40 μF across a supply voltage of 50 V. Determine the resonant frequency and the circuit current at resonance.

16 A coil with a resistance of 10 Ω and an inductance of 50 mH is connected in series with a capacitor of capacitance 12 μF across a supply voltage of 20 V. Determine the resonant frequency and the voltage across the capacitor at resonance.

8.6.2 *Q*-factor

In a series *RLC* circuit, the voltage across the inductance at resonance is IX_L and thus can be very large, likewise the voltage across the capacitor IX_C. A factor known as the *Q-factor* or 'quality factor' is used to indicate the voltage magnification across either the inductor or capacitor at resonance compared with the voltage across the resistance:

$$Q\text{-factor} = \frac{\text{voltage across } L \text{ or } C \text{ at resonance}}{\text{voltage across } R \text{ at resonance}} \quad [38]$$

Thus, in terms of the voltage across the inductance:

$$Q\text{-factor} = \frac{IX_L}{IR} = \frac{2\pi f_0 L}{R} \quad [39]$$

or, in terms of the voltage across the capacitance:

$$Q\text{-factor} = \frac{IX_C}{IR} = \frac{X_C}{R} = \frac{1}{2\pi f_0 CR} \quad [40]$$

Since $f_0 = 1/2\pi\sqrt{(LC)}$ we can write, using either equation [39] or [40]:

$$Q\text{-factor} = \frac{1}{R}\sqrt{\frac{L}{C}} \quad [41]$$

The *Q*-factor in giving a measure of the voltage magnification at resonance, also gives a measure of the selectivity of the circuit. The greater the voltage magnification at the resonant frequency the more selective the circuit is of the band of frequencies centred on the resonant frequency from the entire spectrum of frequencies. A high *Q*-factor thus means a high selectivity.

Example

A series *RLC* circuit has a resonant frequency of 50 Hz, a resistance of 20 Ω, an inductance of 300 mH, capacitance and a supply voltage of 24 V. Calculate (a) the value of the capacitance, (b) the circuit current at resonance, (c) the values of the voltages across each component and (d) the *Q*-factor of the circuit.

(a) The resonant frequency $f_0 = 1/2\pi\sqrt{(LC)}$ and so:

$$C = \frac{1}{4\pi^2 f_0^2 L} = \frac{1}{4\pi^2 \times 50^2 \times 0.300} = 33.8\ \mu\text{F}$$

(b) At resonance $I = V/R = 24/20 = 1.2$ A.
(c) At resonance, the voltage across the resistor will be the entire supply voltage of 24 V. The inductive reactance $X_L = 2\pi fL = 2\pi \times 50 \times 0.300 = 94.2\ \Omega$. At resonance $X_L = X_C$ and so the capacitive reactance is also 94.2 Ω. Hence the value of the voltage across the inductor and across the capacitor is $IX = 1.2 \times 94.2 = 113.0$ V.
(d) The *Q*-factor is voltage across *L* or *C* divided by the voltage across *R* and so is 113.0/24 = 4.7.

Revision

17 A series *RLC* circuit has an inductance of 10 mH and a capacitance of 0.1 μF. What will the resistance have to be if a *Q*-factor of 15 is to be obtained?

8.7 Parallel circuits

For a parallel d.c. circuit there is the same voltage drop across each component and the current entering the parallel arrangement is the sum of the currents through each component. For a parallel a.c. circuit there is the same voltage phasor for the voltage across each component and the phasor for the current entering the parallel arrangement is the sum of the phasors for the currents through each component.

For *parallel resistance and inductance* (Figure 8.22(a)), the voltage across each component is the same, being the supply voltage **V**. The current $\mathbf{I_R}$ flowing through the resistance is in phase with the supply voltage; the current $\mathbf{I_L}$ through the inductor lags behind the supply voltage by 90° (Figure 8.22(b)). The phasor sum of the currents $\mathbf{I_R}$ and $\mathbf{I_L}$ must be the supply current **I** entering the parallel arrangement (Figure 8.22(c)). Thus:

$$I^2 = I_R^2 + I_L^2 \qquad [42]$$

$$\tan\phi = \frac{I_L}{I_R} \qquad [43]$$

$$\cos\phi = \frac{I_R}{I} \qquad [44]$$

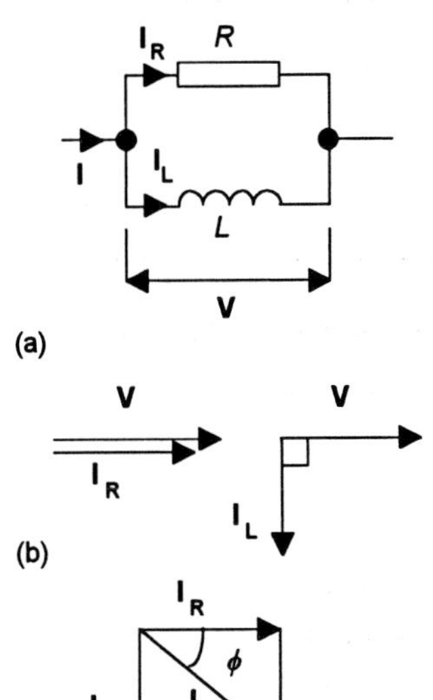

Figure 8.22 *Parallel R and L*

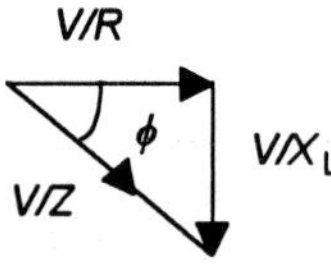

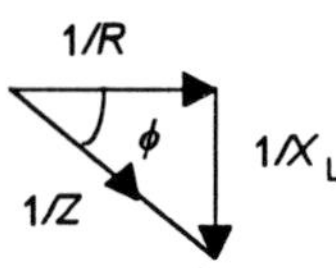

(a)

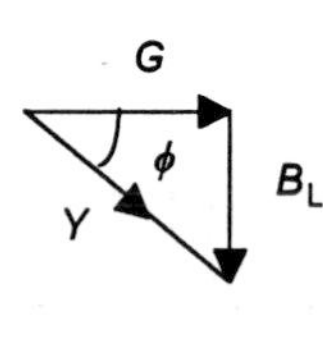

(b)

Figure 8.23 *Parallel R and L*

But $V = IZ$, where Z is the impedance of the parallel arrangement and $V = I_R R$ and $V = I_L X_L$, and so by substitution in equations [42], [43] and [44]:

$$\frac{1}{Z^2} = \frac{1}{R^2} + \frac{1}{X_L^2} \qquad [45]$$

Alternatively we could redraw the current triangle in Figure 8.22(c) as Figure 8.23(a) and use the Pythagoras theorem and trigonometry to obtain the above equations.

The conductance G is the reciprocal of resistance, susceptance B the reciprocal of reactance and admittance the reciprocal of impedance. Thus, equation [45] in terms of admittance is:

$$Y^2 = G^2 + B_L^2 \qquad [46]$$

and equations [43] and [44]:

$$\tan\phi = \frac{R}{X_L} \qquad [47]$$

$$\cos\phi = \frac{Z}{R} \qquad [48]$$

Alternatively we can obtain the above equations from the impedance triangle in Figure 8.23(a); when each side length is divided by V we obtain the admittance triangle shown in Figure 8.23(b).

For *parallel resistance and capacitance* (Figure 8.24(a)), the voltage across each component is the same, being the supply voltage **V**. The current $\mathbf{I_R}$ flowing through the resistance is in phase with the supply voltage; the current $\mathbf{I_C}$ through the inductor leads the supply voltage by 90° (Figure 8.24(b)). The phasor sum of the currents $\mathbf{I_R}$ and $\mathbf{I_C}$ is the current **I** entering the parallel arrangement (Figure 8.24(c)). Thus:

$$I^2 = I_R^2 + I_C^2 \qquad [49]$$

$$\tan\phi = \frac{I_C}{I_R} \qquad [50]$$

$$\cos\phi = \frac{I_R}{I} \qquad [51]$$

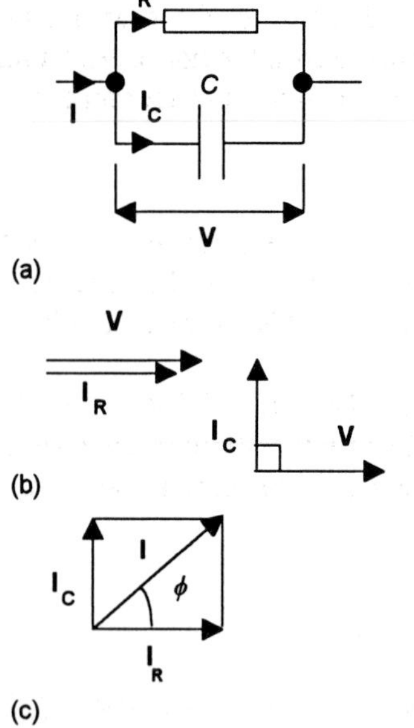

Figure 8.24 *Parallel R and C*

But $V = IZ$, where Z is the impedance of the parallel arrangement and $V = I_R R$ and $V = I_C X_C$, thus by substitution in equations [49], [50] and [51]:

$$\frac{1}{Z^2} = \frac{1}{R^2} + \frac{1}{X_C^2} \qquad [52]$$

Since conductance G is the reciprocal of resistance, susceptance B the reciprocal of reactance and admittance the reciprocal of impedance, equation [52] can be written as:

$$Y^2 = G^2 + B_C^2 \qquad [53]$$

By substitution in equations [50] and [51]:

$$\tan\phi = \frac{R}{X_C} \qquad [54]$$

$$\cos\phi = \frac{Z}{R} \qquad [55]$$

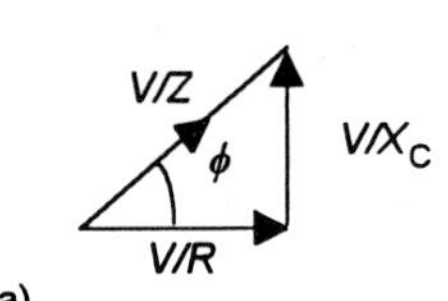

(a)

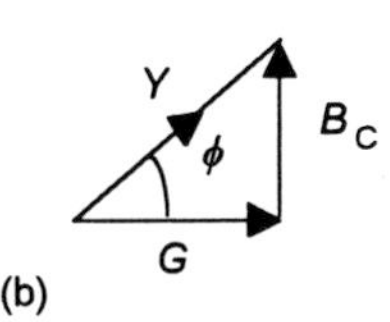

(b)

Figure 8.25 *Parallel R and C*

Alternatively we could redraw the current triangle in Figure 8.24(c) as Figure 8.25(a). When each side length is divided by V, we obtain the admittance triangle shown in Figure 8.25(b). Hence the above equations can be obtained by the use of the Pythagoras theorem and trigonometry.

Example

A resistance of 10 Ω is in parallel with an inductance of 50 mH. What will be the current drawn from the supply and the impedance of the circuit when a voltage of 24 V r.m.s., 50 Hz is applied to it?

The inductive reactance is $X_L = 2\pi fL = 2\pi \times 50 \times 0.050 = 15.7\ \Omega$. Thus $I_L = V/X_L = 24/15.7 = 1.53$ A. The current through the resistor $I_R = V/R = 24/10 = 2.4$ A. The magnitude of the current drawn from the supply is thus:

$$I = \sqrt{I_R^2 + I_L^2} = \sqrt{2.4^2 + 1.53^2} = 2.85 \text{ A}$$

The phase angle ϕ by which the supply current lags the supply voltage is:

$$\phi = \tan^{-1}\frac{R}{X_L} = \tan^{-1}\frac{10}{15.7} = 32.5°$$

The circuit impedance has the value:

$$Z = \frac{V}{I} = \frac{24}{2.85} = 8.42\ \Omega$$

Example

A resistance of 10 Ω is in parallel with a capacitance of 10 μF. What will be the current drawn from the supply and the impedance of the circuit when a voltage of 20 V r.m.s., 1 kHz is applied to it?

The inductive capacitance is $X_C = 1/2\pi fC = 1/2\pi \times 1000 \times 10 \times 10^{-6} = 15.9\ \Omega$. Thus $I_C = V/X_C = 20/15.9 = 1.26$ A. The current through the resistor $I_R = V/R = 20/10 = 2.0$ A. The magnitude of the current drawn from the supply is thus:

$$I = \sqrt{I_R^2 + I_C^2} = \sqrt{2.0^2 + 1.26^2} = 2.36 \text{ A}$$

The phase angle ϕ by which the supply current leads the voltage is:

$$\phi = \tan^{-1}\frac{R}{X_C} = \tan^{-1}\frac{10}{15.9} = 32.2^{\circ}$$

The circuit impedance has the value $Z = V/I = 20/2.36 = 8.47\ \Omega$.

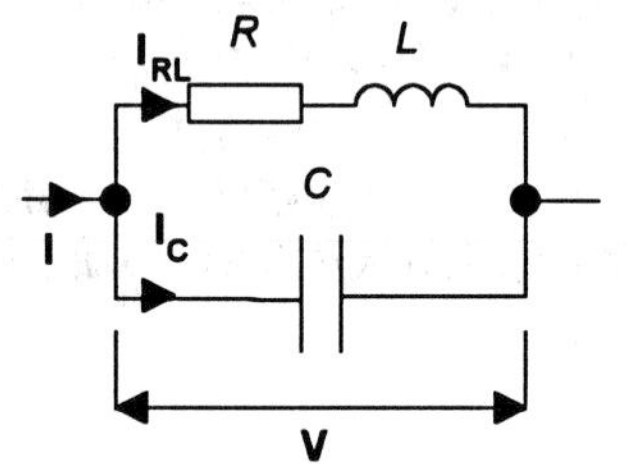

Figure 8.26 *RL in parallel with C*

Revision

18 A resistance of 100 Ω is in parallel with an inductance of 100 mH. What will be the current drawn from the supply and the impedance of the circuit when a voltage of 12 V r.m.s., 50 Hz is applied to it?

19 A resistance of 1 kΩ is in parallel with a capacitance of 2 μF. What will be the current drawn from the supply and the impedance of the circuit when a voltage of 12 V r.m.s., 50 Hz is applied to it?

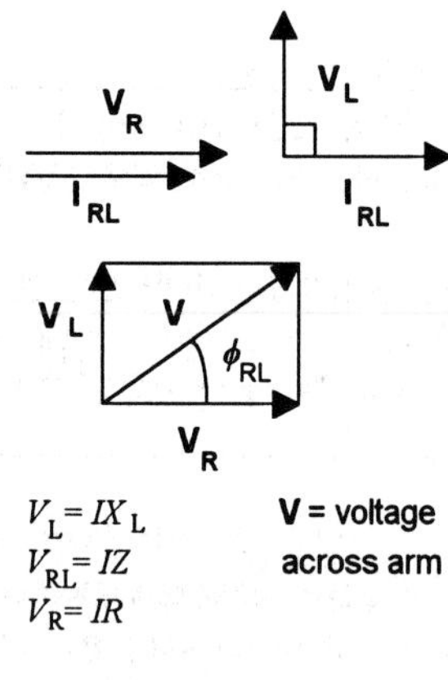

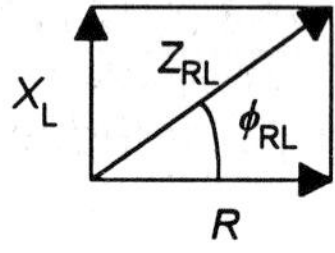

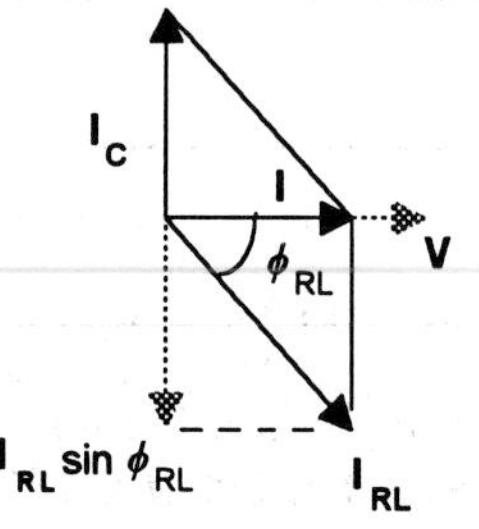

Figure 8.27 *RL arm of circuit*

8.7.1 Resonance

A widely used circuit has a coil in parallel with a capacitor, the coil having both inductance and resistance (Figure 8.26). For the inductive arm, the effect of having a resistance in series with an inductance is to give the phasors shown in Figure 8.27. The outcome is a current of $\mathbf{I}_{RL} = \mathbf{V}/\mathbf{Z}_{RL}$ which is at a phase angle of ϕ_{RL} to the circuit current. Figure 8.28 shows the current phasors for the parallel circuit. Resonance occurs when the magnitude of the current through the capacitor I_C is equal to $I_{RL} \sin \phi_{RL}$; the resonance condition is when the circuit current is in phase with the applied voltage. One way of describing this equality is that the reactive current through the capacitor equals the reactive current through the *RL* arm, the current through the *RL* arm having a reactive component which is at 90° to the voltage and a resistive component which is in phase with the voltage. Since $I_C = V/X_C$, $I_{RL} = V/Z_{RL}$ and $\sin \phi_{RL} = X_L/Z_{RL}$, where X_C is the reactance of the capacitor and Z_{RL} is the impedance of the *RL* combination, then at resonance:

Figure 8.28 *Resonance*

$$I_C = I_{RL} \sin \phi_{RL}$$

and so:

$$\frac{V}{X_C} = \frac{V}{Z_{RL}} \frac{X_L}{Z_{RL}}$$

and hence:

$$Z_{RL}^2 = X_L X_C$$

Since $X_L = \omega_0 L$, $X_C = 1/\omega_0 C$, then:

$$Z_{RL}^2 = \frac{L}{C} \qquad [56]$$

As $Z_{RL} = \sqrt{[R^2 + (\omega_0 L)^2]}$, then equation [56] can be written as:

$$R^2 + (\omega_0 L)^2 = \omega_0 L \times \frac{1}{\omega C} = \frac{L}{C}$$

and so:

$$f_0 = \frac{1}{2\pi}\sqrt{\frac{1}{LC} - \frac{R^2}{L^2}} \qquad [57]$$

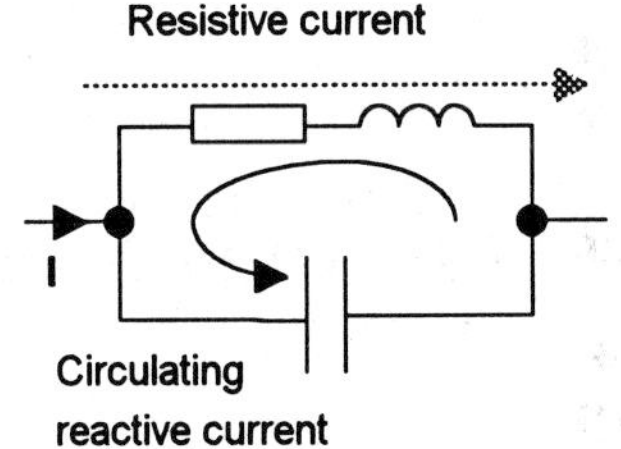

Figure 8.29 *RL in parallel with C at resonance*

Often R^2/L^2 is negligible when compared with $1/LC$ and the equation reduces to the form:

$$f_0 \approx \frac{1}{2\pi\sqrt{LC}} \qquad [58]$$

At resonance the current I_C can be much larger than the current I drawn from the supply. We can think of this reactive current as circulating round the *RLC* circuit with only the resistive component of the current through *RL* passing through the system (Figure 8.29).

The current I drawn from the supply is, at resonance, the resistive component and so $I = I_{RL} \cos \phi_{RL}$. But $I_{RL} = V/Z_{RL}$ and, from Figure 8.27, $\cos \phi_{RL} = R/Z_{RL}$. Thus:

$$I = \frac{V}{Z_{RL}} \frac{R}{Z_{RL}} = \frac{VR}{Z_{RL}^2}$$

Hence, using equation [55]:

$$I = \frac{VCR}{L}$$

Since the current drawn from the supply is, at resonance, in phase with the supply voltage, the impedance of the circuit, i.e. V/I, is a resistance. This resistance is known as the dynamic resistance R_D.

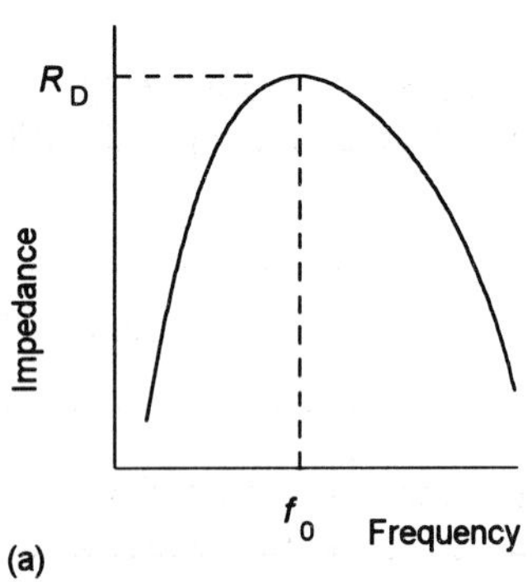

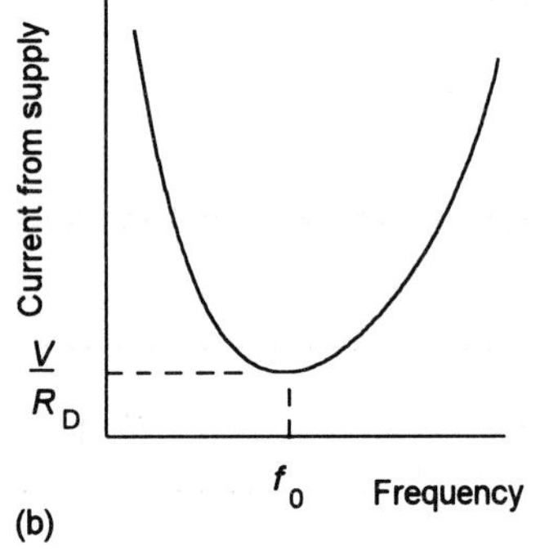

Figure 8.30 *Variation of (a) impedance, (b) current with frequency*

$$R_D = \frac{V}{I} = \frac{L}{CR} \qquad [59]$$

Figure 8.30 shows how (a) the impedance and (b) the current drawn from the supply by the circuit varies with frequency. The impedance is at a maximum value of R_D and the current at a minimum value of V/R_D at the resonant frequency; R_D is termed the *dynamic resistance* and is equal to L/RC. For this reason the parallel resonant circuit is known as the *rejecter circuit* since it 'rejects current' at the resonant frequency.

Example

A coil of resistance 10 Ω and inductance 50 mH is connected in parallel with a capacitor of capacitance 10 μF. Determine (a) the resonant frequency, (b) the impedance at resonance and (c) the current taken by the circuit at resonance when an alternating voltage of 5 V is applied.

(a) Using equation [56]:

$$f_0 = \frac{1}{2\pi}\sqrt{\frac{1}{LC} - \frac{R^2}{L^2}} = \frac{1}{2\pi}\sqrt{\frac{1}{0.050 \times 10 \times 10^{-6}} - \frac{10^2}{0.050^2}}$$

$$= 223 \text{ Hz}$$

Neglecting the R^2/L^2 term gives the answer 225 Hz.

(b) At resonance the impedance is the dynamic resistance:

$$R_D = \frac{L}{RC} = \frac{0.050}{10 \times 10 \times 10^{-6}} = 500\ \Omega$$

(c) The current at resonance is given by:

$$\mathbf{I} = \frac{\mathbf{V}}{Z} = \frac{5}{500} = 10 \text{ mA}$$

This current is in phase with the applied voltage.

Revision

20 A coil of resistance 10 Ω and inductance 5 mH is connected in parallel with a capacitor of capacitance 0.25 μF. Determine (a) the resonant frequency, (b) the impedance at resonance and (c) the current taken by the circuit at resonance when an alternating voltage of 10 V is applied.

21 Determine the resonant frequency when a 120 mH pure inductor is connected in parallel with a capacitance of 60 μF.

22 Determine the resonant frequency when a coil of inductance 10 mH and resistance 6 Ω is in parallel with a capacitance of 10 μF.

8.7.2 *Q*-factor

The *Q-factor* of a parallel resonant circuit can be defined as being the circulating current magnification in the parallel circuit compared with the supply current and so:

$$Q\text{-factor} = \frac{I_c}{I} \qquad [60]$$

where I_c is the size of the circulating current and I that taken from the supply by the circuit. For a coil, with resistance R and inductance L, in parallel with capacitance C, the circulating current is the reactive current $I_l \sin \phi_{RL}$ (see Figure 8.29) and for a supply current i we thus have:

$$Q\text{-factor} = \frac{I_l \sin \phi_{RL}}{I}$$

But, as Figure 8.28 indicates, this ratio is tan ϕ_{RL}. But Figure 8.29 shows that tan $\phi_{RL} = X_L/R$ and so we can write:

$$Q\text{-factor} = \frac{2\pi f_0 L}{R} \qquad [61]$$

Example

Determine the *Q*-factor for a circuit involving a coil, of inductance 120 mH and resistance 20 Ω, in parallel with a 60 μF capacitor.

The resonant frequency is given by equation [56] as:

$$f_0 = \frac{1}{2\pi}\sqrt{\frac{1}{LC} - \frac{R^2}{L^2}} = \frac{1}{2\pi}\sqrt{\frac{1}{0.12 \times 60 \times 10^{-6}} - \frac{20^2}{0.12^2}}$$

$$= 53.1 \text{ Hz}$$

Using equation [61]:

$$Q\text{-factor} = \frac{2\pi f_0 L}{R} = \frac{2\pi \times 53.1 \times 0.120}{20} = 2.0$$

Revision

23 Determine the *Q*-factor for a circuit involving a coil, of inductance 100 mH and resistance 1 Ω, in parallel with a 1 μF capacitor.

8.8 Power

The power developed in a d.c. circuit is the product of the current and the resistance. With an a.c. circuit the voltage and current are varying with time, thus we refer to the instantaneous power p as the product of the instantaneous current i and instantaneous voltage v:

$$p = iv \qquad [62]$$

Since the current and voltage are changing with time, the product varies. For this reason we quote the average power dissipated over a cycle.

8.8.1 Power for a purely resistive circuit

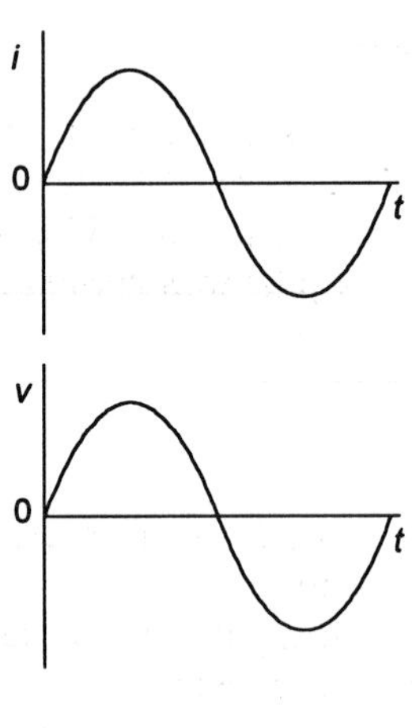

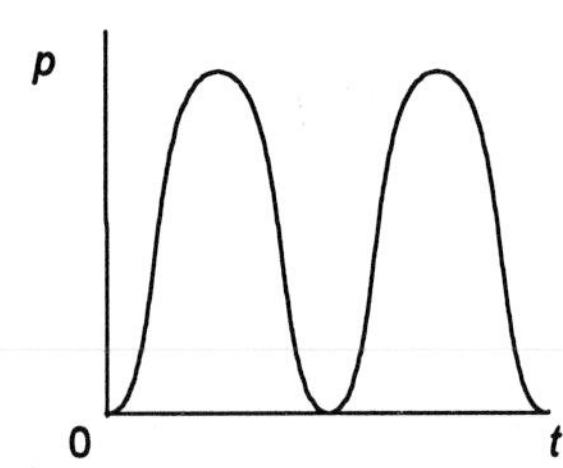

Figure 8.31 *Current, voltage and power as functions of time*

For a purely resistive circuit the current and voltage are in phase and we have $i = I_m \sin \omega t$ and $v = V_m \sin \omega t$, where i and v are the instantaneous values and I_m and V_m the maximum values. Hence the instantaneous power p is:

$$p = iv = I_m \sin \omega t \times V_m \sin \omega t = I_m V_m \sin^2 \omega t \qquad [63]$$

Figure 8.31 shows the graphs of i, v and p as functions of time t. By inspection of the graph of p varying with time we might deduce that, since the graph is symmetrical about a horizontal axis through its middle, the average value of p over one cycle is:

$$\text{average power } P = \tfrac{1}{2} I_m V_m \qquad [64]$$

Alternatively, we can use the trigonometric relation $\cos 2\theta = 1 - 2 \sin^2 \theta$ to write equation [64] as:

$$p = I_m V_m \tfrac{1}{2}(1 - \cos 2\omega t) = \tfrac{1}{2} I_m V_m - \tfrac{1}{2} I_m V_m \cos 2\omega t \qquad [65]$$

The average value of a cosine function over one cycle is zero. Hence:

$$\text{average power } P = \tfrac{1}{2} I_m V_m \qquad [66]$$

The r.m.s. current $I_{r.m.s.}$ is $I_m/\sqrt{2}$ and the r.m.s. voltage $V_{r.m.s.}$ is $V_m/\sqrt{2}$. Hence we can write the equation for the average power dissipated with a purely resistive circuit as:

$$\text{average power } P = I_{r.m.s.} V_{r.m.s.} \qquad [67]$$

Example

Determine the average power developed when a 240 V r.m.s. alternating current supply is connected across a resistance of 100 Ω.

Since $V_{r.m.s.} = RI_{r.m.s.}$ then equation [67] can be written as:

$$\text{average power} = I_{r.m.s.} V_{r.m.s.} = \frac{V_{r.m.s.}^2}{R} = \frac{240^2}{100} = 576 \text{ W}$$

Revision

24 Calculate the resistance of an electric light bulb that develops a power of 100 W when connected across a 240 V alternating current supply.

8.8.2 Power for a purely reactive circuit

Consider a circuit which is purely reactive. For such a component there is a phase angle of 90° between the current and the voltage (Figure 8.32). For example, we might have $i = I_m \sin \omega t$ and $v = V_m \sin (\omega t - 90°)$. The instantaneous power p is the product of the instantaneous current and voltage:

$$p = I_m \sin \omega t \times V_m \sin (\omega t - 90°) \qquad [68]$$

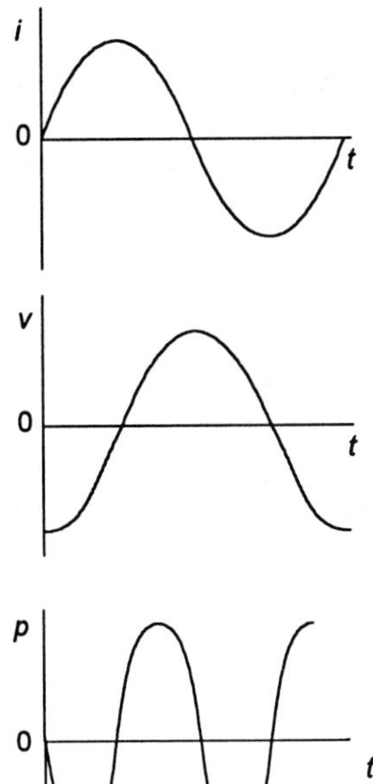

Figure 8.32 *Current, voltage and power as functions of time*

Figure 8.32 shows i, v and p as functions of time. Over one cycle the average power for the purely reactive element is zero.

We can demonstrate that this is the case by using the trigonometric relation $2 \sin A \sin B = \cos (A - B) - \cos (A + B)$, this giving:

$$2 \sin \omega t \times \sin (\omega t - 90°) = \cos 90° - \cos (2\omega t - 90°)$$

$$= 0 - \sin 2\omega t$$

and hence:

$$p = \tfrac{1}{2} I_m V_m \sin 2\omega t \qquad [69]$$

The average value of $\sin 2\omega t$ over a cycle is 0 and thus the average power is 0.

Average power for a purely reactive circuit = 0	[70]

8.8.3 Power for a circuit having resistance and reactance

For a component or circuit having both resistance and reactance, there is a phase difference other than 90° between the current and the voltage. thus we might have $i = I_m \sin \omega t$ and $v = V_m \sin (\omega t - \phi)$ with ϕ being the phase difference (Figure 8.33). If we resolve the voltage phasor into horizontal and vertical components of $V \cos \phi$ and $V \sin \phi$, then the component at 90° to **I** is the situation we would get with a purely reactive circuit and so the average power over a cycle is zero for that component. The component in phase with the current is the situation we get with a purely resistive circuit and so the average power is given by using equation [67]. Thus the average power P over a cycle is given by the in-phase components as:

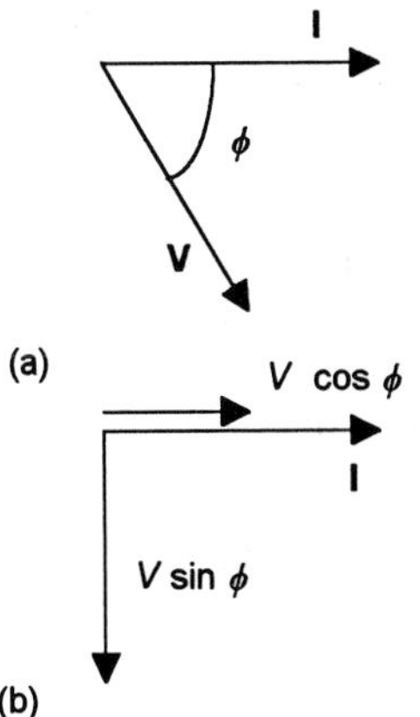

Figure 8.33 *Phasors*

$$P = IV \cos \phi \qquad [71]$$

where I and V are r.m.s. values.

We can obtain the above relationship by using trigonometric relations. Thus for:

$$p = I_m \sin \omega t \times V_m \sin (\omega t - \phi) \qquad [72]$$

we can use the relation $2 \sin A \sin B = \cos (A - B) - \cos (A + B)$ and so $2 \sin \omega t \sin (\omega t - \phi) = \cos \phi - \cos (2\omega t - \phi)$. Hence:

$$p = \tfrac{1}{2} I_m V_m \cos \phi - \tfrac{1}{2} I_m V_m \cos (2\omega t - \phi) \qquad [73]$$

The average value of $\cos (2\omega t - \phi)$ over one cycle is zero and thus:

$$\text{average power } P = \tfrac{1}{2} I_m V_m \cos \phi = I_{r.m.s.} V_{r.m.s.} \cos \phi \qquad [74]$$

where I and V are r.m.s. values, the average power being in watts. The term *power factor* is used for $\cos \phi$, since it is the factor by which the power for a purely resistive circuit is multiplied to give the power when there is both resistance and reactance. See Section 8.8.4 for further discussion of the power factor.

Example

Determine the power dissipated when a coil of inductance 40 mH and resistance 20 Ω is connected to a 240 V r.m.s., 50 Hz supply.

The inductive reactance is $X_L = 2\pi fL = 2\pi \times 50 \times 0.040 = 12.6\ \Omega$. Thus the impedance $Z = \sqrt{(20^2 + 12.6^2)} = 23.6\ \Omega$ and so the current $I = 240/23.6 = 10.2$ A. For a series RL circuit we have $\cos \phi = R/Z = 20/23.6 = 0.85$. Thus the power factor is 0.85. The average power dissipated is thus $IV \cos \phi = 10.2 \times 240 \times 0.85 = 2081$ W. Alternatively we could recognise that power will only be dissipated in the resistance, no power being dissipated in the reactance. Thus the power developed is $I^2R = 10.2^2 \times 20 = 2081$ W.

Example

Determine the resistance and inductance of a series RL circuit if, when connected to a 240 V r.m.s., 50 Hz supply, it takes a power of 0.8 kW and a current of 4 A?

The circuit has a resistive component of R and a reactive component of X_L. The power is the power dissipated in just the resistive component and thus $P = I^2R$ and so $R = 800/4^2 = 50\ \Omega$. The circuit impedance $Z = V/I = 240/4 = 60\ \Omega$. Since $Z = \sqrt{(R^2 + X_L^2)}$ then $X_L = \sqrt{(Z^2 - R^2)} = \sqrt{(60^2 - 50^2)} = 33.2\ \Omega$. Since the reactance $X_L = 2\pi fL$ then $L = 33.2/(2\pi \times 50) = 0.11$ H.

Revision

25 Determine the power dissipated when a coil of inductance 100 mH and resistance 60 Ω is connected across a 12 V r.m.s., 100 Hz supply.

26 Determine the power dissipated with an 8 μF capacitor in series with a 1 kΩ resistor when the series arrangement is connected across a 12 V r.m.s., 50 Hz supply.

8.8.4 Apparent power, power factor and reactive power

The average *active power*, the term *true power* is often used, dissipated in an alternating current circuit is given by equation [74] as:

$$\text{average active power } P = I_{\text{r.m.s.}} V_{\text{r.m.s.}} \cos \phi$$

The product of the r.m.s. values of the current and voltage is called the *apparent power S*, the unit of apparent power being volt amperes (V A). Multiplying the apparent power by cos ϕ gives the true power dissipated. For this reason, cos ϕ is called the *power factor*. It has no units. Thus:

average power = apparent power × power factor	[75]

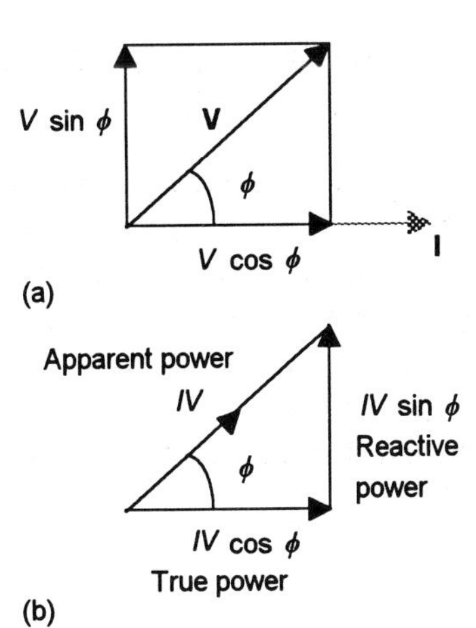

Figure 8.34 *(a) Voltage components, (b) power diagram*

For a purely reactive circuit the power factor is 0, for a purely resistive factor it is 1. A circuit in which the current lags the voltage, i.e. an inductive circuit, is said to have a lagging power factor; a circuit in which the current leads the voltage, i.e. a capacitive circuit, is said to have a leading power factor.

If we consider an alternating current circuit such as a *RL* series circuit, the overall voltage will be leading the current by some phase angle ϕ. We can resolve the voltage phasor into two components, one in phase with the current of magnitude $V \cos \phi$ and one, at a phase of 90° to the current, of magnitude $V \sin \phi$ (Figure 8.34(a)). Multiplying the magnitude of each of these by the magnitude of the current I transforms the phasor diagram into a power triangle (Figure 8.34(b)). For this triangle, the horizontal is the active power $P = IV \cos \phi$, the hypotenuse is the apparent power $S = IV$ and the vertical is $VI \sin \phi$, this being termed the *reactive power Q* and given the unit of volt amperes reactive (V Ar).

Example

A load takes 120 kW at a power factor of 0.6 lagging. Determine the apparent power and the reactive power.

The true power P = 120 kW. Thus $120 = VI \cos \phi = VI \times 0.6$ and so the apparent power $S = IV = 120/0.6 = 200$ kV A.

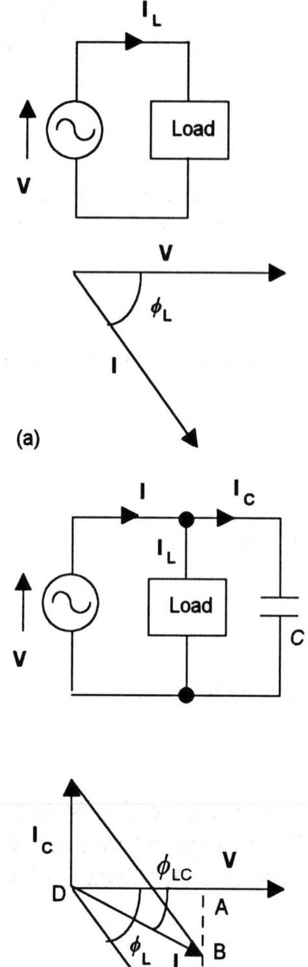

Figure 8.35 *Improving the power factor*

The reactive power $Q = IV \sin \phi$. Since $\cos \phi = 0.6$ then $\phi = 53.1°$ and so $\sin \phi = 0.8$ and $Q = 200 \times 0.8 = 160$ kV Ar.

Revision

27 A coil with an inductance of 1 H and a resistance of 200 Ω is connected to a 240 V r.m.s. 50 Hz supply. Determine the active power and the apparent power.

28 A coil with an inductive reactance of 10 Ω and a resistance of 7.5 Ω is connected to a 240 V r.m.s., 50 Hz supply. Determine the power factor, the apparent power and the active power.

8.8.5 Power factor improvement

The active power is the power content of the apparent power that is delivered by a supply system. The lower the power factor the greater is the current needed to produce the same active power. The higher the current, the greater the size of cable required to deliver the supply. Thus it is more economical in the delivery of power to have a system with a large power factor. Since most residential and industrial loads are inductive and have a lagging power factor, power factors can be improved by the connection of capacitors in parallel with the load.

Figure 8.35(a) shows the situation with an inductive load and Figure 8.35(b) the same load when a capacitor is connected in parallel with it. With no capacitor the current supplied **I** is the same as the current through the load $\mathbf{I_L}$ and we have a lagging phase angle of ϕ_L and thus a lagging power factor of $\cos \phi_L$. When the capacitor is connected in parallel, the current supplied gives a current through the load and a current through the capacitor. **I** is thus the phasor sum of $\mathbf{I_C}$ and $\mathbf{I_L}$. The current **I** taken from the supply is now at a lower lagging phase angle ϕ_{LC} and hence a higher power factor.

Initially, Figure 8.35(b) gives AC = $I_L \sin \phi_L$. After connecting the capacitor, AB = $I \sin \phi_{LC}$. Since BC = I_C then:

$$I_C = AC - AB = I_L \sin \phi_L - I \sin \phi_{LC} \quad [76]$$

Multiplying this equation by V gives:

$$VI_C = VI_L \sin \phi_L - VI \sin \phi_{LC} \quad [77]$$

But $VI_L \sin \phi_L$ is the reactive power with no capacitor Q_L and $VI \sin \phi_{LC}$ is the reactive power Q_{LC} with the capacitor. Thus:

$$VI_C = Q_L - Q_{LC} = \text{change in reactive power} \quad [78]$$

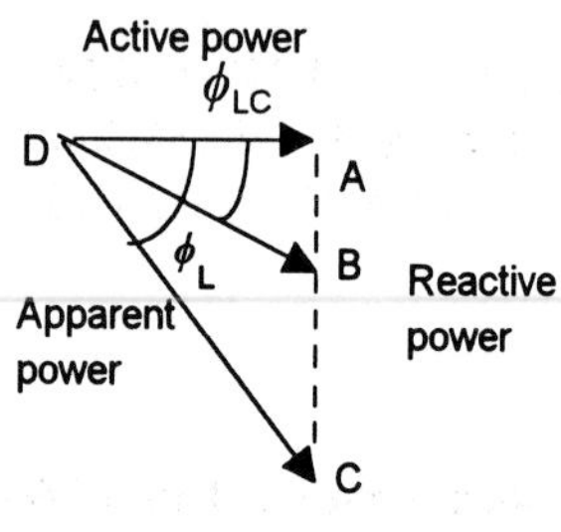

Figure 8.36 *Power triangles*

The triangles ABD and ACD with their sides multiplied by V give the power triangles (Figure 8.36). AB is the reactive power with the capacitor, AD is the reactive power without the capacitor, DC is the apparent power without the capacitor, DB is the apparent power with the capacitor and DA, since it is the apparent power multiplied by the cosine of the phase angle, which is the active power.

Example

A 240 V r.m.s., 50 Hz a.c. motor gives an active power output of 1 kW when operating at a power factor of 0.8 lagging. Calculate the reduction in the current taken by the motor if the power factor is improved to 0.9 lagging.

At a power factor of 0.8 lagging, the current taken by the motor from the supply is given by equation [74] as:

$$I_{\text{r.m.s.}} = \frac{P}{V_{\text{r.m.s.}} \cos\phi} = \frac{1\times 10^3}{240\times 0.8} = 5.21 \text{ A}$$

At a power factor of 0.9 lagging, the current taken is:

$$I_{\text{r.m.s.}} = \frac{P}{V_{\text{r.m.s.}} \cos\phi} = \frac{1\times 10^3}{240\times 0.9} = 4.62 \text{ A}$$

The current is thus reduced by 5.21 – 4.62 = 0.59 A.

Example

Calculate the capacitance required for a capacitor which is to be connected in parallel with a load of 500 W, supplied by a power supply of 240 V, 50 Hz, to change the power factor from 0.75 to 0.85 lagging.

The apparent power with the power factor 0.75 is 500/0.75 = 666.7 V A. With the power factor of 0.75 we have $\cos\phi = 0.75$ and so ϕ = 41.4°. Hence the reactive power is 666.7 × sin 41.4° = 440.9 V Ar. With the power factor of 0.85, the apparent power is 500/0.85 = 588.2 V A, ϕ = 31.8° and the reactive power is 588.2 sin 31.8° = 310.0 V Ar. The change in reactive power required is thus 440.9 – 310.0 = 130.9 V Ar. Using equation [78]:

$$VI_C = \text{change in reactive power} = 130.9 \text{ V Ar}$$

The current taken by the capacitor is thus 130.9/240 = 0.545 A. Since current is the voltage across the capacitor divided by its reactance, the reactance is 240/0.545 = 440.4 Ω. Thus 440.4 = $1/2\pi fC$ and so C = 7.23 μF.

Example

A load takes a current of 4 A r.m.s. when connected to a 240 V r.m.s., 50 Hz supply and has a power factor of 0.75 lagging. What capacitance will a capacitor need to have if, when connected in parallel with the load, it is to increase the power factor to 0.85 lagging?

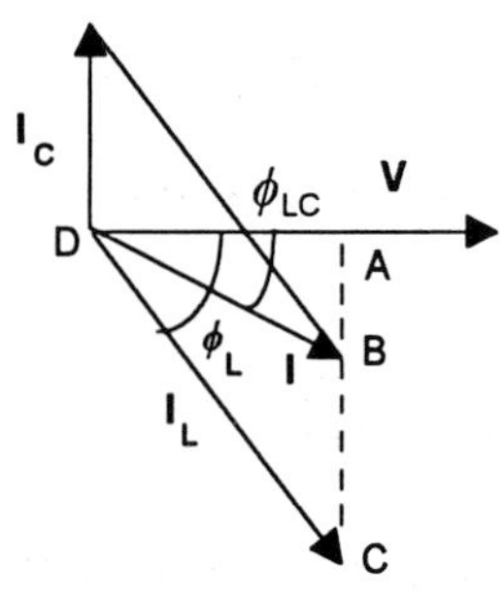

Figure 8.37 *Example*

Since the initial power factor cos ϕ_L is 0.75, the initial phase angle is ϕ_L = 41.4°. The corrected power factor is 0.85 and so the corrected phase angle is ϕ_{LC} = 31.8°. The phasor diagram for the currents is shown in Figure 8.37. For triangle ACD we have $AC = I_L \sin \phi_L$ = 4 × sin 41.4° = 2.65 A. For triangle ABD we have AB = AD tan ϕ_{LC} and since AD = I_L cos ϕ_L = 4 × 0.75 = 3 then AB = 3 tan 31.8° = 1.86 A. Thus I_C = AC – AB = 2.65 – 1.86 = 0.79 A. For the capacitor we have $X_C = V/I_C$ = 240/0.79 = 304 Ω and so $1/2\pi fC$ = 304 and C = 10.5 μF.

Revision

29 Calculate the reactive power required of a capacitor which when connected in parallel with a load of 650 kV A will change its power factor from 0.7 to 0.9 lagging.

30 A load takes a current of 6 A r.m.s. when connected to a 240 V r.m.s., 50 Hz supply and has a power factor of 0.65 lagging. What capacitance will a capacitor need to have if, when connected in parallel with the load, it is to increase the power factor to 0.90 lagging?

Problems

1 A sinusoidal voltage has a maximum value of 10 V and a frequency of 50 Hz. (a) Write an equation describing how the voltage varies with time. (b) Determine the voltages after times from t = 0 of (i) 0.002 s, (ii) 0.006 s and (iii) 0.012s.

2 A sinusoidal current has a maximum value of 50 mA and a frequency of 2 kHz. (a) Write an equation describing how the current varies with time. (b) Determine the currents after times from t = 0 of (i) 0.4 ms, (ii) 0.8 ms, (iii) 1.6 ms.

3 For a sinusoidal voltage described by v = 10 sin 1000t volts, what will be (a) the value of the voltage at time t = 0, (b) the maximum value of the voltage, (c) the voltage after 0.2 ms?

4 Determine the capacitive reactance of:

(a) a capacitor of capacitance 0.1 μF when the frequency of the current through it is 1 kHz,

(b) a capacitor of capacitance 100 μF when the voltage across it is 12 sin 5000 t V,

(c) a capacitor of capacitance 20 pF when the current through it is 100 sin 12000t mA.

5 Determine the inductive reactance of:

(a) an inductor of inductance 20 mH when the frequency of the current through it is 20 kHz,
(b) an inductor of inductance 1 H when the voltage across it is 10 sin 4000t V,
(c) an inductor of inductance 100 mH when the current through it is 20 sin 100t mA.

6 Determine the current through a pure capacitor of capacitance 0.01 μF when the voltage across it is 40 sin 1000t V.

7 At what frequency will the inductive reactance of an inductor of inductance 15 mH be 2.5 kΩ?

8 At what frequency will the capacitance reactance of a capacitor of capacitance 20 μF be 100 Ω?

9 A coil of inductance 50 mH and resistance 15 Ω is connected across a 240 V r.m.s., 50 Hz supply. Determine the circuit impedance, the current taken and the phase angle between the current and voltage.

10 A coil has a resistance of 15 Ω and an inductive reactance of 8 Ω. Determine the impedance of the coil and the phase angle between the voltage across the coil and the current through it.

11 A voltage of 100 V r.m.s., 50 Hz is connected to a series *RL* circuit, the resistance being 45 Ω and the inductance having an inductive reactance of 60 Ω. Determine (a) the circuit impedance, (b) the current through the circuit, (c) the voltages across each component.

12 A voltage of 100 V r.m.s., 50 Hz is connected to a series *RL* circuit, the resistance being 15 Ω and the inductance being 60 mH. Determine (a) the circuit impedance, (b) the magnitude and phase angle of the current through the circuit.

13 A voltage of 50 V r.m.s., 50 Hz is applied to a series *RC* circuit. If the current taken is 1.7 A when the voltage drop across the resistor is 34 V, what is the capacitance of the capacitor?

14 A voltage of 240 V r.m.s., 50 Hz is applied to a series *RC* circuit. If the resistance is 40 Ω and the capacitance 50 μF, determine (a) the circuit impedance, (b) the magnitude and phase angle of the current.

15 A voltage of 24 V r.m.s., 400 Hz is applied to a series *RC* circuit. If the resistance is 40 Ω and the capacitance 15 μF, determine (a) the circuit impedance, (b) the magnitude and phase angle of the current.

16 A series *RLC* circuit takes a current of 5 A r.m.s. If the resistance is 10 Ω, the inductive reactance 20 Ω and the capacitive reactance 10 Ω, what will be (a) the magnitude and phase of the voltage applied to the circuit, (b) the voltages across each component?

17 A *RLC* circuit has a coil having a resistance of 75 Ω and an inductance of 0.15 H in series with a capacitance of 8 μF. If a voltage of 100 V r.m.s., 200 Hz is applied to it, determine (a) the magnitude and phase angle of the current, (b) the voltage across the coil, (c) the voltage across the capacitor.

18 A series *RLC* circuit has a resistance of 10 Ω, an inductance of 100 mH and a capacitance of 0.2 μF. Determine the resonant frequency and the current at resonance when the alternating supply voltage is 12 V.

19 A series *RLC* circuit has a resistance of 10 Ω, an inductance of 100 mH and a capacitance of 2 μF. What is the resonant frequency of the circuit?

20 A series *RLC* circuit has a resistance of 50 Ω, an inductance of 200 mH and a capacitance of 0.2 μF. Determine the voltages across each component at resonance if the supply voltage is $15\angle 0°$.

21 A series *RLC* circuit resonates at a frequency of 100 Hz. If the resistance is 20 Ω and the inductance 300 mH, what is (a) the circuit capacitance, (b) the *Q*-factor of the circuit?

22 A series *RLC* circuit has a resistance of 4 Ω, an inductance of 60 mH and a capacitance of 30 μF. Determine (a) the resonant frequency and (b) the *Q*-factor of the circuit.

23 A circuit consists of a resistance of 10 Ω in parallel with an inductance of 50 mH. What is the magnitude and phase angle of the current taken from the supply when a voltage of 240 V r.m.s., 50 Hz is applied?

24 A circuit consists of a resistance of 40 Ω in parallel with an inductance of reactance 30 Ω. What is the magnitude and phase angle of the current taken from the supply when a voltage of 120 V r.m.s., 50 Hz is applied?

25 A circuit consists of a resistance in parallel with capacitance. What are the values of the resistance and capacitance if, when a voltage of 240 V r.m.s., 200 Hz is applied, the current taken from the supply is 2 A with a phase angle of 53.1° leading?

26 A resistance is connected in parallel with a capacitance. If, when a voltage of 100 V r.m.s. is applied to the circuit, the current drawn from the supply is 2 A at a phase angle of 30° leading, what are (a) the magnitudes of the currents through the resistance and the capacitance, and (b) the values of the resistance and the capacitive reactance?

27 A coil with a resistance of 10 Ω and an inductance of 50 mH is in parallel with a capacitance of 0.01 μF. Determine the resonant frequency.

28 A coil with a resistance of 5 Ω and an inductance of 50 mH is in parallel with a capacitance of 0.1 μF and a voltage supply of 100 V r.m.s., variable frequency, is applied. Determine the resonant frequency, the dynamic resistance, the current drawn at resonance and the *Q*-factor.

29 A coil with a resistance of 10 Ω and an inductance of 120 mH is in parallel with a capacitance of 60 μF and a voltage supply of 100 V r.m.s., variable frequency, is applied. Determine the resonant frequency, the dynamic resistance, the current drawn at resonance and the *Q*-factor.

30 Determine the *Q*-factor for a circuit involving a coil, of inductance 10 μH and resistance 3 Ω, in parallel with a 100 pF capacitor.

31 Determine the power dissipated when a coil of inductance 100 mH and resistance 40 Ω is connected across a 10 V, 500 Hz supply.

32 A coil has an inductance of 10 H and resistance of 250 Ω and is connected in series with a 5 μF capacitor across a 240 V, 50 Hz supply. Determine the power factor, the apparent power, the reactive power and the active power.

33 A circuit consists of a coil, resistance 200 Ω and inductance 10 H, in parallel with a capacitance of 5 μF. A voltage of 240 V r.m.s., 50 Hz is applied to the circuit. Determine (a) the current drawn from the supply, (b) the power factor of the circuit, (c) the active power, (d) the apparent power, (e) the reactive power.

34 A series *RC* circuit has a resistance of 2 kΩ and a capacitance of 1 μF and a 240 V r.m.s., 50 Hz supply is applied to the circuit. Determine (a) the power factor, (b) the active power, (c) the apparent power, (d) the reactive power.

35 A load takes 80 kW at a power factor of 0.6 lagging. Determine the apparent power and the reactive power.

36 Calculate the reactive power required of a capacitor which when connected in parallel with a load of 300 kV A will change its power factor from 0.7 to 0.9 lagging.

37 A load takes a current of 5 A r.m.s. when connected to a 240 V r.m.s., 50 Hz supply and has a power factor of 0.7 lagging. What capacitance will a capacitor need to have if, when connected in parallel with the load, it is to increase the power factor to 0.9?

9 Complex waveforms

9.1 Introduction

The alternating waveforms considered in Chapter 8 were all assumed to be sinusoidal. This will not always be the case. For example, many voltages which might initially have been sinusoidal have their waveforms 'distorted' by being applied to some non-linear device and thus we need to be able to considered the behaviour of this waveform with an electrical circuit. In other cases we might have a rectangular waveform rather than a sinusoidal one. A waveform which is not sinusoidal is termed a *complex waveform*. This chapter is a basic consideration of such non-sinusoidal waveforms.

9.1.1 Principle of superposition

The basic principle we will be using with circuit analysis using complex waveforms is the *principle of superposition*. When this is applied to circuits termed *linear circuits*, then because the current is proportional to the voltage, if a voltage v_1 gives a current i_1 and a voltage v_2 gives a current i_2 then a voltage equal to the sum of these voltages, i.e. $(v_1 + v_2)$, will give a current equal to the sum of the currents, i.e. $(i_1 + i_2)$. For example, consider the situation where we first apply a d.c. voltage of 2 V to a circuit and determine the current, then apply a d.c. voltage of 3 V and determine the current. If we then applied a voltage of 5 V to the circuit, the current would be the same as the sum of the currents due to the 2 V and the 3 V when separately applied. As another example, if we first apply a sinusoidal voltage of 2 sin ωt V to a circuit and determine the current and then apply a sinusoidal voltage of 3 sin ωt V to the circuit and determine the current, then the current we obtain when applying a voltage of 5 sin ωt V is the same as the sum of the current obtained when 2 sin ωt V and 3 sin ωt V were separately applied.

9.2 The Fourier series

A waveform which varies sinusoidally with time can be represented by an equation of the form:

$$v_1 = A \sin \omega t \qquad [1]$$

where ω is the angular frequency and equal to $2\pi f$, f being the frequency, v_1 is the value of this waveform at time t and A is the maximum value of the waveform, i.e. its amplitude. Such a waveform repeats itself every 2π radians, i.e. 360°. If the frequency is doubled, with the maximum value unchanged, then equation [1] becomes:

$$v_2 = A \sin 2\omega t \qquad [2]$$

If the frequency is trebled, with the maximum value unchanged, we have:

$$v_3 = A \sin 3\omega t \qquad [3]$$

Suppose that, in addition to doubling and trebling the frequency, we also change the maximum values of the waveforms. Equations [1], [2] and [3] can then be written as:

$$v_1 = A_1 \sin 1\omega t \qquad [4]$$

$$v_2 = A_2 \sin 2\omega t \qquad [5]$$

$$v_3 = A_3 \sin 3\omega t \qquad [6]$$

The above equations all describe waveforms that have started off with v being zero at time $t = 0$. When this is not the case then equations [4], [5] and [6] become:

$$v_1 = A_1 \sin(1\omega t + \phi_1) \qquad [7]$$

$$v_2 = A_2 \sin(2\omega t + \phi_2) \qquad [8]$$

$$v_3 = A_3 \sin(3\omega t + \phi_3) \qquad [9]$$

How does this help with describing a complex waveform? Well, Jean Baptiste Fourier in 1822 proposed that any periodic waveform can be made up of a combination of sinusoidal waveforms, i.e.

$$v = A_1 \sin(1\omega t + \phi_1) + A_2 \sin(2\omega t + \phi_2) + A_3 \sin(3\omega t + \phi_3) + \dots \qquad [10]$$

We can also include a d.c. component A_0 to give:

$$v = A_0 + A_1 \sin(1\omega t + \phi_1) + A_2 \sin(2\omega t + \phi_2) + A_3 \sin(3\omega t + \phi_3) + \dots \qquad [11]$$

This is termed the *Fourier* series. The waveform element with the $1\omega t$ frequency is called the *fundamental frequency* or the *first harmonic*, the element with the $2\omega t$ frequency the *second harmonic*, the element with the $3\omega t$ the third harmonic, and so on.

There is an alternative, simpler, way of writing equation [11]. Since $\sin(A + B) = \sin A \cos B + \cos A \sin B$ we can write:

$$A_1 \sin(1\omega t + \phi_1) = A_1 \sin \phi_1 \cos 1\omega t + A_1 \cos \phi_1 \sin 1\omega t$$

If we represent the non-time varying terms $A_1 \sin \phi_1$ by a constant a_1 and $A_1 \cos \phi_1$ by b_1, then:

$$A_1 \sin(1\omega t + \phi_1) = a_1 \cos 1\omega t + b_1 \sin 1\omega t$$

Likewise we can write:

$$A_2 \sin(2\omega t + \phi_2) = a_2 \cos 2\omega t + b_2 \sin 2\omega t$$

$$A_3 \sin(3\omega t + \phi_3) = a_3 \cos 3\omega t + b_3 \sin 3\omega t$$

and so on. Thus equation [11], for convenience we choose to write $\frac{1}{2}a_0$ for A_0, can be written as:

$$v = \tfrac{1}{2}a_0 + a_1 \cos 1\omega t + a_2 \cos 2\omega t + a_3 \cos 3\omega t + \dots + b_1 \sin 1\omega t + b_2 \sin 2\omega t + b_3 \sin 3\omega t + \dots \quad [12]$$

Fourier series: any periodic waveform can be represented by a constant d.c. signal term plus terms involving sinusoidal waveforms of multiples of a basic frequency.

Revision

1 What harmonics are present in the waveform given by $v = 1.0 - 0.67 \cos 2\omega t - 0.13 \cos 4\omega t$?

9.2.1 Graphical synthesis of a complex waveform

When we have two waveforms we can obtain the waveform describing their sum by adding the values of the two, ordinate by ordinate. Figure 9.1 illustrates this with the addition of a d.c. signal $v = 1$ and a sinusoidal waveform $v = 1 \sin \omega t$ to give the waveform for $1 + 1 \sin \omega t$.

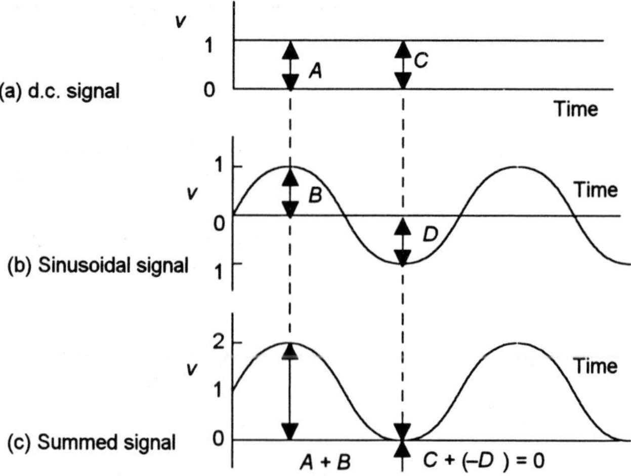

Figure 9.1 *Addition of two waveforms*

Consider the waveform produced by having just sine terms with the fundamental and the third harmonic and $a_3 = a_1/3$, i.e.

$$v = a_1 \sin 1\omega t + \tfrac{1}{3} a_1 \sin 3\omega t \qquad [13]$$

Figure 9.2 shows graphs of the two terms and the waveform obtained by adding the two, ordinate by ordinate.

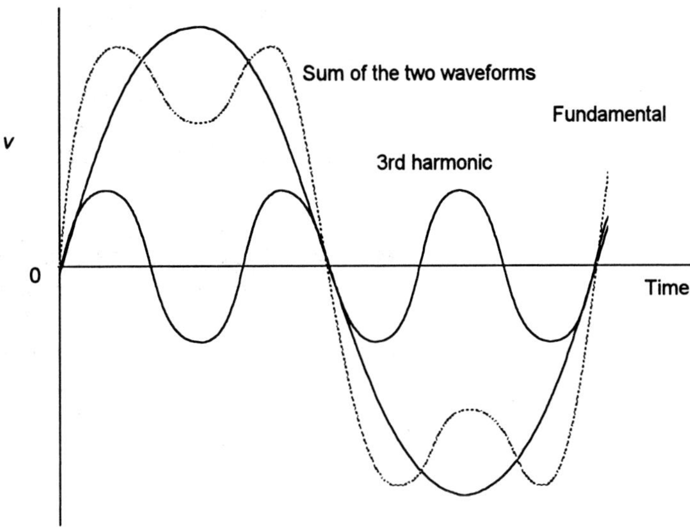

Figure 9.2 *Adding two waveforms*

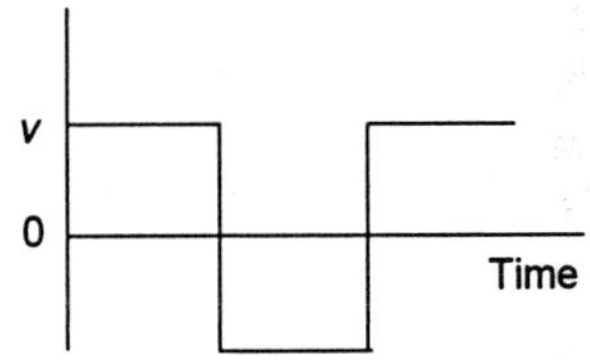

Figure 9.3 *Square waveform*

The result of adding the two waveforms is something that begins to look a bit like a square waveform. A better approximation to a square waveform (Figure 9.3) is given by adding more terms:

$$v = a_1 \sin 1\omega t + \tfrac{1}{3} a_1 \sin 3\omega t + \tfrac{1}{5} a_1 \sin 5\omega t + \tfrac{1}{7} a_1 \sin 7\omega t + \dots \qquad [14]$$

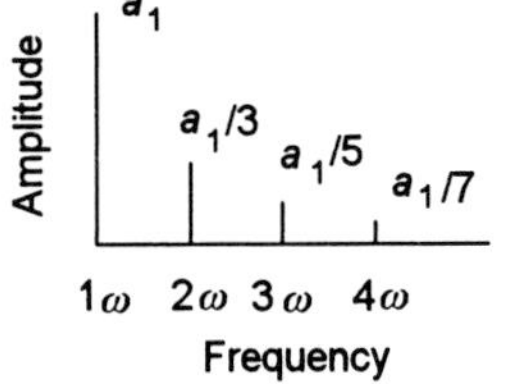

Figure 9.4 *Line spectrum*

We can describe the above representation of a square waveform by its *line spectrum*. This is a graph showing the amplitudes of the various frequencies (Figure 9.4).

If we add to the waveform a d.c. term then we can shift the sum graph up or down accordingly. Thus if we add $0.79a_1$, i.e. we have:

$$v = 0.79a_1 + a_1 \sin 1\omega t + \tfrac{1}{3} a_1 \sin 3\omega t + \tfrac{1}{5} a_1 \sin 5\omega t + \tfrac{1}{7} a_1 \sin 7\omega t + \dots \qquad [15]$$

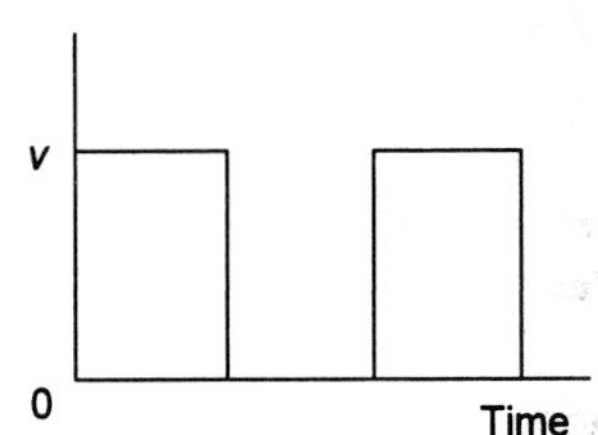

Figure 9.5 *Rectangular waveform*

and obtain a rectangular waveform which approximates to a periodic sequence of pulses (Figure 9.5). Figure 9.6 shows the addition of such a term to the waveform summation in Figure 9.2. Note that the average value of this waveform over one cycle is 0.79. The term $\frac{1}{2}a_0$ in the

Fourier series (equation [12]) thus represents the average value of the waveform over a cycle. Figure 9.7 shows the line spectrum.

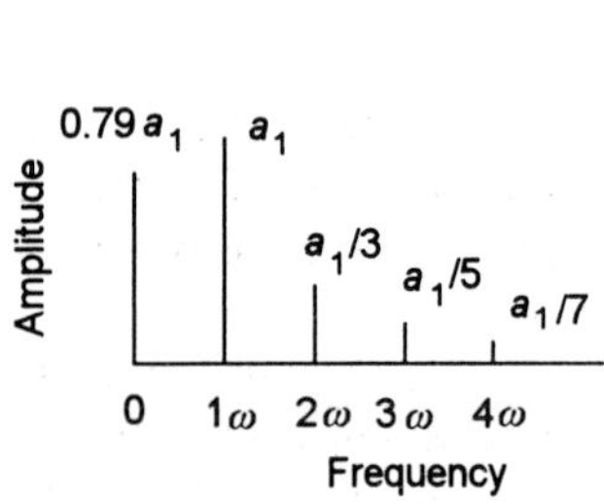

Figure 9.7 *Line spectrum*

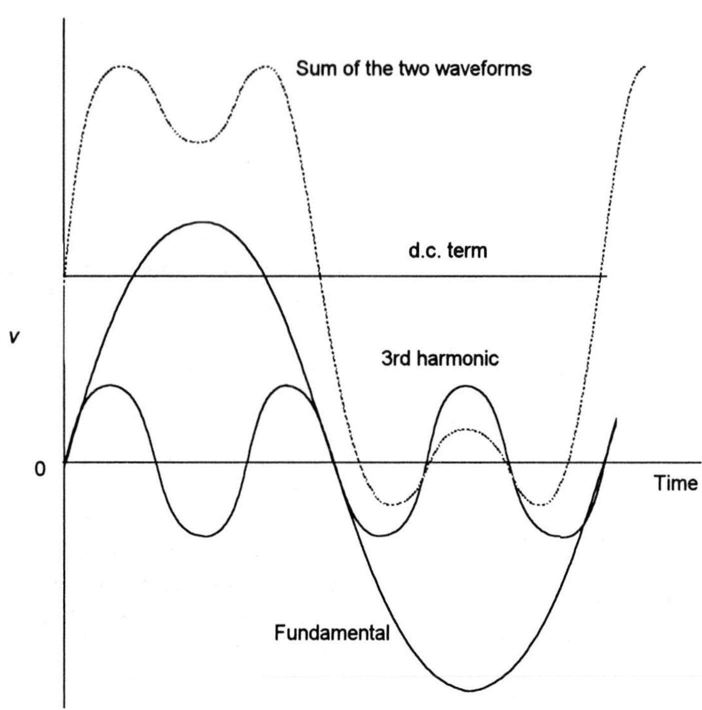

Figure 9.6 *Adding waveforms*

Revision

2 By constructing a graph, determine the waveform for:

(a) $v = 10 \sin \omega t + 3 \cos 3\omega t$, (b) $v = 5 + 10 \sin \omega t + 3 \cos 3\omega t$.

9.3 Circuit analysis with complex waveforms

Often in considering electrical systems the input is not a simple d.c. or sinusoidal a.c. signal but perhaps a square wave periodic signal or a distorted sinusoidal signal or a half-wave rectified sinusoid. Such problems can be tackled by representing the waveform as a Fourier series and using the *principle of superposition*; we find the overall effect of the waveform by summing the effects due to each term in the Fourier series considered alone. Thus if we have a voltage waveform:

$$v = V_0 + V_1 \sin \omega t + V_2 \sin 2\omega t + V_3 \sin 3\omega t + \ldots + V_n \sin n\omega t \qquad [16]$$

then we can consider the effects of each element taken alone. Thus we can calculate the current due to the voltage V_0, that due to $V_1 \sin \omega t$, that due to $V_2 \sin 2\omega t$, that due to $V_3 \sin 3\omega t$, and so on for all the terms in the series. We then add these currents to obtain the overall current due to the waveform.

As an illustration, consider the application to a pure resistance R of such a voltage. Since $i = v/R$ and resistance R is independent of frequency, then the current due to the V_0 term is V_0/R, that due to the first harmonic term is $(V_1 \sin \omega t)/R$, that due to the second harmonic term is $(V_2 \sin 2\omega t)/R$ and so on. Thus the resulting current waveform is:

$$i = \frac{V_0}{R} + \frac{V_1}{R}\sin \omega t + \frac{V_2}{R}\sin 2\omega t + \ldots + \frac{V_n}{R}\sin n\omega t \quad [17]$$

Because the resistance is the same for each harmonic, the amplitude of each voltage harmonic is reduced by the same factor, i.e. the resistance. The phases of each harmonic are not changed. The current waveform is thus the same shape as the voltage waveform.

The situation is different with the application to a pure inductance L of the voltage described by equation [16]. The impedance of a pure inductance depends on the frequency, i.e. its reactance $X_L = \omega L$. Also the current lags the voltage by 90°. The impedance is 0 when the frequency is 0 and thus the current due to the V_0 term will be 0. The current due to the first harmonic will be the voltage of that harmonic divided by the impedance at that frequency and so $V_1 \sin (\omega t - 90°)/\omega L$. The current due to the second harmonic will be the voltage of that harmonic divided by the impedance at that frequency and so $V_1 \sin (2\omega t - 90°)/2\omega L$. Thus the current waveform will be

$$i = \frac{V_1}{\omega L}\sin(\omega L - 90°) + \frac{V_2}{2\omega L}\sin(2\omega L - 90°) + \ldots$$
$$+ \frac{V_n}{n\omega L}\sin(n\omega L - 90°) \quad [18]$$

Each of the voltage terms has its amplitude altered by a different amount; the phase, however, is changed by the same amount. The result is that the shape of the current waveform is different to that of the voltage waveform.

Consider a pure capacitor capacitance C when the voltage applied across it is as described by equation [16]. The impedance of a pure capacitor depends on the frequency, i.e. its reactance $X_C = 1/\omega C$, and the current leads the voltage by 90°. The impedance is 0 when the frequency is 0 and thus the current due to the V_0 term will be 0. The current due to the first harmonic will be the voltage of that harmonic divided by the impedance at that frequency and so $V_1 \sin (\omega t + 90°)/(1/\omega C)$. For the second harmonic the current will be the voltage of that harmonic divided by the impedance at that frequency and so $V_1 \sin (2\omega t + 90°)/(1/2\omega C)$. Thus the current waveform will be:

$$i = \omega C V_1 \sin(\omega t + 90°) + 2\omega C V_2 \sin(2\omega t + 90°) + \ldots$$
$$+ n\omega C V_n \sin(n\omega t + 90°) \quad [19]$$

Application

Consider the effect of using a rectifier, or rectifiers, on a sinusoidal current. The result with half-wave rectification is of the form shown in Figure 9.8(a) and with full-wave rectification as shown in Figure 9.8(b). With the half-wave rectification the output can be expressed as:

$$i = \frac{A}{\pi}\left[1 + \frac{\pi}{2}\sin \omega t - \frac{2}{3}\cos 2\omega t + \frac{2}{15}\cos 4\omega t + \ldots\right]$$

and with full-wave rectification as:

$$i = \frac{2A}{\pi}\left[1 - \frac{2}{3}\cos 2\omega t + \frac{2}{15}\cos 4\omega t + \ldots\right]$$

We can thus carry out the analysis of circuits involving such signals in terms of their Fourier series.

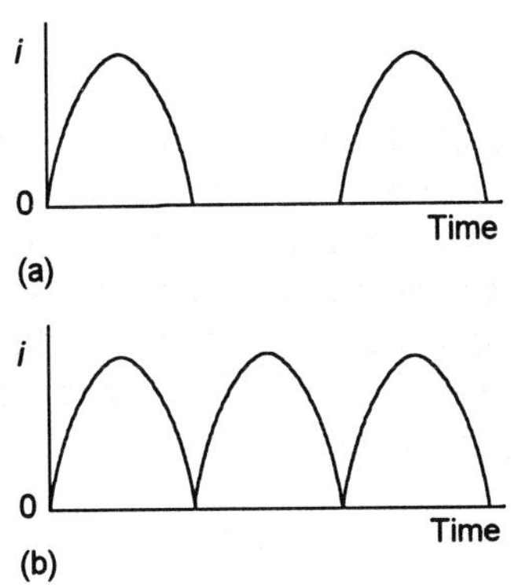

Figure 9.8 *(a) Half-wave rectification, (b) full-wave rectification*

Each of the voltage terms has had their amplitude altered by a different amount but the phase changed by the same amount. The result of this is that the shape of the current waveform is different to that of the voltage waveform.

Example

A complex voltage of $2.5 + 3.2 \sin 100t + 1.6 \sin 200t$ V is applied across a resistor having a resistance of 100 Ω. Determine the current through the resistor.

The complex current will be the sum of the currents due to each of the voltage terms in the complex voltage. Since the resistance is the same at all frequencies, the complex current will be:

$$i = 0.025 + 0.032 \sin 100t + 0.016 \sin 200t \text{ A}$$

Example

A complex voltage of $2.5 + 3.2 \sin 100t + 1.6 \sin 200t$ V is applied across a pure inductor having a inductance of 100 mH. Determine the current through the inductor.

The impedance is 0 when the frequency is 0 and thus the current due to the 2.5 V term will be 0. For the second term, the reactance is $100 \times 0.100 = 10$ Ω and the current lags the voltage by 90° and so the current due to this harmonic is $0.32 \sin (100t - 90°)$ A. For the third term, the reactance is $200 \times 0.100 = 20$ Ω and the current lags the voltage by 90° and so the current due to this harmonic is $0.08 \sin (100t - 90°)$ A. Thus the current waveform is:

$$i = 0.32 \sin (100t - 90°) + 0.08 \sin (100t - 90°) \text{ A}$$

Example

Determine the waveform of the current occurring when a 2μF capacitor has connected across it the half-wave rectified sinusoidal voltage $v = 0.32 + 0.5 \cos 100t + 0.21 \cos 200t$ V.

There will be no current arising from the d.c. term. For the first harmonic the reactance is $1/(2 \times 10^{-6} \times 100)$ Ω and so we have a current of $0.5 \times 2 \times 10^{-6} \times 100 \cos (100t + 90°)$ A. For the second harmonic the reactance is $1/(2 \times 10^{-6} \times 200)$ Ω and so the current is $0.21 \times 2 \times 10^{-6} \times 200 \cos (200t + 90°)$. Thus the resulting current is:

$$i = 2 \times 10^{-6} \times 0.5 \times 100 \cos (100t + 90°) + 2 \times 10^{-6} \times 0.21 \times 200 \cos (200t + 90°) \text{ A}$$

Revision

3 Determine the waveform of the current occurring when a resistor of resistance 1 kΩ has connected across it the half-wave rectified sinusoidal voltage $v = 0.32 + 0.5 \cos 100t + 0.21 \cos 200t$ V

4 Determine the waveform of the current when a pure inductor of inductance 10 mH has connected across it the half-wave rectified sinusoidal voltage $v = 0.32 + 0.5 \cos 100t + 0.21 \cos 200t$ V.

5 A voltage of $2.5 + 3.2 \sin 100t + 1.6 \sin 200t$ V is applied across a 10 μF capacitor. Determine the current.

9.4 Production of harmonics

The waveforms produced by a.c. generators is usually very nearly perfectly sinusoidal and thus the amount of harmonics is small. Most of the harmonic content of waveforms in circuits is, however, produced by non-linear circuit elements. A *non-linear circuit element* is one for which the current flowing through it is not proportional to the voltage across it. Examples of such devices are semiconductor diodes, transistors and ferromagnetic-cored coils. As an illustration, consider a transistor with the relationship between current and voltage shown in Figure 9.9. A sinusoidal voltage input gives rise to a distorted sinusoidal current.

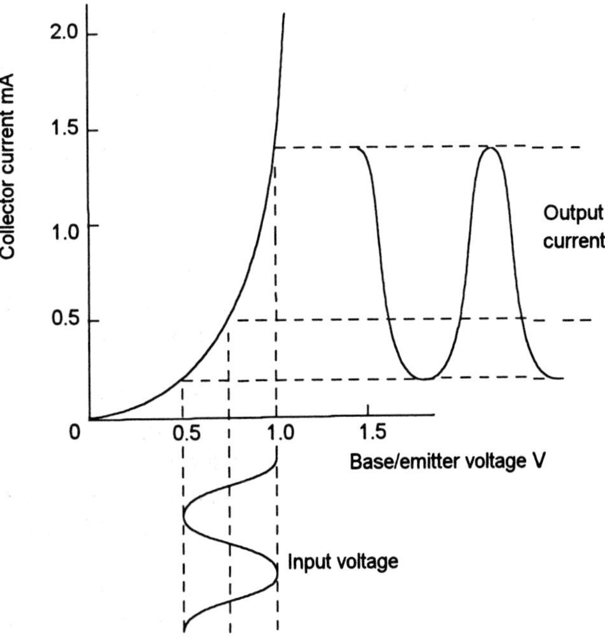

Figure 9.9 *Characteristic of a transistor*

Another example of non-linearity is that given by an iron-cored circuit element such as a transformer. When a sinusoidal alternating voltage is applied to a coil wound on an iron core, the current through the coil produces magnetic flux in the core. The relationship between the current and the magnetic flux is not linear. Indeed, not only is it not linear but an increasing value of current produces a different flow density to the same value of current when it is decreasing, this effect being termed hysteresis. Figure 9.10 illustrates this and shows how a sinusoidal current produces magnetic flux which varies with time in a non-sinusoidal manner. Because the core flux is non-sinusoidal, the e.m.f. induced in the secondary winding of a transformer by the magnetic flux will be non-sinusoidal.

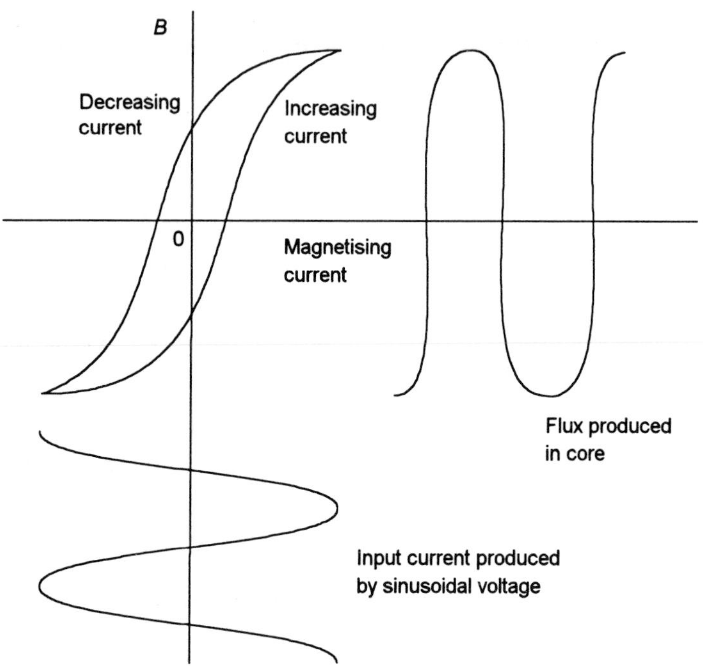

Figure 9.10 *Effect of iron core on magnetic flux waveform*

The current–voltage relationship for non-linear circuit elements can, in general, be represented by an equation of the form:

$$i = a + bv + cv^2 + dv^3 + \dots \qquad [20]$$

where a, b, c, d, etc. are constants. Suppose we have an applied voltage of $v = V \sin \omega t$, then if we consider just the a, b, c terms:

$$i = a + bV \sin \omega t + c(V \sin \omega t)^2$$

But $\sin^2 \omega t = \frac{1}{2}(1 - \cos 2\omega t)$ and so:

$$i = a + bV \sin \omega t + \tfrac{1}{2}cV^2(1 - \cos 2\omega t) \qquad [21]$$

Thus a second harmonic has been introduced. The amount of distortion resulting from this can be expressed as the amplitude of the second harmonic as a percentage of the fundamental.

$$\% \text{ second harmonic} = \frac{\frac{1}{2}cV^2}{bV} \times 100\% = \frac{cV}{2b} \times 100\% \qquad [22]$$

A typical amplifier stage response has a current–voltage characteristic of the form:

$$i = a + bv + cv^2 \qquad [23]$$

The input v is generally made up of a bias voltage V_b plus the signal to be amplified. If this is a sinusoidal signal $V_s \sin \omega t$ then equation [23] gives:

$$i = a + b(V_b + V_s \sin \omega t) + c(V_b + V_s \sin \omega t)^2$$

$$= a + bV_b + bV_s \sin \omega t + cV_b^2 + V_s^2 \sin^2 \omega t + 2cV_bV_s \sin \omega t$$

$$= a + bV_b + cV_b^2 + (bV_s + 2cV_bV_s) \sin \omega t + cV_s^2 \sin^2 \omega t$$

$$= a + bV_b + cV_b^2 + (bV_s + 2cV_bV_s) \sin \omega t + \tfrac{1}{2}cV_s^2(1 - \cos 2\omega t)$$

When the alternating signal is zero then $i = a + bV_b + cV_b^2$. Let this current be represented by d. If we represent $(bV_s + 2cV_bV_s)$ by e and $\frac{1}{2}cV_s^2$ by f then:

$$i = d + e \sin \omega t + f - f \cos 2\omega t \qquad [24]$$

$d + f$ is the steady direct current, e is the peak value of the fundamental and f the peak value of the second harmonic. The percentage second harmonic distortion is thus $(f/e) \times 100\%$. We can write this in terms of current values occurring with the current waveform. The alternating current output will alternate between a maximum value, which occurs when $\omega t = 90°$, of $i_{max} = d + e + f + f$ and a minimum value, which occurs when $\omega t = 270°$, of $i_{min} = d - e + f + f$. Thus $i_{max} + i_{min} = 2d + 4f$. But d is the current occurring when the signal has a zero value. If we designate this as i_0 then $i_{max} + i_{min} - 2i_0 = 4f$. The maximum current value minus the minimum current value is $(d + e + f + f) - (d - e + f + f) = 2e$. Hence:

$$\% \text{ 2nd harmonic} = \frac{f}{e} \times 100\% = \frac{i_{max} + i_{min} - 2i_0}{2(i_{max} - i_{min})} \times 100\% \qquad [25]$$

Example

A non-linear circuit element gave the current–voltage characteristic $i = 2.0 + 1.2v + 0.6v^2$ mA. Determine the percentage second harmonic content when $v = 1.0 \sin \omega t$ V.

Using $\sin^2 \omega t = ½(1 - \cos 2\omega t)$:

$$i = 2.0 + 1.2 \times 1.0 \sin \omega t + ½ \times 0.6 \times 1.0^2(1 - \cos 2\omega t)$$

$$= 2.3 + 1.2 \sin \omega t - 0.3 \cos \omega t$$

The percentage second harmonic content is $(0.3/1.2) \times 100 = 25\%$.

Example

The current waveform produced by an amplifier stage is found to alternate between 60 mA and 85 mA when there is an input of a sinusoidal voltage with a constant current of 72 mA being given when the signal voltage is zero. If the amplifier characteristic is of the form $i = a + bv + cv^2$, determine the percentage second harmonic content.

Using equation [25]:

$$\% \text{ 2nd harmonic} = \frac{i_{max} + i_{min} - 2i_0}{2(i_{max} - i_{min})} \times 100\%$$

$$= \frac{85 + 60 - 2 \times 72}{2(85 - 72)} \times 100 = 3.8\%$$

Revision

6 A non-linear circuit element gave the current–voltage characteristic $i = 2.6 + 1.0v + 0.3v^2$ mA. Determine the percentage second harmonic content when $v = 1.0 \sin \omega t$ V.

7 A transistor amplifier stage has the current–voltage characteristic $i = 0.5 - 20v + 200v^2$ mA when v is in volts. Determine (a) the steady direct current and (b) the percentage second harmonic content when $v = 0.65 + 0.05 \sin \omega t$ V.

Problems

1 Plot the graph of the waveform $v = 3 \sin \omega t + \sin 2\omega t$.

2 Plot the graph of the waveform $v = 2 \sin \omega t + \cos 3\omega t$.

3 A voltage of $2 \sin 500t + 1 \sin 1000t$ V is applied across a resistor having a resistance of 10 Ω. Determine the circuit current.

4 Determine the waveform of the current occurring when a 2 μF capacitor has connected across it the half-wave rectified sinusoidal voltage $v = 0.32 + 0.5 \cos 100t + 0.21 \cos 200t$ V.

5 Determine the waveform of the current occurring when a 10 μF capacitor has connected across it the voltage $v = 2.5 + 3.2 \cos 100t + 1.6 \cos 200t$ V.

6 A non-linear circuit element gave the current–voltage characteristic $i = 2.6 + 1.4v + 0.6v^2$ mA. Determine the percentage second harmonic content when $v = 2.0 \sin \omega t$ V.

7 The current waveform produced by an amplifier stage is found to alternate between 2 mA and 12 mA when there is an input of a sinusoidal voltage with a constant current of 6 mA being given when the signal voltage is zero. If the amplifier characteristic is of the form $i = a + bv + cv^2$, determine the percentage second harmonic content.

10 Transformers

10.1 Introduction

A transformer is a device for taking an alternating current at one voltage and transforming it to alternating current at another voltage; it basically consists of two coils, called the primary and secondary coils, linked by a common magnetic circuit. The chapter deals with the basic principles of transformers, their basic constructional form and materials and their use for impedance matching.

Figure 10.1 shows the symbols that are used for transformers in circuit diagrams.

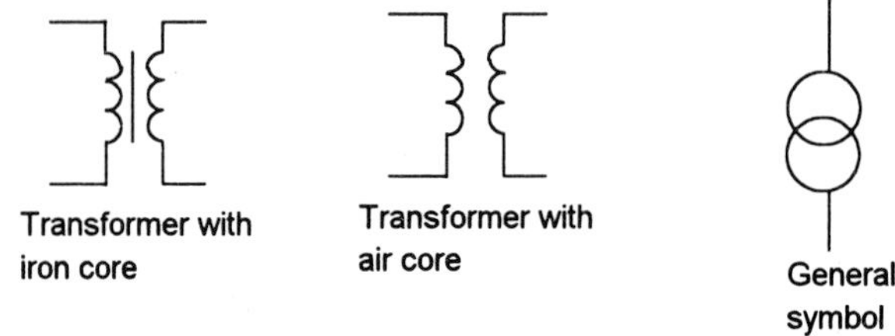

Figure 10.1 *Symbols*

10.2 Basic principles

Basically a transformer consists of two coils wound on a magnetic material core (Figure 10.2). One coil, called the primary coil, is supplied with the input alternating current and the output is taken from the other coil, termed the secondary coil. The transformer transfers electrical energy from the primary circuit to the secondary circuit by means of a magnetic field that links both circuits. No electrical connection between the primary and secondary is necessary; the transfomer may thus be used to isolate one circuit from another but still permit the exchange of energy between them.

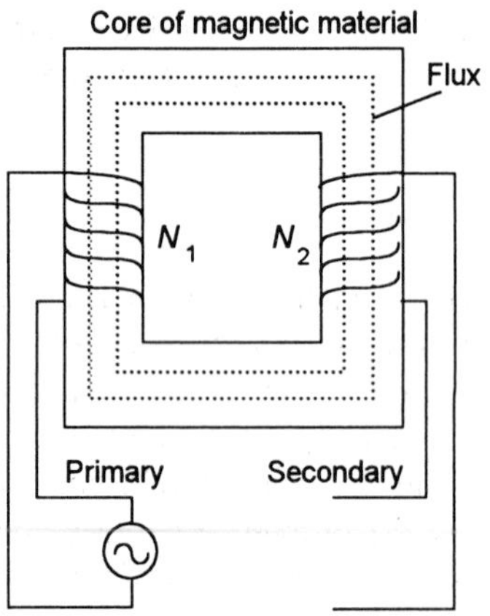

Figure 10.2 *Basic transformer on no-load*

10.2.1 Transformer on no-load

For an *ideal transformer*, i.e. one where all the magnetic flux is confined to the core and links both the primary and secondary coils and the exciting current needed to establish the magnetic flux in the core is negligible, an a.c. input to the primary coil produces an alternating magnetic flux and induces the same e.m.f. in each turn of both coils. Thus, if N_1 is the number of turns in the primary coil and N_2 the number of turns in the secondary coil, then (induced e.m.f. in primary coil)/(induced e.m.f. in secondary coil) = N_1/N_2. When there is no load connected to the secondary coil, the voltage V_2 between the terminals of the coil is the same as the induced e.m.f. If there is no load then there is

Application
The a.c. supply voltage has often to be stepped down to a lower voltage for use with electronic equipment and this is achieved by the use of a step-down transformer. It might then be rectified to give a low voltage d.c. supply.

Step-down voltage transformers are also used with electrical power transmission. The power in a single-phase circuit is given by $VI \cos \phi$. Thus for the transmission of a given power and power factor, VI is constant and so doubling V means halving I. Thus the higher the voltage of transmission the lower the power losses. For this reason, power in Britain is transmitted over long distances at a voltage of 400 kV. Step-down transformers are used to bring the voltage down to levels at which it is used in industry and in homes (Figure 10.3).

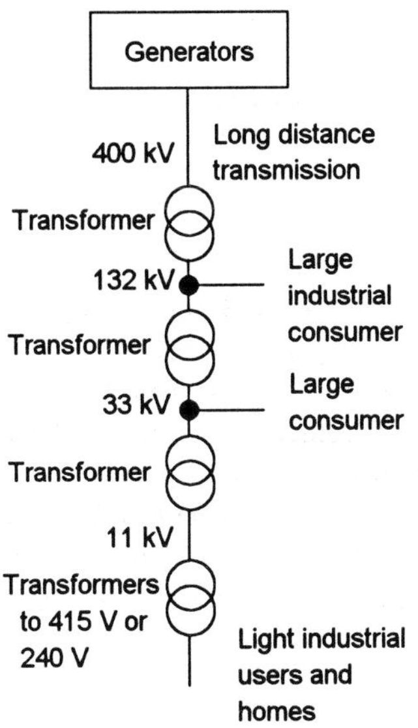

Figure 10.3 *Power transmission*

no current and so no energy taken from the secondary coil and so no energy is taken from the primary coil. This can only be the case if the induced e.m.f. is equal to and opposing the input voltage V_1. Thus:

$$\frac{V_1}{V_2} = \frac{N_1}{N_2} = \text{turns ratio} \qquad [1]$$

If the secondary voltage is greater than the primary voltage, the transformer is said to have a *step-up voltage ratio* and if the secondary voltage is less than the primary voltage a *step-down voltage ratio.*

Example

An ideal transformer has a primary with 900 turns. How many turns will there need to be on the secondary coil if a 240 V a.c. input is to give rise to a no-load a.c. output of 80 V?

Using equation [1]:

$$\frac{V_1}{V_2} = \frac{N_1}{N_2} = \text{turns ratio}$$

$$\frac{240}{80} = \frac{900}{N_2}$$

Hence $N_2 = 300$.

Revision

1 An ideal transformer has 800 primary turns and 200 secondary turns. What will be the secondary voltage when the primary is supplied from a 240 V a.c. supply?

2 An ideal transformer has a primary coil with 500 turns. How many secondary turns will it have if a 240 V a.c. input gives rise to no-load 48 V a.c. output?

10.2.2 Transformer with resistive load

If there is a resistive load connected to the secondary (Figure 10.4), a secondary current flows and hence power is dissipated. This current in the secondary coil produces its own alternating flux in the core and results in an alternating e.m.f. being induced in the primary coil and so a current in the primary coil. If the power losses in a transformer are negligible, then when there is a current in the primary coil the power supplied to the primary coil must equal the power taken from the secondary coil:

$$I_1V_1 = I_2V_2 \qquad [2]$$

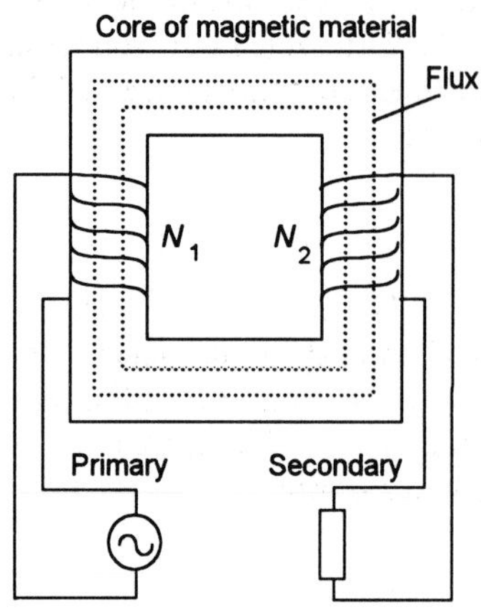

Figure 10.4 *Basic transformer with resistive load*

where I_1 is the current in the primary coil and I_2 that in the secondary coil. We can state the above equation as:

input volt-amperes = output volt-amperes [3]

Equation [2] can be rearranged as $V_1/V_2 = I_2/I_1$ and so we have the general equations:

$$\frac{V_1}{V_2} = \frac{I_2}{I_1} = \frac{N_1}{N_2} = \text{turns ratio} \quad [4]$$

and:

$$I_1N_1 = I_2N_2 \quad [5]$$

The product of the current through a coil and its number of turns is called its *ampere-turns*. Thus the input ampere-turns = the output ampere-turns.

Example

An ideal transformer is used to light a 12 V, 15 W lamp from the 240 V mains supply. Determine the turns ratio required and the current taken from the mains supply.

Turns ratio $V_1/V_2 = 240/12 = 20$. The current taken by the lamp $I_2 = 15/12 = 1.25$ A. Thus, since I_2/I_1 = turns ratio, we have $I_1 = 1.25/20 = 0.0625$ A.

Example

An ideal transformer has 20 primary turns and 80 secondary turns and supplies a secondary current of 1.5 A. What is (a) the value of the ampere-turns of the transformer, (b) the primary current?

(a) Value of ampere-turns = $I_2N_2 = 80 \times 1.5 = 120$ ampere-turns.
(b) $I_1N_1 = I_2N_2$ and so $I_1 = (N_2/N_1)I_2 = (80/20) \times 1.5 = 6.0$ A.

Revision

3 An ideal transformer is used to light a 12 V, 150 W lamp from the 240 V mains supply. Determine the turns ratio required and the current taken from the mains supply.

4 An ideal transformer has a step-up voltage ratio of 6:1. If it has a secondary coil voltage of 120 V and a secondary current of 2 A, what is (a) the primary coil voltage, (b) the primary coil current?

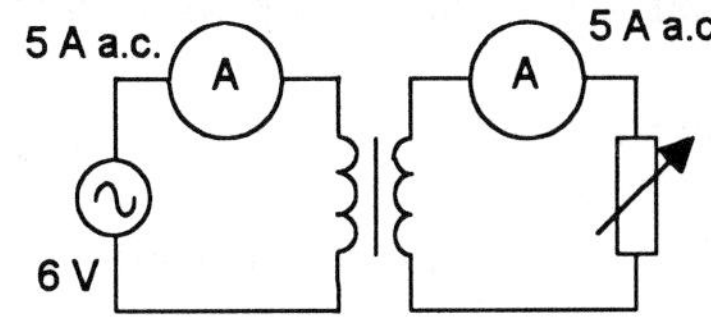

Figure 10.5 *Activity*

5 An ideal transformer has a turns ratio of 5, a primary coil of 200 turns and a primary coil current of 2 A. What is (a) the number of turns on the primary coil, (b) the value of the ampere-turns of the transformer, (c) the secondary coil current?

Activity

Use two C-cores and coils, e.g. two 120 + 120 turn coils, to form a transformer and hence investigate the behaviour of the transformer using the circuit shown in Figure 10.5. Examine, for different turn ratios, e.g. 1:1, 2:1, 1:2, how the primary current depends on the current in the secondary coil by varying the secondary load resistance.

10.3 Transformer construction

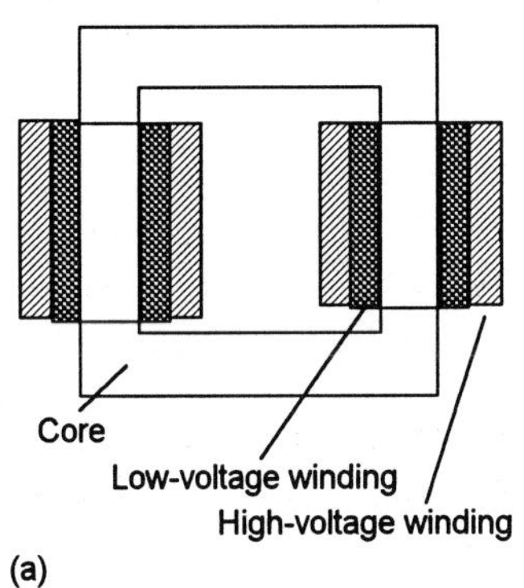

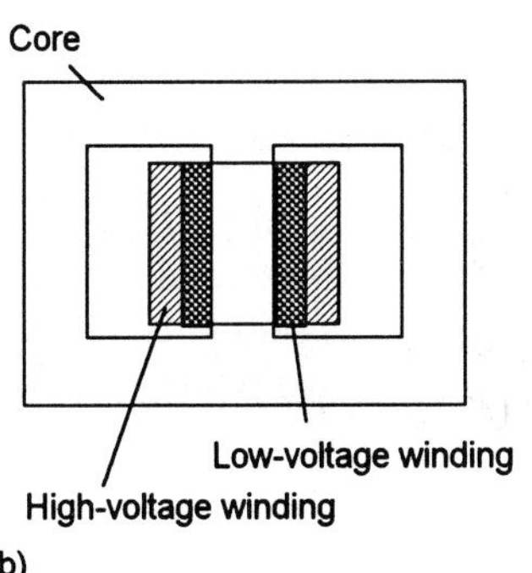

Figure 10.6 *(a) Core, (b) shell*

For power transformers, i.e. transformers used on the fixed power frequency of 50 Hz, it is important that all the magnetic flux generated by the primary coil links with the secondary coil and so they are both wound on a closed iron core. Two basic types of core construction are used. In the *core type* the windings are wound around two legs of a rectangular-shaped magnetic core, one winding on top of the other (Figure 10.6(a)), so that both the primary and the secondary have one half of each winding on each limb. The low-voltage winding is the innermost winding. In the *shell-type* the windings are wound around the centre leg of a three-legged core (Figure 10.6(b)). The low-voltage winding is the innermost winding.

10.3.1 Core material

When an alternating current is applied to the primary coil of a transformer, the magnetic state of the transformer core is taken through a complete magnetisation cycle, i.e. a *hysteresis loop* (Figure 10.7). When the magnetic field strength is increased from zero for an unmagnetised material, the magnetic flux increases from point 0 to A on Figure 10.7. When the field is reduced back to zero the magnetic flux does not go back to zero but to point C. When the magnetic field is then reversed in direction, because we are starting the second-half of the alternating current cycle, the magnetic flux goes from C to D to E. As the magnetic field is then decreased back to zero to complete the cycle, the magnetic flux goes from E to F. The next cycle takes the magnetic flux from F to G to A to C to D to E and then back to F. The energy consumed in this cycle is proportional to the area enclosed by the hysteresis loop. Hence, to minimise such losses, a material is required which has a hysteresis loop with a small enclosed area.

Another source of loss is *eddy currents*. The changing magnetic field produced by the alternating current in the primary coil will, as well as inducing e.m.f.s in the coils, induce e.m.f.s in the core and hence give rise to currents in the core. Such currents are termed *eddy currents*. A solid core can have a very low electrical resistance and so these eddy currents can dissipate large amounts of power. To minimise such losses, power transformer cores are made of a stack of thin laminations, each

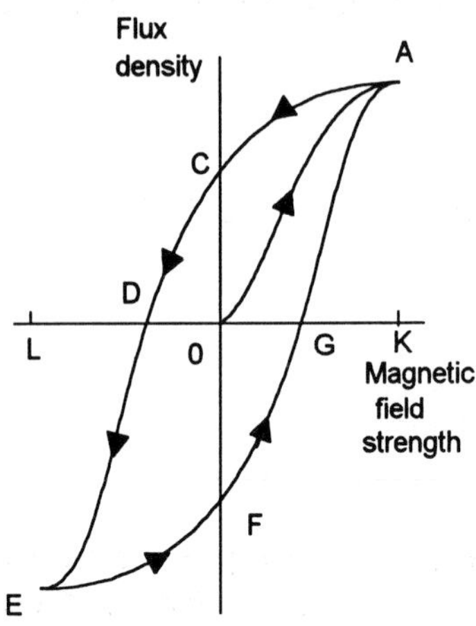

Figure 10.7 *Hysteresis loop showing how the magnetic flux density in the core changes as the magnetic field strength produced by the primary current alternates between K and L*

electrically insulated from its neighbours. This increases the electrical resistance of the core and so reduces the size of eddy currents and hence eddy current losses. The material used for power transformer cores is generally silicon steel, this material having a high value of permeability at low magnetic field strengths. The relative permeability of a material is the factor by which the flux density in a material is greater than that which would occur in a vacuum for the same magnetic field, thus the higher the permeability the greater the magnetic flux density produced for a given primary current.

The transformers used in audio-frequency applications involving matching impedances for maximum power transfer (see Section 10.4), e.g. connecting the output stage of an amplifier to a loudspeaker, have to handle a range of frequencies, e.g. 20 to 15 000 Hz, and for such transformers a compressed powdered ferromagnetic alloy, known as permalloy, might be used for the core. This is because the higher frequencies require laminations to be made thinner in order to reduce eddy currents. This is because the rate of change of current with time is being increased and so the induced e.m.f.s increase. With a powder, the binding material interface between powder grains provides the required high resistance.

Application
As an illustration, a typical 230-240 V mains transformer might have a twin, tapped, secondary winding and have the specification of: 9 VA rating, secondary windings 0-12 V, 0-12V, maximum current for each winding of 330 mA.

10.3.2 Power rating

The power rating for a transformer is defined as:

$$\text{power rating in VA} = V_2 I_{2FL} \qquad [6]$$

where V_2 is the secondary voltage and I_{2FL} is the secondary current on full-load. Full-load is the load which allows the maximum power output. This maximum power output is determined by the rate at which the heat generated by internal losses can be dissipated. The major part of this is generated by the current through the primary and secondary windings, i.e. the power loss = I^2R; this loss is termed the *copper loss*. Thus there is a maximum current which can be obtained from the secondary without the transformer overheating. When a transformer is said to be operating at half load, it means that it is delivering half its rated power output and hence its secondary current is half the full load value.

Example

A 2.75 kVA single-phase transformer has a primary with 600 turns and a secondary with 100 turns. If the alternating primary voltage is 3300 V, what will be the minimum resistance which can be connected as the load?

The secondary voltage = V_1N_2/N_1 = 3300 × (100/600) = 550 V. Since V_2I_{2FL} = 3000, then I_{2FL} = 2750/550 = 5 A. The value of the resistance to give this current is $R = V_2/I_{2FL}$ = 550/5 = 110 Ω. This is the minimum value of resistance since any smaller value would give rise to a larger current.

Revision

6 A single-phase transformer is specified as 15 kVA 600/240 V. What are the full-load primary and secondary currents.

10.3.3 Efficiency

The power losses that occur with a transformer are:

1 *Copper loss*
This is generated by the current through the primary and secondary windings and is thus power loss = $I_1^2R_1 + I_2^2R_2$, where I_1 is the current through a primary of resistance R_1 and I_2 is the current through a primary of resistance R_2.

2 *Iron loss*
This is the loss in the core resulting from hysteresis and eddy currents.

The efficiency of a transformer is defined as:

$$\text{efficiency} = \frac{\text{output power}}{\text{input power}} \times 100\% \qquad [7]$$

The output power is $V_2I_2 \cos \phi$ and equals the input power minus losses.

Example

A transformer delivers a power of 5 kW when on-load. What will be its efficiency if the copper loss is then 100 W and the iron loss 75 W?

The input power = output power + losses = 5000 + (100 + 75) = 5175 W. Hence the efficiency = (5000/5175) × 100 = 96.6%.

Example

A 40 kVA single-phase transformer has an iron loss of 400 W and a full load copper loss of 600 W. What is the efficiency of the transformer for a power factor of 0.8 at (a) full load, (b) half load?

(a) The total loss at full load = 400 + 600 = 1000 W. At a power factor of 0.8, output power = $V_2I_2 \cos \phi$ = 40 000 × 0.8 = 32 000 W. The input power is the output power plus losses = 32 000 + 1000 = 33 000 W. Hence the efficiency = (32 000/33 000) × 100% = 97.0%. (b) At half load, the current through the secondary will be half the full load value. Since the copper loss is proportional to I_2^2 then, at half load, the copper loss is reduced to $(\frac{1}{2})^2$ of 600 W, i.e. 150 W. The iron loss is not affected by a change in current. The total loss is now 400 + 150 = 550 W. The output power = $V_2 I_2 \cos\phi = \frac{1}{2} \times 40\,000 \times 0.8 = 16\,000$ W. The input power is output power plus losses = 16 000 + 550 = 16 550 W. Hence the efficiency = (16 000/16 550) × 100% = 96.7%.

Revision

7 What is the efficiency of a transformer which supplies 2 kW when on-load if the copper loss is then 50 W and the iron loss 35 W?

8 A 100 kVA single-phase transformer has an iron loss of 800 W and, at full load, a copper loss of 1000 W. If the power factor is 0.8, what will be (a) the efficiency at full load, (b) the efficiency at half-load?

10.4 Impedance matching

One use of a transformer is to alter the apparent value of an impedance. Consider the situation where a transformer has a secondary load of impedance Z_2 (Figure 10.8), then $Z_2 = V_2/I_2$. The primary has a current I_1 for a voltage V_1 and so the transformer has an input impedance $Z_1 = V_1/I_1$. Thus Z_1 is the impedance into which the source generator is working but Z_2 is the impedance seen from the secondary.

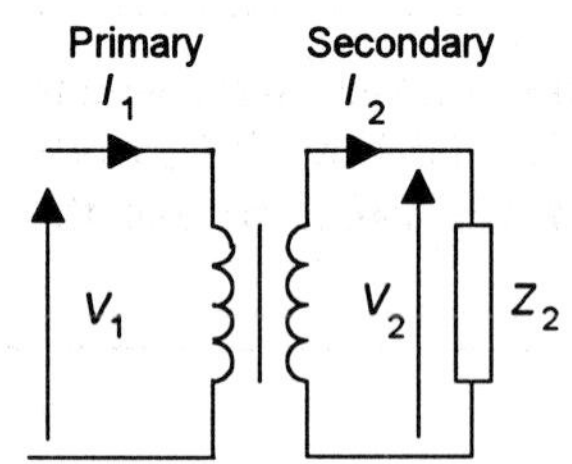

Figure 10.8 *Impedance transforming*

For a transformer we have $V_1/V_2 = N_1/N_2$ and $I_2/I_1 = N_1/N_2$. Thus:

$$V_1 = \frac{N_1}{N_2}V_2 \quad \text{and} \quad I_1 = \frac{N_2}{N_1}I_2$$

$$\frac{V_1}{I_1} = \left(\frac{N_1}{N_2}\right)^2 \frac{V_2}{I_2}$$

$$Z_1 = \left(\frac{N_1}{N_2}\right)^2 Z_2 \qquad [8]$$

Example

A step-down transformer with a turns ratio of 3:1 is connected to a load of 100 Ω. What will be the input resistance of the transformer?

$R_1 = (N_1/N_2)^2 R_2 = 3^2 \times 100 = 900\ \Omega$.

Revision

9 A step-up transformer has a turns ratio of 1:4 and is connected to a load of 40 Ω. What will be the input resistance of the transformer?

10.4.1 Maximum power transfer theorem

A use of such impedance transforming is to match, for example, the impedance of a loudspeaker to the output impedance of an amplifier, so

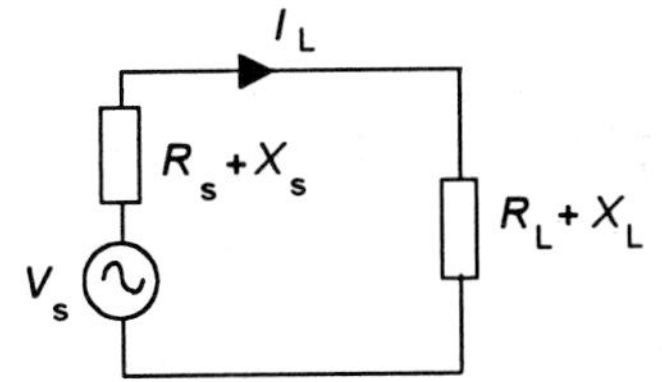

Figure 10.9 *Power transfer*

that maximum power is transferred. Such a process is known as *impedance matching*.

Consider a circuit having an a.c. source of voltage V_s and internal impedance Z_s and a load of impedance Z (Figure 10.9). What is the condition for the maximum load power? For an a.c. generator the internal impedance is $R_s + X_s$. The load will have the impedance $R_L + X_L$. The total impedance Z is given by:

$$Z = \sqrt{(R_s + R_L)^2 + (X_s + X_L)^2}$$

The current I_L through the load will be:

$$I_L = \frac{V_s}{\sqrt{(R_s + R_L)^2 + (X_s + X_L)^2}}$$

The load power is thus:

$$P_L = I_L^2 R_L = \frac{V_s^2 R_L}{(R_s + R_L)^2 + (X_s + X_L)^2} \qquad [9]$$

> The first requirement for maximum power transfer to the load is that we have $X_s = -X_L$, i.e. reactance of the load should be made equal and opposite to the reactance of the source so that the total reactance of the circuit is zero.

With this condition, equation [9] becomes:

$$P_L = \frac{V_s^2 R_L}{(R_s + R_L)^2} = V_s^2[R_L(R_s + R_L)^2]$$

For maximum power we must have $dP_L/dR_L = 0$, hence:

$$\frac{dP_L}{dR_L} = V_s^2[(R_s + R_L)^{-2} - 2R_L(R_s + R_L)^{-3}] = 0$$

and so:

$$2R_L = R_s + R_L$$

$$R_L = R_s \qquad [10]$$

> The second requirement for maximum power transfer is thus that the resistance of the load is equal to the resistance of the source.

The above requirements are known as the *maximum power transfer theorem*.

Example

The output stage of an amplifier has an impedance of 135 Ω and is to be used to drive a loudspeaker. Determine the turns ratio of the transformer which will match this impedance to that of a loudspeaker of impedance 15 Ω to give maximum power transfer.

$Z_1 = (N_1/N_2)^2 Z_2 = 135 = (N_1/N_2)^2 \times 15$, so $(N_1/N_2) = \sqrt{(135/15)} = 3$.

Revision

10 A generator and its transmission lines have together a resistance of 10 Ω and a reactance of +5 Ω. What should the resistance and reactance of the load be for maximum power transfer?

Problems

1 An ideal transformer has 1000 primary turns and 3500 secondary turns. If the a.c. mains voltage of 240 V is connected to the primary, what voltage will be produced at the secondary?

2 An ideal transformer has a primary a.c. input of 100 V and gives a secondary output of 75 V across a 25 Ω resistor. What is (a) the turns ratio, (b) the primary current?

3 An ideal transformer has a turns ratio of 10:3. If there is an a.c. input of 250 V to the primary and a resistance of 15 Ω across the secondary, determine (a) the secondary e.m.f., (b) the secondary current, (c) the primary current, (d) the power input required.

4 A 5 kVA ideal transformer has a primary with 500 turns and a secondary with 40 turns. When there is a primary input of an a.c. voltage of 3 kV what will be (a) the primary and secondary currents when the transformer is on full load, (b) the secondary e.m.f.?

5 A step-down ideal transformer has a secondary output of 240 V. When the primary current is 3.0 A, the secondary load is 10 Ω. What will be the primary voltage?

6 An ideal transformer has a primary coil with 1200 turns and the current through it is 0.5 A when the input a.c. voltage is 100 V. The secondary has 300 turns. What will be (a) the secondary voltage, (b) the ampere-turns?

7 A 100 VA ideal transformer transforms a primary voltage of 240 V to a secondary voltage of 12 V. Determine the primary and secondary currents when the transformer is on full load.

8 A 500 VA single-phase transformer has an iron loss of 20 W and a full load copper loss of 30 W. Determine the efficiency when it has a resistive load at (a) full load, (b) half load.

9 A 100 kVA single-phase transformer has an iron loss of 800 W and a full load copper loss of 1000 W. Determine the efficiency at a power factor of 0.8 when it is operating at (a) full load, (b) half load.

10 A 50 kVA single-phase transformer operating at a power factor of 0.8 has an efficiency of 98.0% on full load and 96.9% on quarter load. Determine the iron loss and the full load copper loss.

11 A 5 kVA single-phase transformer has an iron loss of 50 W and a full load copper loss of 120 W. What is the efficiency when it is operating with a resistive load at half load?

12 The output stage of an amplifier has an impedance of 50 Ω and is to be connected to a circuit of impedance 1000 Ω. Determine the turns ratio of the transformer which will match the amplifier impedance to that of the circuit to give maximum power transfer.

13 The output stage of an amplifier has an impedance of 50 Ω and is to be connected to a circuit of impedance 800 Ω. Determine the turns ratio of the transformer which will match the amplifier impedance to that of the circuit to give maximum power transfer.

14 An electronic circuit with an output impedance of 1.2 kΩ is to be connected to a circuit of impedance 10 Ω. Determine the turns ratio of the transformer which will enable maximum power transfer between the circuits.

15 A load of impedance 10 Ω is supplied with power via a step down transformer having a turns ratio of 5. What will be the impedance perceived in the primary?

16 What will be the power delivered to a load of resistance 150 Ω when (a) it is directly connected to an alternator of 10 V and internal resistance 50 Ω, (b) a transformer is used to match the impedances? What turns ratio will be required for the transformer?

17 An electronic circuit has an output impedance of 10 kΩ and delivers an a.c. voltage of 10 V. If it is matched to a load by a step-down single-phase transformer of turns ratio 20, what will be the impedance of the load and the power dissipated in it?

11 Systems

11.1 Introduction

The systems approach to engineering involves attention being focused on the functions of elements rather than how an effect is achieved. In these days of integrated circuits the question is generally which circuit to use, based on a consideration of the function required, rather than a consideration of the circuitry involved in the design of the integrated circuit or the workings of the constituent elements of that circuit. Thus, for example, we might require an integrated circuit to take a number of inputs and give an output when all of them are high signals. This consideration of systems enables the same basic approach to be used for a wide variety of engineering processes. For example, a measurement system consists of three basic system elements: a sensor, signal conditioning and display. We can apply this model to a simple measurement system such as a Bourdon gauge used for pressure measurement where the entire system is 'mechanical' and involves a Bourdon tube which is used to rotate, via gearing, a pointer across a scale or to an electronic measurement system for pressure measurement involving a semiconductor pressure element with a microprocessor used to process the signal and give a display.

11.2 Basic principles

The term system can be defined as:

> A *system* is a set of components which work together to provide some form of output from a specified input or inputs.

Thus an amplifier can be considered to be a system, consisting of components such as transistors and resistors connected together to accomplish the task of taking an input signal and providing a bigger output. All systems are considered in terms of their inputs and outputs and you can think of a system as being rather like a machine into which you feed an input, the machine then processes the input and spews out its output.

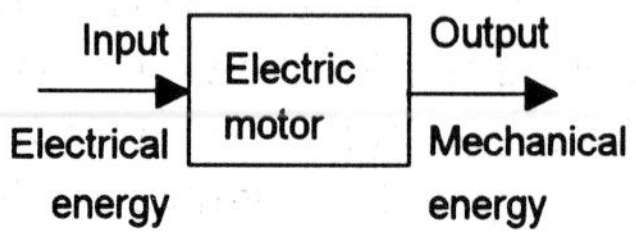

Input → Electric motor → Output

Electrical energy → Mechanical energy

Figure 11.1 *Electric motor system*

11.2.1 Block diagrams

A useful way of representing a system is as a *block diagram*. Within the boundary described by the box outline is the system and inputs to the system are shown by arrows entering the box and outputs by arrows leaving the box. Figure 11.1 illustrates this for an electric motor system;

there is an input of electrical energy and an output of mechanical energy, though you might consider there is also an output of waste heat. The interest is in the relationship between the output and the input rather than the internal science of the motor and how it operates.

11.2.2 Connected systems

We can have a number of systems connected together to achieve some function. For example, we can consider a power station as a system which has an input of fuel and an output of electrical power. However, it can be more useful to consider the power station as a number of linked systems (Figure 11.2). Thus we can have the boiler system which has an input of fuel and an output of steam pressure. The steam pressure then becomes the input to the turbine to give an output of rotational mechanical power. This in turn becomes the input to the electrical generator system which gives an output of electrical power.

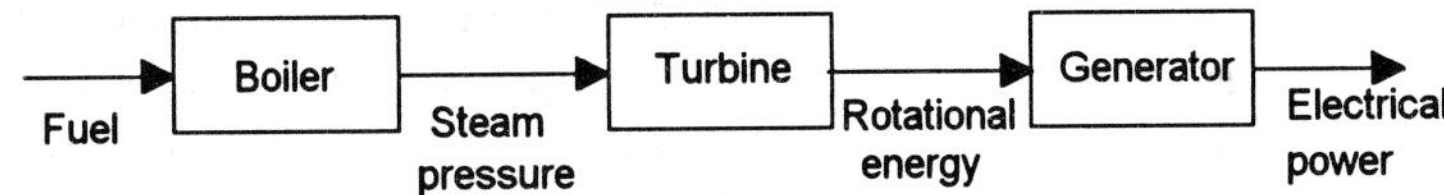

Figure 11.2 *Power station*

As another example a hi-fi music system might involve a CD player, a cassette player, a record player, an amplifier and loudspeakers. It is convenient to think of the systems as being interconnected by means of their inputs and outputs. Thus, for the hi-fi system, the output of the CD player becomes the input to the amplifier, the output of the amplifier becomes the input to the loudspeakers.

As illustrated above, a convenient way of displaying such interconnections between system elements is in terms of interconnected *block diagrams*. In drawing a system as a series of interconnected blocks, it is necessary to recognise that the lines drawn to connect boxes indicate a flow of information in the direction indicated by the arrow and not necessarily physical connections.

As a further illustration, Figure 11.3 shows the basic form of a radio communication involving analogue signals. The input signal is the input to the modulator system which puts it into a suitable form for transmission. The signal is then transmitted before becoming the input to the receiver where it passes through the demodulator system to be put into a suitable form for reception. The input to the demodulator system is likely to be not only the transmitted signal but also noise and interference, so the output signal from the modulator has added to it, at a *summing junction*, the noise and interference signal. A summing junction is represented by a circle with the inputs to quadrants of the circle given + or – signs to indicate whether we are summing two positive quantities or summing a positive quantity and a negative quantity and so subtracting signals.

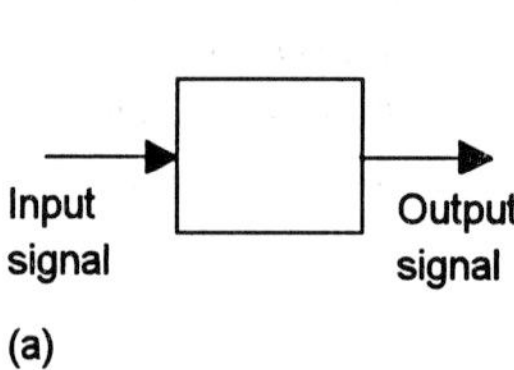

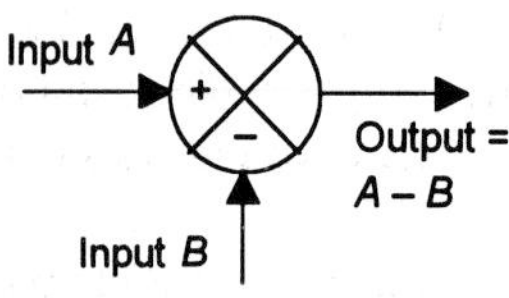

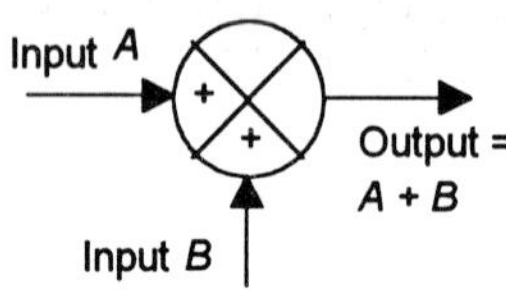

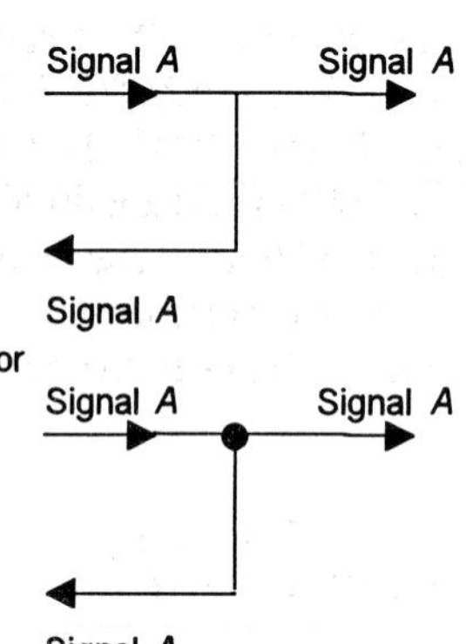

Figure 11.4 *Block diagram elements*

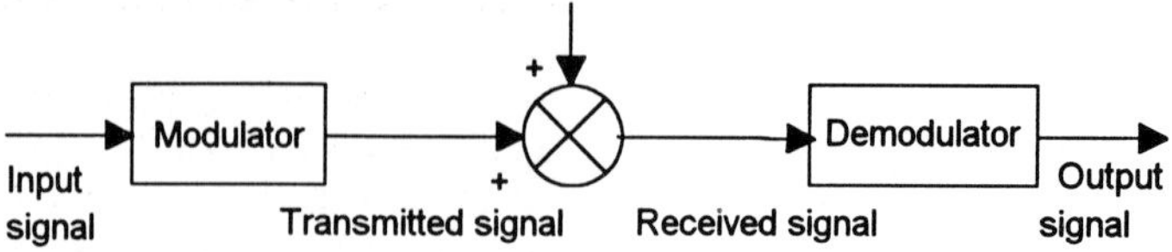

Figure 11.3 *A radio system*

To summarise:

> With *block diagrams*, a rectangular box is used to represent a system, and inputs and outputs are represented by lines with arrows (Figure 11.4(a)). The system represented by the box operates on the input signal to produce the output signal. A summing junction is represented by a circle (Figure 11.4(b)) with one or more arrowed lines for inputs coming in and an arrowed line for an output going out, plus or minus signs being placed against the input arrow heads to indicate whether the signals have to be added or subtracted. A take-off point (Figure 11.4(c)) allows a signal to be tapped and used elsewhere, the assumption being made that the signal is not affected by introducing the take-off point.

As an illustration of a block diagram involving all the elements shown in Figure 11.4, consider a domestic central heating system (Figure 11.5). It consists of a thermostat element which has inputs of the required temperature and the actual temperature and an output of a signal representing the difference between the required temperature and the actual temperature. This difference signal is used to control the central heating boiler and switch it on or off. The output from the boiler is used to give an input of heat to the room and so control its temperature. The output signal from the room system element is thus a temperature signal. A take-off point is used to indicate that the temperature signal is fed back to a summing junction where it is summed, i.e. in this case subtracted, with the required temperature signal to give the difference signal.

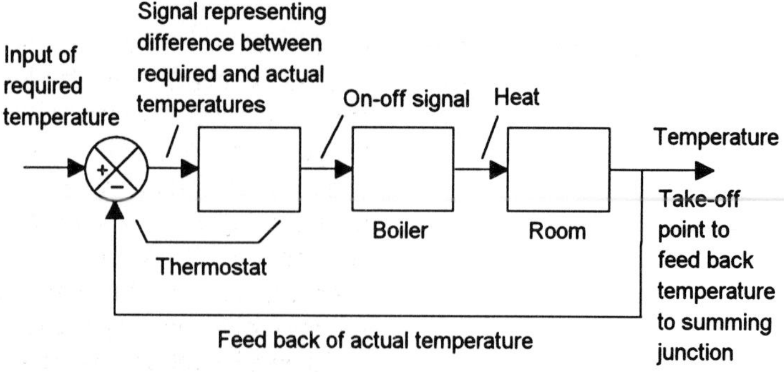

Figure 11.5 *Central heating system*

11.2.3 Input–output relationships

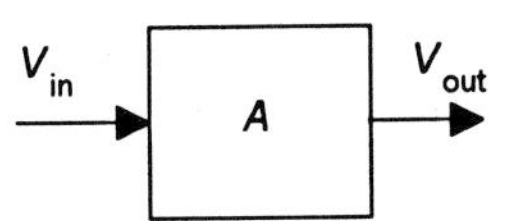

Figure 11.6 *Amplifier system*

A system takes an input, performs some action on it and then gives an output which is somehow related to the input. The output is thus some function of the input. In the case of an amplifier system we might have an input voltage V_{in} and an output voltage V_{out} which is proportional to the input and so:

$$V_{out} = Av_{in} \qquad [1]$$

where A is the factor by which the amplifier multiplies the input, i.e. the amplifier gain. We can represent such a system by the block diagram shown in Figure 11.6. The A in the box is the factor which operates on the input to give the output.

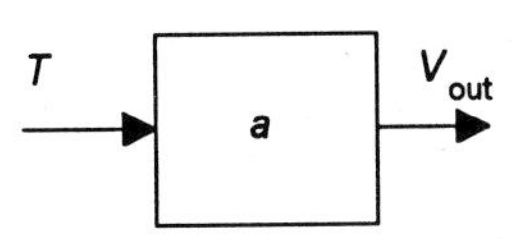

Figure 11.7 *Thermocouple system*

As another example, consider a sensor such as a thermocouple which has an input of temperature and an output of a voltage. When the voltage V_{out} is proportional to the temperature T we have (Figure 11.7):

$$V_{out} = aT \qquad [2]$$

with a being a constant and having the unit of V/°C.

A temperature measurement system might consist of a thermocouple connected to an amplifier and hence to a display (Figure 11.8). Suppose, for a steady temperature input, the thermocouple gives 4 mV/°C. When there is a temperature input to the system of, say, 10°C then the thermocouple gives an output of 40 mV. If the amplifier has a gain of 20 then the input to it of 40 mV becomes an output of 800 mV. The display might be a moving-coil voltmeter and give a pointer rotation of 1° per 50 mV. Thus the pointer rotates through 16° for the input to the system of 10°C.

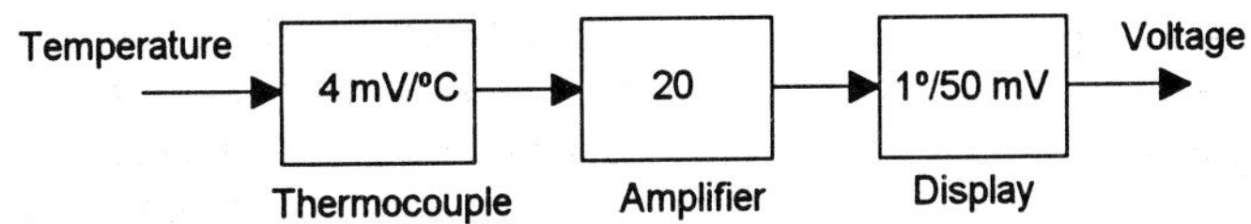

Figure 11.8 *Temperature measurement system*

11.3 Information and signals

Electronic systems react to *information*. This may take many forms. Thus there might be an electronic measurement system which is used for the measurement of temperature and so the system reacts to an input of temperature information. Another form of electronic system might be a digital camera which reacts to an input of the detailed information in the light coming from some scene to give a picture. Yet another form of electronic system is one used to count the number of items passing along a production line, the information then being in the form of on–off signals indicating the presence or not of items.

As an illustration, Figure 11.9 shows the form the information might take for a system when the input is suddenly increased, e.g. a switch

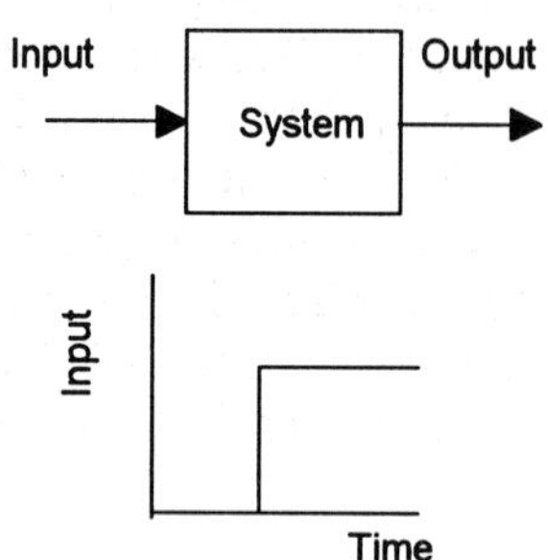

Figure 11.9 *Step input to a system*

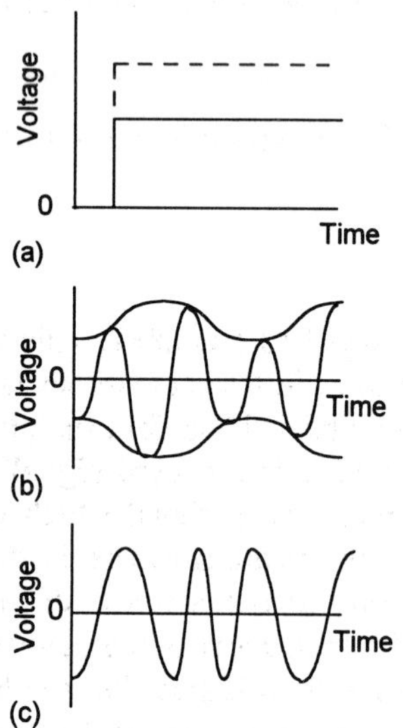

Figure 11.10 *Analogue signals*

being activated to supply a voltage signal. The input is then basically a step form of signal showing a sudden change in signal size.

The input signals to an electronic system will be converted by the system to time-varying voltages and currents which can then be processed by the system to give the system output. Between the input and output the signals can take a number of forms. Thus in a microprocessor-based system we might have the input signal from the sensor converted into a voltage signal which has an amplitude which varies with time, this then being converted into a digital signal where the pulse sequence is a measure of the amplitude variation, and then, after processing by the microprocessor, the digital output from the microprocessor is converted back into an output voltage which has an amplitude which varies with time.

11.3.1 Analogue signals

With an *analogue* signal the information is carried by the size, amplitude or frequency of the signal varying with time. This might involve the size of the signal being changed, as in Figure 11.10(a). For an alternating signal the amplitude (Figure 11.10(b)) or frequency (Figure 11.10(c)) might be changed, the term used being modulated. With *amplitude modulation* (Figure 11.10(b)) the amplitude of the alternating carrier signal is varied according to how the information transmitted varies with time. With *frequency modulation* (Figure 11.10(c)), the modulating signal varies the frequency of the carrier signal as opposed to the amplitude in the case of amplitude modulation.

As an illustration, Figure 11.11 shows a block diagram of the system used for an AM (amplitude modulation) receiver.

1 The aerial picks up transmitted radio signals. At this point the required frequency is just one of many that has been received.

2 The radio frequency (RF) amplifier amplifies the very small signals. This is a type of amplifier specifically designed for amplifying such high frequency signals.

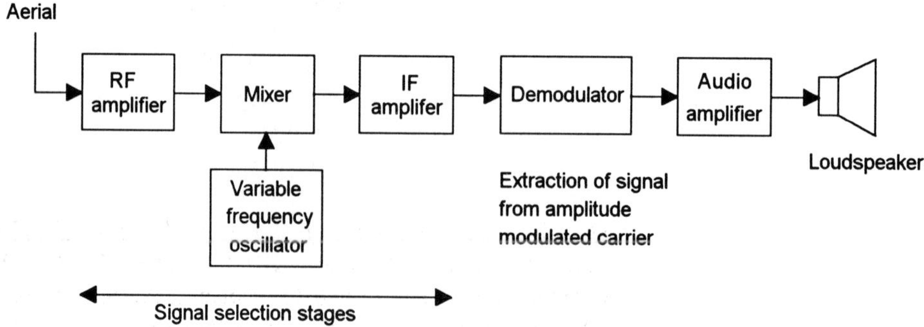

Figure 11.11 *An AM receiver*

3 The required carrier frequency is selected by first shifting the high frequency down to an intermediate (IF) frequency. This is achieved by mixing (heterodyning) it with the output from a variable frequency oscillator to give an output which is the sum and difference frequencies of the oscillator and the carrier frequency. The result is that the carrier frequencies can be shifted down to the standard intermediate frequency for AM receivers of about 460 Hz and tuning can be achieved by shifting just the required carrier frequency down to this frequency.

4 A further stage of amplification is then carried out, the IF amplifier being just needed to amplify signals at this 460 Hz frequency.

5 Demodulation is then used to extract the amplitude modulation signal from the carrier frequency.

6 An audio amplifier is then used to amplify the signals for conversion to sound by the loudspeakers.

11.3.2 Digital signals

With a *digital* signal the information is conveyed as a number. This is generally by means of a signal which can assume just one of two levels, these levels representing the binary numbers 0 and 1. Thus, for example, information about an analogue signal of size 15 V is conveyed digitally by a signal representing the number 15 rather than a signal which is made a size which is related to 15 V. The numbers are conveyed in the binary system. The *binary system* is based on just two states 0 and 1, these being termed binary digits or *bits*. When a number is represented by this system, the digit position in the number indicates the weight attached to each digit with the weight increasing by a factor of 2 as we move from right to left in a number. The digit at the right-hand end is called the least significant bit (LSB) and the digit at the left-hand end the most significant bit (MSB). For example, the 4-bit binary number 1111 is:

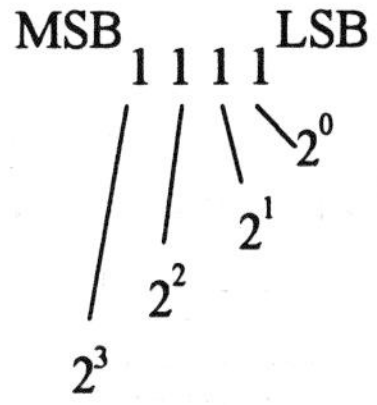

MSB 1 1 1 1 LSB

2^3 2^2 2^1 2^0

and so is the decimal number $2^0 + 2^1 + 2^2 + 2^3 = 15$. The 8-bit binary number 10111011 is $2^0 + 2^1 + 0 + 2^3 + 2^4 + 2^5 + 0 + 2^7 = 187$. The combination of bits to represent a number is termed a *word*.

Digital signals can be conveyed either in parallel or serial. *Parallel transmission* with, say, an 8-bit word involves eight parallel wires with each bit of the word being transmitted along its own wire. *Serial transmission* involves the bits of the word being transmitted one after the other along a single wire.

As an illustration of a system involving digital signals, consider the mobile telephone system. Figure 11.12 shows the basic elements of such a system.

1 *For transmission*
The speech input to the microphone gives an analogue signal which is converted into a digital signal by the analogue-to-digital converter. This is then coded in the vocoder, i.e. the binary representation of the signal has its format and bit ordering organised, to give a suitable serial binary signal which is used to modulate a radio frequency carrier for transmission.

2 *For reception*
The received signal from the aerial is a binary modulated radio frequency carrier and is amplified, filtered, demodulated and decoded to provide the wanted signal in digital form. A digital-to-analogue converter (DAC) is used to provide an analogue output for the loudspeaker.

3 *Dialling*
The microcontroller in the SIM card has a digital input from a keypad for the dialled number. This is then transformed into a suitable digital signal by the analogue-to-digital converter (ADC) in the processing module. It also provides an output to a display to give a visual display of the dialled number and other messages. The SIM card stores, in digital form, such data as the directory number of the user, subscription information and data to determine access to the network. A mobile telephone cannot access the network service unless it can provide a valid access code to enable the network to authenticate the user.

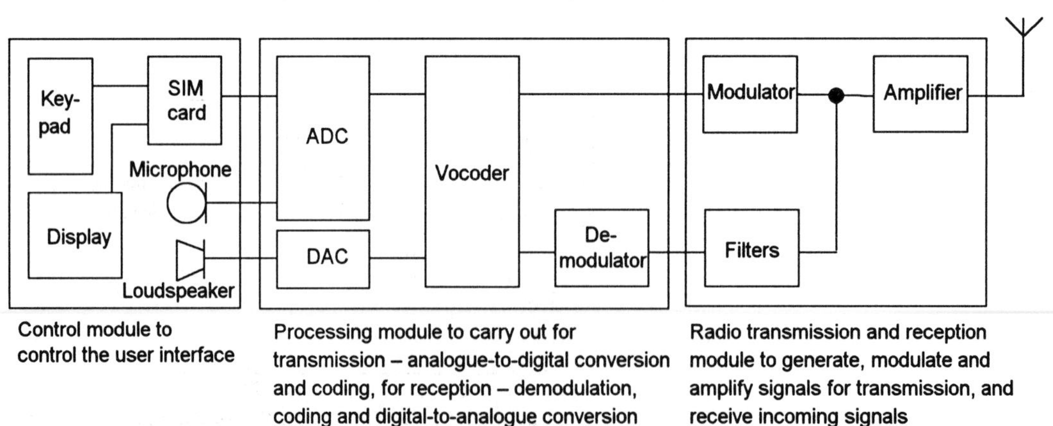

Figure 11.12 *Mobile telephone system*

11.4 Signal processing

Electronic circuits are used to process signals and the following are some of the types of processes involved.

11.4.1 Analogue to digital conversion

Signals from sensors are generally analogue and need to be converted to digital signals if the electronic system uses a microprocessor. An analogue-to-digital converter (ADC) has an input of an analogue signal and an output of a digital word which represent the analogue signal. The basic principles of analogue to digital conversion are that:

1 The analogue signal is sampled.

2 The sampled signal is then transformed into a digital signal.

3 Steps 1 and 2 are then repeated at regular intervals of time, the process being controlled by a clock.

A consequence of the sampling is that the ADC gives digital signals which represent the size of the analogue input signal at regular intervals of time rather than a continuous representation of the signal.

The relationship between the sampled analogue signal and the digital output is illustrated by Figure 11.13 where the output is in the form of a 3-bit word. With a 3 bit word there are just $2^3 = 8$ possible binary signals and so we divide the sampled analogue signal into 8 levels and represent each level by a word. The possible levels are called *quantisation levels* and the difference in analogue voltage between two adjacent levels is termed the *quantisation level.* Thus for the analogue to digital conversion represented by Figure 11.13, the quantisation interval is 1 V. When the analogue input is at a value which is centred over the quantisation interval the output word accurately represents the analogue signal. However, there will be an error when the analogue input is at other values, the maximum error being equal to one-half of the interval or ±½ bit. This is termed the *quantisation error.*

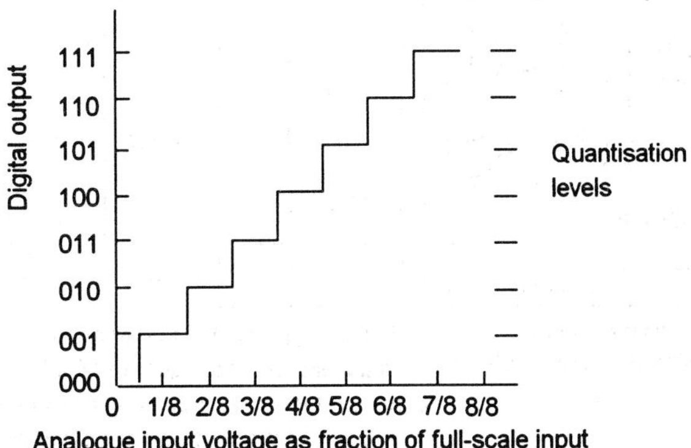

Figure 11.13 *Analogue to digital conversion*

Application
As an illustration of commercially available analogue-to-digital converters, the following is part of a specification:

Resolution 12 bits
Input voltage range ±10 V
Conversion cycle time 10 μs

The word length determines the *resolution*, i.e. the smallest changes in input which will result in a change in digital output, of the analogue-to-digital converter. The smallest change in digital output that can be registered is one bit in the least significant bit position in the word. Thus with a word length of n bits, since there are 2^n levels, the full-scale analogue input V_{FS} is divided into 2^n segments and so the smallest change in input that can be registered is $V_{FS}/2^n$. Thus:

$$\text{resolution} = \frac{V_{FS}}{2^n} \qquad [3]$$

Example

An analogue-to-digital converter has a word length of 8 bits and the analogue signal input varies between 0 and 10 V, what is the resolution?

The resolution is $10/2^8 = 0.039$ V. Any change less than this will fail to change the output. Thus an input of 0.030 V will give an output of 0000 0000, an input of 0.039 V an output of 0000 0001, an input of 0.060 V an output of 0000 0001, an input of 0.078 V an output of 0000 0010.

Revision

1 An electronic temperature measurement system employs a temperature sensor which gives an output which is converted to a digital signal by an analogue-to-digital converter and the digital signal is then displayed. If the range of temperatures to be measured is 0 to 200°C and a resolution of 0.5°C is required, what should be the word length of the ADC?

11.4.2 Digital to analogue conversion

Application
As an illustration of commercially available digital-to-analogue converters, the following is part of a specification:

Resolution 10 bits
Relative accuracy ±1.75 LSB max
Output voltage span 0 to 10 V
Settling time 100 ns

The output from a microprocessor is digital and often needs to be converted into an analogue signal to operate devices such as control valves or motors. This is carried out by a digital-to-analogue converter (DAC). The input to a digital-to-analogue converter is a binary word and the output is an analogue signal that represents the word. Thus, for example, an input to a 3-bit DAC of 010 must give an analogue output which is twice that given by an input of 001 and an input of 011 must give an analogue output which is three times that of 001. Thus we might have 001 giving 1 V, 010 giving 2 V, 011 giving 3 V. Figure 11.14 illustrates this for an input to a 3-bit DAC for which each additional bit increases the output voltage by 1 V. The word length that can be handled by a DAC determines its resolution. Thus a 4-bit DAC with a full-scale voltage of V_{FS} gives an analogue output which changes in steps of $V_{FS}/2^4$ while an 8-bit DAC gives an output which changes in steps of $V_{FS}/2^8$.

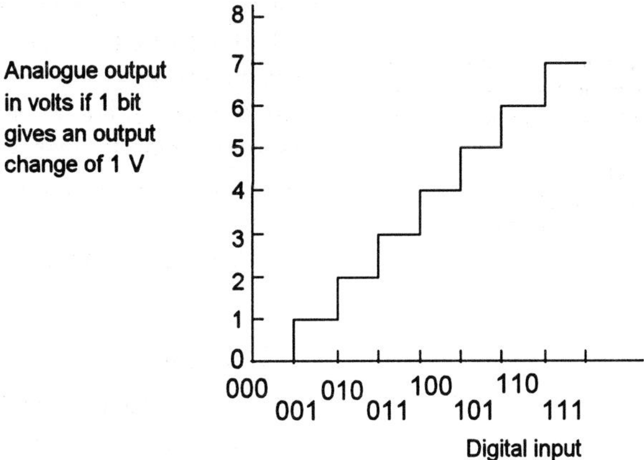

Figure 11.14 *Digital to analogue conversion*

Example

A microprocessor gives an output of an 8-bit word which is fed through an 8-bit digital-to-analogue converter to a control valve. The control valve requires the voltage to change from 0 to 10 V to go from fully closed to fully open. If the fully closed state is represented by 0000 0000 and the fully open state by 1111 1111 what will be the change in output to the valve when there is a change of 1 bit in the output from the microprocessor?

The full-scale output voltage of 10 V will be divided into 2^8 intervals. A change of 1 bit is thus a change in the output voltage of $10/2^8 = 0.039$ V.

11.4.3 Amplification

Amplifiers can be purchased as complete integrated circuits, the circuits containing many transistors and other components. The concern is then with the overall properties of an amplifier, i.e. the behaviour of the amplifier system, rather than how the circuit works.

Amplification is ideally the process of increasing in size or amplitude a signal without changing its waveform. We can consider the voltage gain, current gain or power gain. The *voltage gain* is the ratio of the input to output voltage, the *current gain* is the ratio of the input to output current and the *power gain* is the ratio of the input to output power.

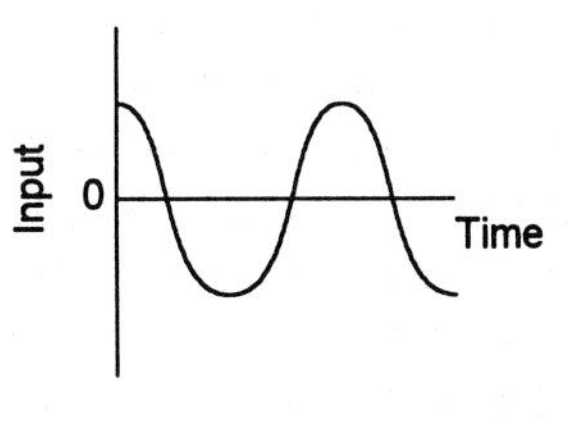

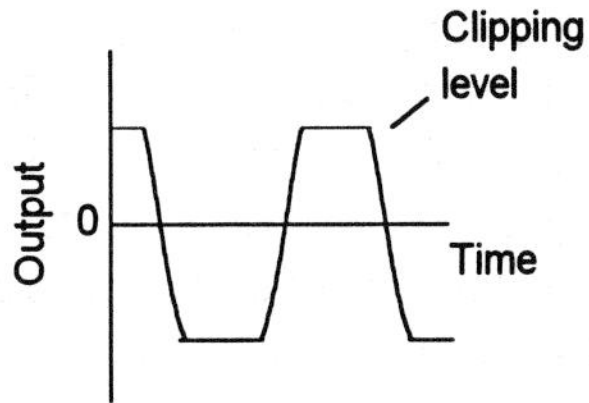

Figure 11.15 *Clipping*

We cannot, however, in practice assume that the gain applies to all input values or to all frequencies. The range within which the output signal can vary is invariably limited; this means that there is a maximum value of input that will produce an output signal for which the waveform is essentially unchanged from that of the input. Increasing the input signal beyond this value will produce *clipping* of the higher levels of the output (Figure 11.15). The gain of an amplifier does not remain constant with the frequency of the input signal. *Direct coupled amplifiers* have

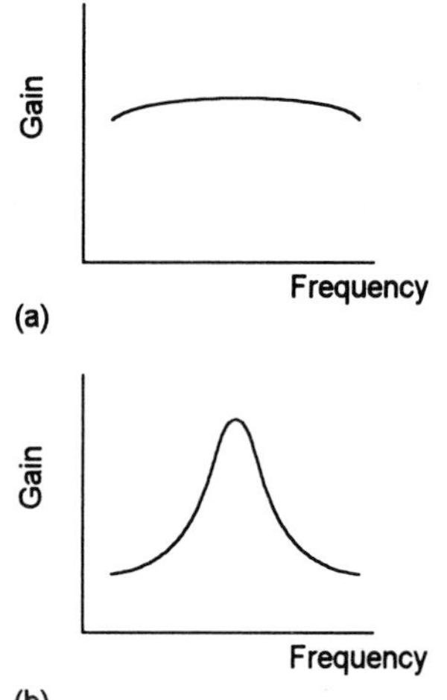

Figure 11.16 *Effect on gain of frequency: (a) direct coupled amplifier, (b) frequency selective amplifier*

gains that remain essentially constant from zero frequency to a relatively high frequency (Figure 11.16(b)); *frequency selective amplifiers* (Figure 11.16(b)) are designed to amplify only over particular frequency ranges. The effect of the gain changing with frequency with a complex waveform is to produce different amplification of some of the harmonics that constitute it and so give an output waveform different from that of the input.

Gains are often expressed in decibels. The *bel* is a logarithmic unit of power ratio:

$$\text{power gain in bels} = \lg\left(\frac{\text{output power}}{\text{input power}}\right) \qquad [4]$$

Because the bel is a rather large unit, decibels (dB) are used:

$$\text{power gain in dB} = 10\lg\left(\frac{\text{output power}}{\text{input power}}\right) \qquad [5]$$

Decibels can also be used to express ratios of d.c. or a.c. voltages (r.m.s. or maximum values). Because the power dissipated in a resistance R when there is a voltage V across it is V^2/R it is convenient to assume that the input and output voltages are across the same resistance and so we can write equation [5] as:

$$\text{voltage gain in db} = 10\lg\left(\frac{\text{output voltage}^2}{\text{input voltage}^2}\right) \qquad [6]$$

or

$$\text{voltage gain in db} = 20\lg\left(\frac{\text{output voltage}}{\text{input voltage}}\right) \qquad [7]$$

Likewise, since the power dissipated in a resistance by a current I through it is I^2R:

$$\text{current gain in db} = 20\lg\left(\frac{\text{output current}}{\text{input current}}\right) \qquad [8]$$

An amplifier might thus be specified as having a power gain of 15 dB.

An amplifier can be represented by the circuit shown in Figure 11.17. The input signal is connected across the amplifier input resistance R_{in} and the output signal is a voltage source, which has a voltage of the product of the amplifier voltage gain A_V and the input voltage V_{in}, in series with the amplifier output resistance R_{out}.

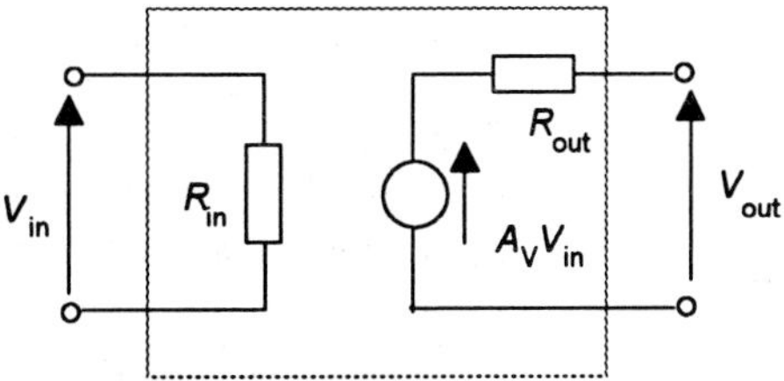

Figure 11.17 *Equivalent circuit for an amplifier*

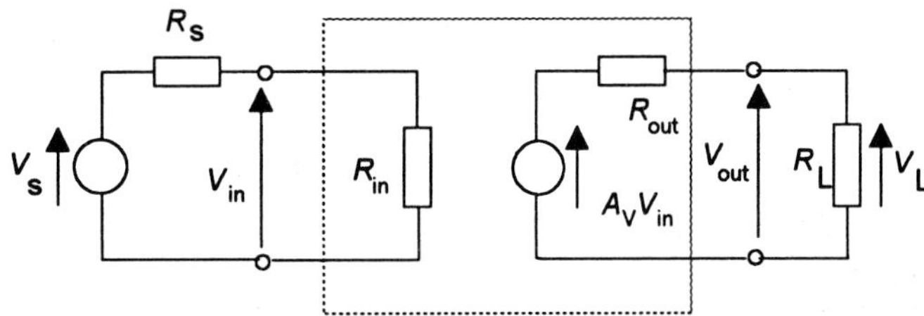

Figure 11.18 *An amplifier inserted between a source and load*

When in use, an amplifier is essentially connected to a voltage source V_s with an internal resistance R_s and with a load of resistance R_L across the amplifier output. (Figure 11.18). The two resistors R_{in} and R_s form a voltage divider so the input voltage to the amplifier is:

$$V_{in} = V_s \frac{R_{in}}{R_s + R_{in}} \quad [9]$$

The larger the value of the amplifier input resistance the closer is the value of V_{in} to V_s. The voltage value obtained by multiplying the input voltage by the voltage gain of the amplifier, i.e. $A_V V_{in}$, is applied across the output resistance of the amplifier and the load resistance R_L. The voltage appearing across the load V_L is thus:

$$V_L = A_V V_{in} \frac{R_L}{R_L + R_{out}} \quad [10]$$

Note that the $A_V V_{in}$ is the voltage produced when there is an open-circuit and is *not* the voltage produced across a load. The voltage gain A_V is thus the gain of an amplifier which is defined for the case of an open circuit load.

As an illustration, an amplifier might be specified as having a gain of 34 dB, an input impedance of 50 kΩ and an output impedance of 5 kΩ.

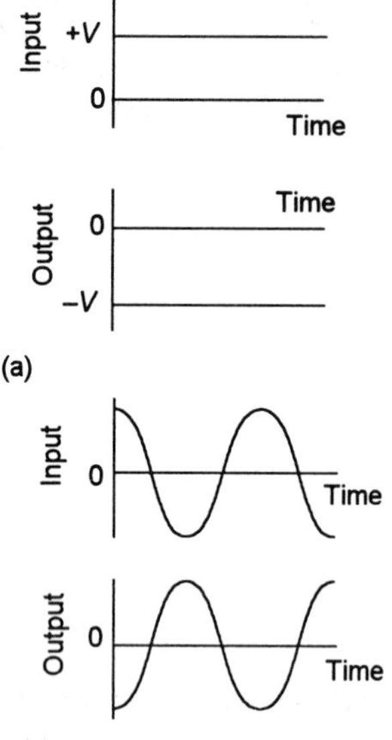

Figure 11.19 *Inversion with: (a) d.c. and (b) a.c. signals*

Application
As an illustration, a commercially available audio amplifier includes as part of its specification the following information:

12 W audio amplifier
Supply voltage range: 6 to 18 V d.c.
Frequency response: 50 Hz to 25 kHz
Voltage gain:
pre-amp 17.7 dB
power amp 29.5 dB
Total 47 dB
Input impedance 40 kΩ

The following are some of the basic types of amplifiers that are encountered.

1 *Inverting and non-inverting amplifiers*
Amplification is often accompanied by inversion when a positive input gives rise to a negative output; Figure 11.19 illustrates this for d.c. and a.c. signals. Amplifiers giving inversion are termed inverting amplifiers and those giving no inversion are termed non-inverting amplifiers.

2 *Voltage and current amplifiers*
The term voltage amplifier is used for an amplifier designed to amplify the voltage of the input signal. Its voltage gain is specified but its current gain is not; it may give an increase in signal current if the load impedance is low enough. The term current amplifier is used for an amplifier designed to amplify the current of the input signal. Both types of amplifier may be inverting or non-inverting.

3 *Power amplifiers*
The term power amplifier is used where both the current and the voltage of the output are significant and the amplifier is required to supply signal power to a particular load. A typical specification might thus include: output power 40 W into a load of 8 Ω, output impedance less that 0.4 Ω, input resistance greater than 10 kΩ.

4 *Differential amplifiers*
A differential amplifier has two input signals and is required to amplify the difference between them.

5 *Buffer amplifiers*
A buffer amplifier is used as the interface between a high impedance signal source and a low impedance load. The gain is not significant and indeed is often 1.

Example

A sinusoidal voltage supply of 2 mV has an internal resistance of 9 kΩ and is connected to an amplifier of open-circuit voltage gain 100, input resistance 91 kΩ and output resistance 100 Ω. Determine the voltage appearing across a load of 1 kΩ connected across the output terminals of the amplifier.

The input voltage to the amplifier is given by equation [9] as:

$$V_{in} = V_s \frac{R_{in}}{R_s + R_{in}} = 2 \times \frac{91}{91+9} = 1.82 \text{ mV}$$

The voltage across the load is given by equation [10] as:

$$V_L = A_V V_{in} \frac{R_L}{R_L + R_{out}} = 100 \times 1.82 \times \frac{1000}{1000+100} = 165 \text{ mV}$$

Example

An amplifier is specified as having a voltage gain of 12 dB. Express this as a voltage ratio.

Using equation [7]:

$$12 = 20 \lg \text{(voltage ratio)}$$

We can write this as:

$$10^{12} = \text{(voltage ratio)}^{29}$$

and so

$$\text{voltage ratio} = 10^{12/20} = 3.98$$

Example

Two amplifiers with voltage gains of 12 dB and 5 dB are connected in cascade with the output of one feeding the input of the other. What is the overall voltage gain if it can be assumed that the second amplifier does not affect the gain of the first and vice versa?

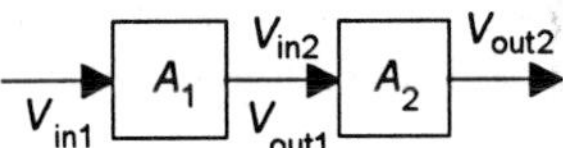

Figure 11.20 *Example*

For two systems connected in cascade (Figure 11.20) we have the output for the first system becoming the input of the second system. Thus if the gain of the first system is $A_1 = V_{out1}/V_{in1}$ then for the second system we have $A_2 = V_{out2}/V_{in2} = V_{out2}/V_{out1}$. The overall gain of the system is $V_{out2}/V_{in1} = (V_{out1}/V_{in1})(V_{out2}/V_{out1}) = A_1 \times A_2$. If we take logarithms then:

$$\lg \text{overall gain} = \lg A_1 + \lg A_2$$

and hence:

$$20 \lg \text{overall gain} = 20 \lg A_1 + 20 \lg A_2$$

Thus the overall gain in dB is equal to the sum of the gains of the two systems when expressed in dB. Hence the overall gain is 12 + 5 = 17 dB.

Example

Two amplifiers with voltage gains of 12 dB and 8 dB are connected in cascade by a cable of attenuation 3 dB with the output of one feeding the input of the other. What is the overall voltage gain if it can be assumed that the second amplifier does not affect the gain of the first and vice versa?

An attenuator is a network with a negative gain, in this case –3 dB. Thus we have three systems in cascade, the two amplifiers and the cable and so the overall gain is 12 – 3 + 8 = 17 dB.

Revision

2 A sinusoidal voltage supply of 10 mV has an internal resistance of 300 Ω and is connected to an amplifier with an open-circuit voltage gain of 100, input resistance 10 kΩ and output resistance 100 Ω. Determine the overall voltage gain and the power gain when the amplifier has a load resistance of 50 Ω connected across it.

3 An amplifier is specified as having a voltage gain of 18 dB. Express this as a voltage ratio.

4 Two amplifiers with voltage gains of 18 dB and 12 dB are connected in cascade with the output of one feeding the input of the other. What is the overall voltage gain if it can be assumed that the second amplifier does not affect the gain of the first and vice versa?

5 Two amplifiers with voltage gains of 10 dB and 8 dB are to be connected in cascade by cable of attenuation 1 dB/metre with the output of one feeding the input of the other. What is the overall voltage gain if it can be assumed that the second amplifier does not affect the gain of the first and vice versa and 1 m of cable is used?

11.4.4 Conversion of a resistance change to a voltage change

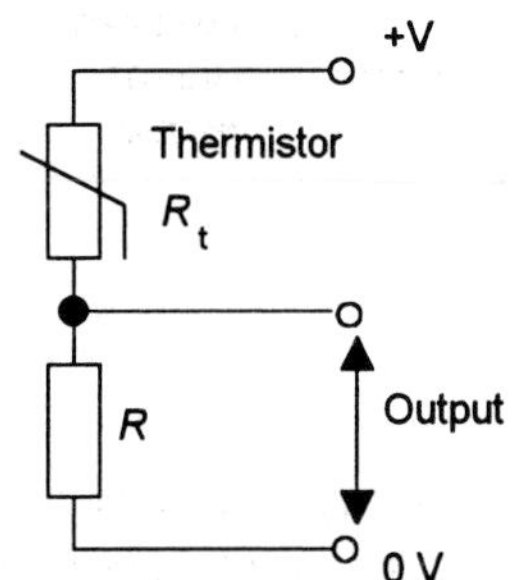

Figure 11.21 *Resistance to voltage conversion for a thermistor*

In measurement systems we often have sensors which give rise to a resistance change as a measure of the variable being sensed, e.g. a thermistor which has a resistance related to the temperature. As a consequence, measurement systems are often concerned with converting resistance changes into voltage changes.

Figure 11.21 shows how a *potential divider circuit* can be used to convert the resistance change produced by a thermistor when subject to a temperature change into a voltage change. A constant voltage V is applied across the thermistor and another resistor in series. When the resistance of the thermistor changes, the fraction of the voltage across the fixed resistor changes. The output voltage is proportional to the fraction of the total resistance which is between the output terminals. Thus:

$$\text{output} = \frac{R}{R+R_t}V \qquad [11]$$

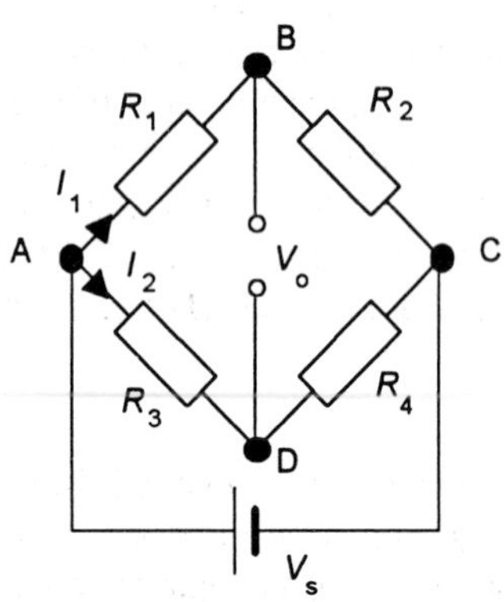

Figure 11.22 *Wheatstone bridge*

Another method is the *Wheatstone bridge* Figure 11.22, this being more suitable when very small resistance changes are involved. The resistance element being monitored forms one of the arms of the bridge. When the output voltage V_o is zero, then there is no potential difference between B and D and so the potential at B must equal that at D. The potential difference across R_1, i.e. V_{AB}, must then equal that across R_3, i.e. V_{AD}, and so $I_1R_1 = I_2R_2$. We also must have the potential difference across R_2, i.e. V_{BC}, equal to that across R_4, i.e. V_{DC}. Since there is no current through BD then the current through R_2 must be the same as that through R_1 and the current through R_4 the same as that through R_3 and so $I_1R_2 = I_2R_4$. Dividing these two equations gives:

$$\frac{R_1}{R_2} = \frac{R_3}{R_4} \qquad [12]$$

If R_1 is the sensor and the others are fixed resistances, then when R_1 changes so the bridge goes out-of-balance and produces an output voltage V_0 which is, for small changes, proportional to the change in resistance.

11.5 Measurement systems

A measurement system (Figure 11.23) consists of three basic elemental systems, these being sensor, signal processor and display.

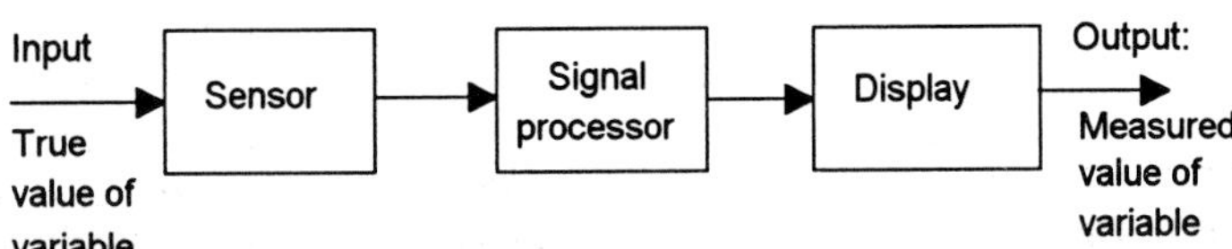

Figure 11.23 *Measurement system elements*

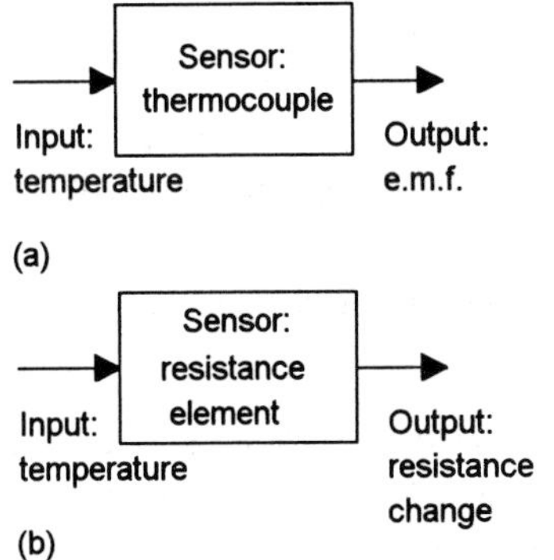

Figure 11.24 *Sensors: (a) thermocouple, (b) resistance thermometer element*

1 *Sensor*
This is the element of the system which is effectively in contact with the process for which a variable is being measured and gives an output which depends on its value, taking information about the variable being measured and changing it into some form which enables the rest of the measurement system to give a value to it. For example, a thermocouple is a sensor which has an input of temperature and an output of a small e.m.f. (Figure 11.24(a)). Another example is a resistance thermometer element which has an input of temperature and an output of a resistance change (Figure 11.24(b)). The term *transducer* is often used for an element which transforms input signals from some form into an electrically equivalent form. Sometimes the term transducer is used for a combination of a sensor and signal processing which takes an input and gives an electrical signal.

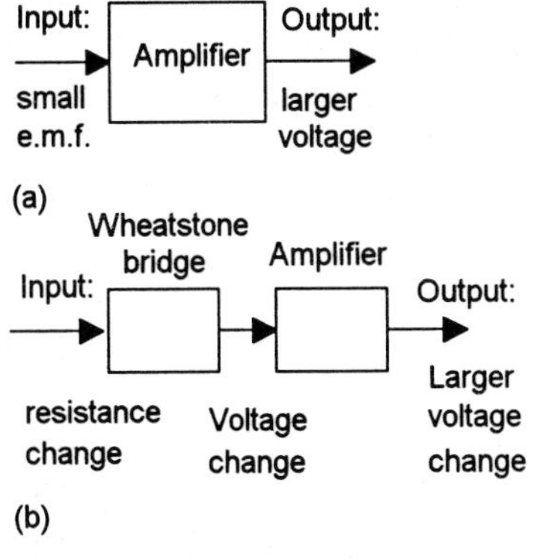

Figure 11.25 *Signal processing*

2 *Signal processor*
This element takes the output from the sensor and converts it into a form which is suitable for display or onward transmission. In the case of the thermocouple this may be an amplifier to make the e.m.f. big enough to register on a meter (Figure 11.25(a)). There often may be more than one item, perhaps an element which puts the output

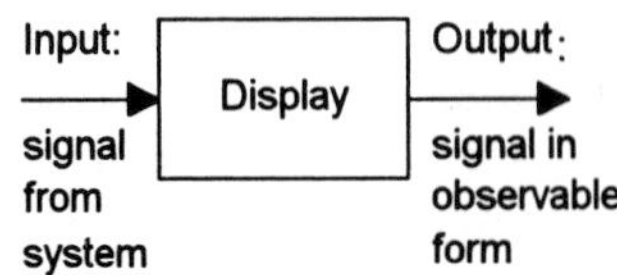

Figure 11.26 *Data presentation element*

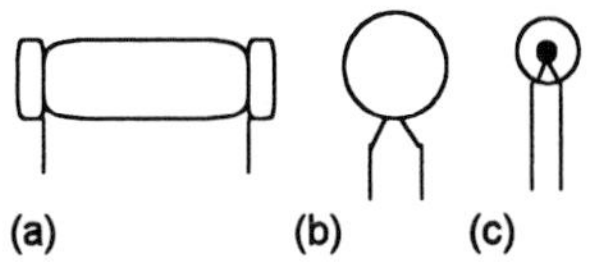

Figure 11.27 *Thermistors: (a) rod, (b) disc, (c) bead*

Application

Consider the determination of the temperature of a liquid in the range 0°C to 100°C where only rough accuracy is required, e.g. the determination of the temperature of the cooling water for a car engine and its display as a pointer moving across a scale marked in different colours to indicate safe and unsafe operating temperatures. A solution might be to use a *thermistor* as a sensor. The resistance change of the thermistor has to be converted into a voltage which can then be applied across a meter and so converted to a current through it and hence a reading on the meter related to the temperature. Figure 11.29 shows a possible solution involving a potential divider circuit to convert the resistance change into a voltage change. We might use a 4.7 kΩ bead thermistor. This has a resistance of 4.7 kΩ at 25°C, 15.28 kΩ at 0°C and 0.33 kΩ at 100°C. The variable resistor might be 0 to 10 kΩ. It enables the sensitivity of the arrangement to be altered. However, if the variable resistor was set to zero resistance then, without a protective resistor, we could possibly have a large current passed through the thermistor. The protective resistor is there to prevent this occurring.

from the sensor into a suitable condition for further processing and then an element which processes the signal so that it can be displayed. Thus in the case of the resistance thermometer there might be a Wheatstone bridge, which transforms the resistance change into a voltage change, then an amplifier to make the voltage big enough for display (Figure 11.25(b)).

3 *Data presentation*
This presents the measured value in a form which enables an observer to recognise it (Figure 11.26). This may be via a display, e.g. a pointer moving across the scale of a meter or perhaps information on a visual display unit (VDU). Alternatively, or additionally, the signal may be recorded, e.g. on the paper of a chart recorder or perhaps on magnetic disc.

11.5.1 Sensors

The following are examples of commonly used sensors and the principles involved in their operation:

1 *Resistive sensors*
The input being measured is transformed into a resistance change. Such sensors include thermistors (Figure 11.27) and resistance temperature detectors (RTDs) where a change in temperature results in a change in resistance, strain gauges (Figure 11.28) where a change in strain results in a change in resistance and photo conductive cells where a change in the intensity of illumination results in a change in resistance.

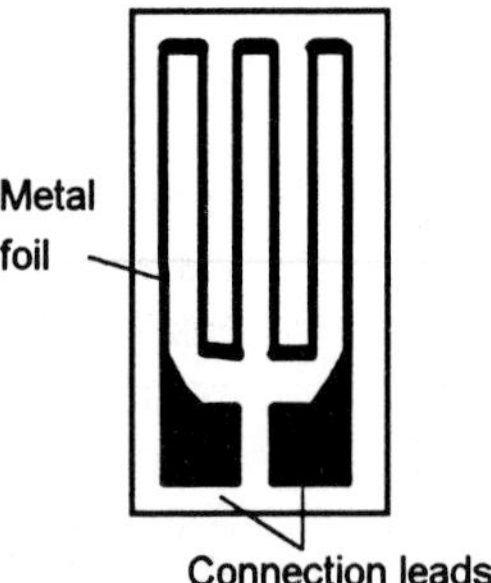

Figure 11.28 *Strain gauge*

2 *Capacitive sensors*
The input being measured is transformed into a capacitance change. Such sensors give displacement and pressure sensors where movement of one plate of a parallel plate capacitor results in a change in capacitance, liquid level sensors where the rise in the level of a liquid into the space between two concentric capacitor plates results in a change in capacitance and humidity sensors where a change in humidity results in a change in the amount of water vapour absorbed by the dielectric and hence a change in dielectric constant and consequently a change in capacitance.

Application

The traditional method of measuring relative humidity involves two thermometers, one with its bulb directly exposed to the air and giving the 'dry temperature' and the other with its bulb covered with muslin which dips into water. The rate of evaporation from the wet muslin depends on the amount of water vapour present in the air; when the air is far from being saturated then the water evaporates quickly, when saturated there is not net evaporation. This rate of evaporation affects the temperature indicated by the thermometer, so giving the 'wet temperature'. Tables are then used to convert these readings into the humidity. Consider the design of a measurement system which will automatically display the humidity. Rather than use a 'wet' thermometer element, a capacitive humidity sensor can be used. The sensor (Figure 11.31(a)) consists of an aluminium substrate with its top surface oxidised to form a porous layer of aluminium oxide. On top of the oxide a very thin gold layer is deposited, this being permeable to water vapour. Electrical connections are made to the gold layer and the aluminium substrate, the arrangement being a capacitor with an aluminium oxide dielectric. Water vapour enters the pores of the aluminium oxide and changes its dielectric constant and hence the capacitance of the capacitor. The capacitance thus gives a measure of the amount of water vapour present in the air. Figure 11.31(b) shows the type of system that might be used with such a sensor. A temperature sensor is also required since the maximum amount of water vapour that air can hold depends on the temperature and thus to compute the humidity the microprocessor needs to know the temperature.

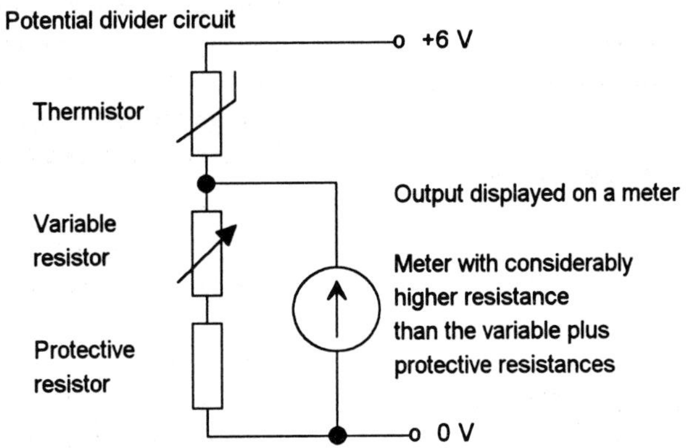

Figure 11.29 *Temperature measurement*

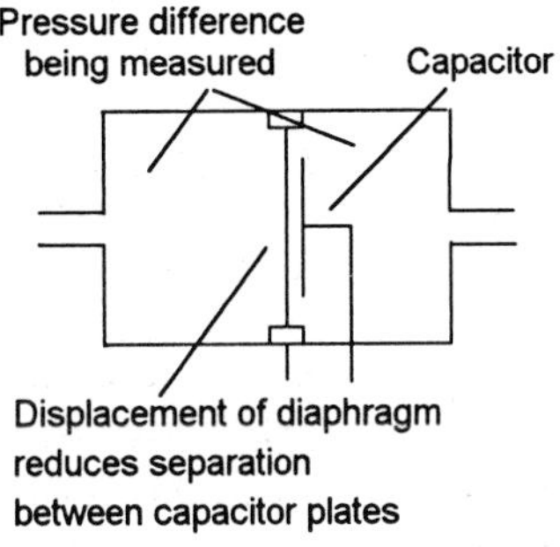

Figure 11.30 *Capacitor pressure gauge*

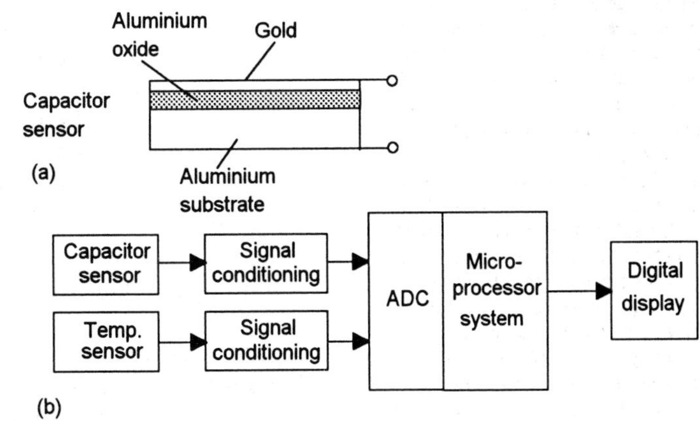

Figure 11.31 *Humidity measurement*

3 *Inductive sensors*

The input being measured is transformed into a change in inductance. A particularly useful form is the linear variable differential transformer (LVDT) (Figure 11.32). This is a

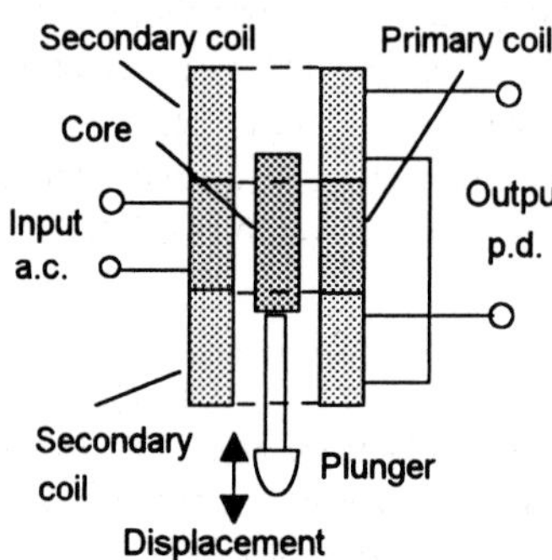

Figure 11.32 *LVDT*

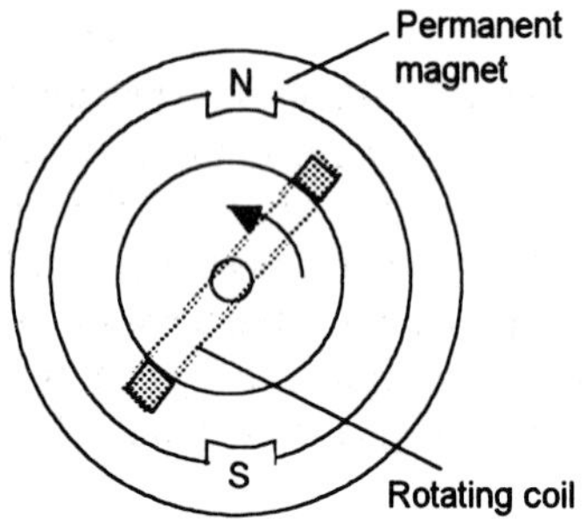

Figure 11.33 *The tachogenerator*

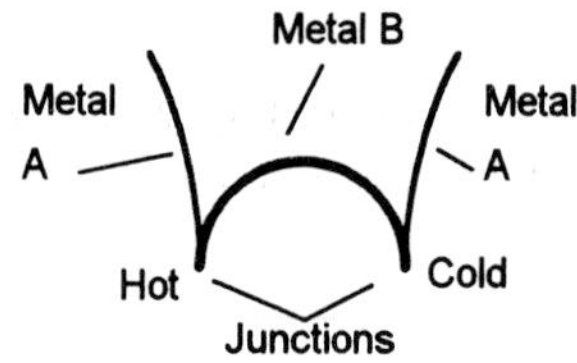

Figure 11.34 *Thermocouple*

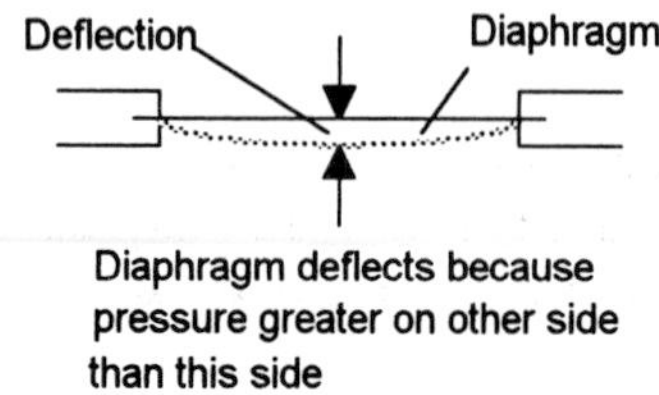

Figure 11.35 *Bending sensor*

transformer with a primary coil and two identical secondary coils wound on the same former. The displacement of a ferromagnetic core inside the former results in the amount inside one secondary coil increasing while that in the other decreases, as a consequence the inductance of one coil increases and that of the other and the difference is thus a measure of the displacement.

4 *Electromagnetic sensors*
These are based on Faraday's laws of electromagnetic induction with the input being measured giving rise to an induced e.m.f. The tachogenerator for the measurement of rotational speed is an example of such a sensor. One form is essentially an a.c. generator with a coil rotating in a magnetic field (Figure 11.33). When the coil rotates electromagnetic induction results in an alternating e.m.f. being induced in the coil. The faster the coil rotates the greater the size of the alternating e.m.f. Thus the size of the alternating e.m.f. is a measure of the angular speed.

5 *Thermoelectric sensors*
When two different metals are joined together, a potential difference occurs across the junction. The potential difference depends on the two metals used and the temperature of the junction. A *thermocouple* involves two such junctions, as illustrated in Figure 11.34. If both junctions are at the same temperature, the potential differences across the two junctions cancel each other out and there is no net e.m.f. If, however, there is a difference in temperature between the two junctions, there is an e.m.f. The value of this e.m.f. E depends on the two metals concerned and the temperatures t of both junctions. Usually one junction is held at 0°C. The input to a thermocouple is thus temperature and the output an e.m.f.

6 *Elastic sensors*
The input being measured is transformed into a displacement. Examples of such sensors are springs for the measurement of force and diaphragm pressure gauges where the pressure causes a diaphragm to deform (Figure 11.35), the amount of deformation being used as a measure of the pressure. This diaphragm may be used as one plate of a parallel plate capacitor and so the deformation is transformed into a change in capacitance, as shown in Figure 11.30. The movement of the centre of a diaphragm can be monitored by some form of displacement sensor. Figure 11.36 shows the form that might be taken when strain gauges are used to monitor the displacement, the strain gauges being stuck to the diaphragm and changing resistance as a result of the diaphragm movement. Typically such sensors are used for pressures over the range 100 kPa to 100 MPa, with an accuracy up to about ±0.1%. Another form of diaphragm pressure gauge uses strain gauge elements integrated within a silicon diaphragm and supplied, together with a resistive network for signal processing, on a single silicon chip as the Motorola MPX pressure sensor. With a voltage supply connected to

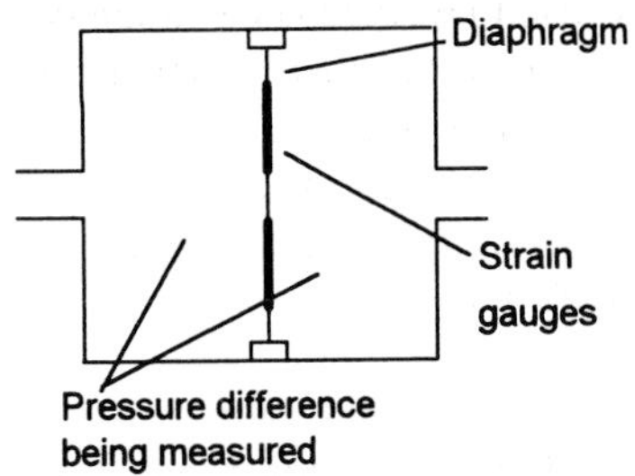

Figure 11.36 *Pressure gauge with strain gauges to sense movement of the diaphragm*

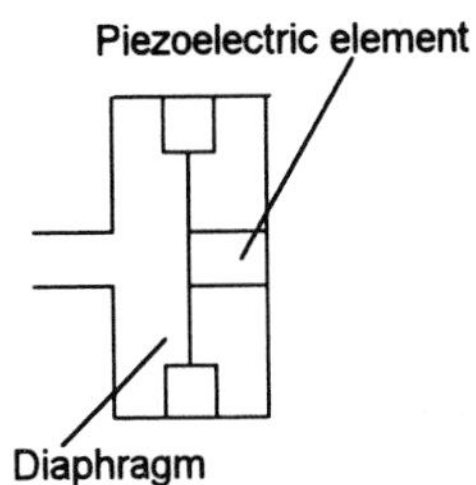

Figure 11.37 *Basic form of a piezoelectric sensor*

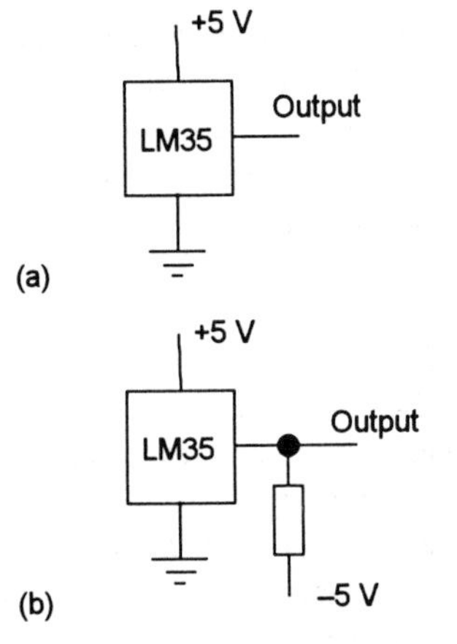

Figure 11.38 *LM35 connections*

the sensor, it gives an output voltage directly proportional to the pressure. Such sensors are available for use for the measurement of absolute pressure, differential pressure or gauge pressure, e.g. MPX2100 has a pressure range of 100 kPa and with a supply voltage of 16 V d.c. gives a voltage output over the full range of 40 mV.

7 *Piezoelectric sensors*
Forces applied to a crystal displace the atoms in the crystal and result in the crystal acquiring a surface charge. Such sensors are used for the measurement of transient pressures, acceleration and vibration. Figure 11.37 shows how such a sensor might be used in a pressure gauge.

8 *Semiconductor sensors*
There is a change in the current across the junction of *semiconductor diodes* and *transistors* when the temperature changes. For use as temperature sensors they are supplied, together with the necessary signal processing circuitry, as integrated circuits. An integrated circuit temperature sensor using transistors is LM35. This gives an output, which is a linear function of temperature, of 10 mV/°C when the supply voltage is 5 V. Figure 11.38(a) shows the connections for the range 12°C to 110° and (b) for –40° to 110°.

9 *Pyroelectric sensors*
Temperature changes give rise to changes in surface charges. Such sensors are widely used for burglar alarm systems to detect the presence of people by their body heat.

10 *Hall effect sensors*
The action of a magnetic field on a flat plate carrying an electric current generates a potential difference which is a measure of the strength of the field.

11.5.2 Terms used to define performance

The following are some of the basic terms used to describe the performance of sensors and measurement systems.

1 *Range*
The range defines the limits between which the input can vary.

2 *Accuracy*
Accuracy is the extent to which the value indicated by a measurement system might be wrong. Accuracy is often expressed as a percentage of the full range output or full-scale deflection.

3 *Sensitivity*
The sensitivity is how much output you get per unit input, e.g. a resistance temperature detector might have a sensitivity of 0.2 Ω/°C. This term is also used to indicate the sensitivity to inputs other than

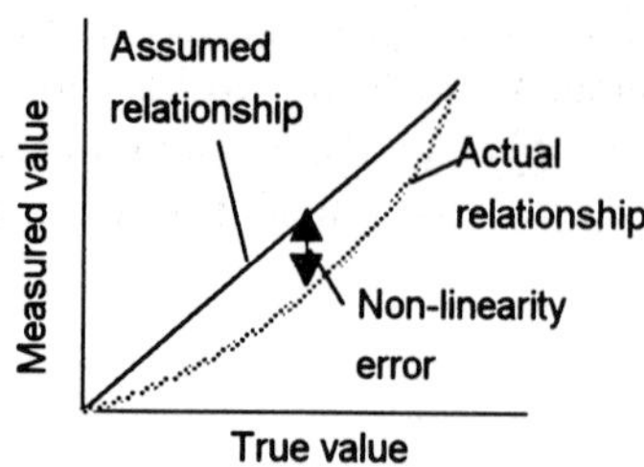

Figure 11.39 *Non-linearity error*

Application
As illustrations, the following are specifications for two temperature sensors:

A bead thermistor
Accuracy ±5%
Maximum power 250 mW
Dissipation factor 7 mW/°C
Response time 1.2 s
Thermal time constant 11 s
Temperature range −40°C to +125°C

An integrated circuit temperature sensor LM35
Accuracy at 25°C ±0.4%
Non-linearity 0.2°C
Sensitivity 10 mV/°C
Quiescent current 65 mA at supply voltage 5 V
Temperature sensitivity of quiescent current +0.39 mA/°C
Output impedance 0.1 mΩ with 1 mA load

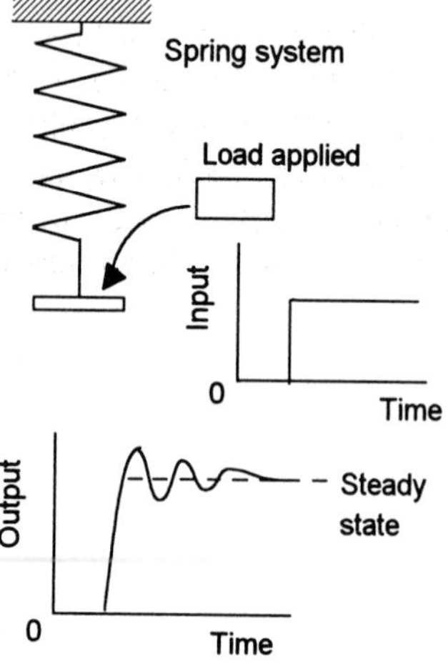

Figure 11.40 *A spring system with a step input takes time to reach the steady state value*

that being measured, i.e. environmental changes. Thus there can be the sensitivity of the transducer to temperature changes in the environment or perhaps fluctuations in the mains voltage supply, e.g. a pressure sensor might have a temperature sensitivity of ±0.1% of the reading per °C change in temperature.

4 *Non-linearity error*
A linear relationship between the input and output might be assumed over the working range for some sensor when it is not perfectly linear and thus errors occur as a result of the assumption (Figure 11.39). The error is defined as the maximum difference from the straight line, e.g. a pressure sensor might be quoted as having a non-linearity error of ±0.5% of the full range.

5 *Stability*
The stability of a sensor is its ability to give the same output when used to measure a constant input over a period of time. The term *drift* is often used to describe the change in output that occurs over time.

6 *Output impedance*
When a sensor giving an electrical output is interfaced with an electronic circuit it is necessary to know the output impedance since this impedance is being connected in either series or parallel with that circuit and so the inclusion of the sensor can significantly modify the behaviour of the system to which it is connected.

7 *Dynamic characteristics*
The *static characteristics* are the values given when steady-state conditions occur, i.e. the values given when the system has settled down after having received some input, the *dynamic characteristics* being the behaviour between the time that the input value changes and the time that the value given by the transducer settles down to the steady-state value. This is illustrated by Figure 11.40 which shows the response of a spring system to a step input signal. Dynamic characteristics are stated in terms of the response of the sensor to inputs in particular forms. Thus the *response time* is the time which elapses after a constant input, a step input, is applied to the sensor up to the point at which the sensor gives an output corresponding to some specified percentage, e.g. 95%, of the value of the input. The *time constant* is the 63.2% response time. A thermocouple in air might have a time constant of perhaps 40 to 100 s. The *rise time* is the time taken for the output to rise to some specified percentage of the steady-state output. Often the rise time refers to the time taken for the output to rise from 10% of the steady-state value to 90 or 95% of the steady-state value.

11.6 Feedback systems

A *feedback system* is one for which the input signal to the system is modified in some way by the system output signal being fed back to the input. The feedback is said to be *negative* when the signal which is fed back subtracts from the input value and *positive* when it adds to the input value.

11.6.1 Amplifiers with feedback

Consider the amplifier system shown in Figure 11.41 in which the output voltage is fed back so that it subtracts from the input voltage, the fed back voltage being in series with the input voltage. It is *negative feedback.*

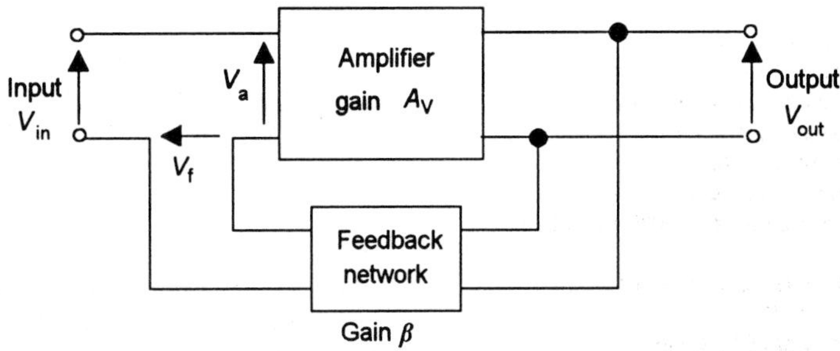

Figure 11.41 *Amplifier with feedback*

The input signal to the amplifier with negative feedback is:

$$V_a = V_{in} - V_f \qquad [13]$$

The feedback signal is the output signal V_{out} after having passed through the feedback circuit. This circuit in its simplest form could be two resistors in series across the output so that the feedback signal is taken as the voltage across one of them, i.e. a voltage divider circuit. The result is that some proportion of the output signal is fed back. If this feedback circuit has a gain β then $V_f = \beta V_{out}$. Equation [13] can thus be written as:

$$V_a = V_{in} - \beta V_{out} \qquad [14]$$

For the amplifier we have $V_{out} = A_V V_a$ and so:

$$V_{out} = A_V(V_{in} - \beta V_{out})$$

$$V_{out}(1 + \beta A_V) = A_V V_{in} \qquad [15]$$

Thus the overall voltage gain with feedback is:

Application

The operational amplifier (op-amp) has a very high gain, usually 100 000 or more, when used without any feedback. As a consequence, a very small input voltage can drive the amplifier into saturation, e.g. a 1 mV input would theoretically produce an output of 100 V but this is way beyond the maximum output level that can be attained. The usefulness of the op-amp when operated in this way is thus very limited. However, when used with negative feedback, the overall voltage gain can be reduced and controlled.

Figure 11.43 shows an op-amp with a negative feedback path. The op-amp has two inputs, termed the inverting input (–) and the non-inverting input (+) and amplifies the difference in voltage levels between the two inputs. With the arrangement shown in the figure, the input is connected via the resistor R_2 to the inverting input, the non-inverting input being connected to earth. The op-amp has virtually an infinite input impedance. Thus the current I through R_1 and the current through R_2 must be equal. Since there is no current between the inverting and non-inverting inputs then there is no potential difference between the two inputs and so point X must be at earth potential. For this reason, X is termed a virtual earth. Thus the voltage across R_1 is the input voltage V_{in} and so we have $I = V_{in}/R_1$. The voltage across R_2 is $-V_{out}$ and so we have $I = -V_{out}/R_2$. Hence we can write $I = V_{in}/R_1 = -V_{out}/R_2$. Thus the gain of the circuit is $V_{out}/I_{in} = -R_2/R_1$. The gain is thus determined solely by the value of the two resistors and so is quite stable and controllable.

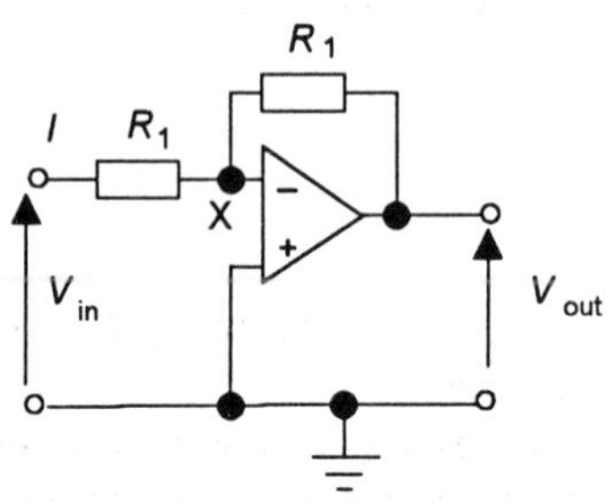

Figure 11.43 *Op-amp with negative feedback*

$$\text{overall gain} = \frac{V_{out}}{V_{in}} = \frac{A_V}{1+\beta A_V} \qquad [16]$$

The overall gain is less than A_V. If βA_V is much greater than 1 then the overall gain is effectively $1/\beta$ and so independent of the voltage gain of the basic amplifier.

The voltage gain of the basic amplifier without feedback depends on the load connected across its output and so is not stable. The voltage gain with feedback can, however, be obtained with a high degree of stability by selection of the component values of the feedback network.

Example

An amplifier has an open-circuit voltage gain of 1000 and an output resistance of 100 Ω. It is connected across a load of resistance 900 Ω. Negative feedback is provided by connection of a voltage divider across the output so that one tenth of the output voltage is fed back to be in series with the input voltage and subtract from it. Determine the overall voltage gain of the system.

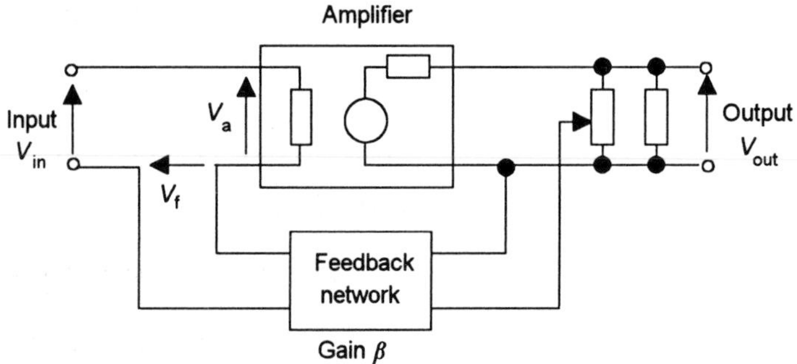

Figure 11.42 *Example*

Figure 11.42 shows the basic system. The gain of the amplifier without feedback is, when the loading effect of the feedback network is neglected:

$$\text{gain without feedback} = 1000 \times \frac{900}{100 \times 900} = 900$$

With negative feedback the gain, using equation [16], becomes:

$$\text{overall gain} = \frac{V_{out}}{V_{in}} = \frac{A_V}{1+\beta A_V} = \frac{900}{1+0.1 \times 900} = 9.89$$

Example

An amplifier is required which will have an overall voltage gain of 100 and for which the gain will not vary by more than 1.0%. What gain and feedback attenuation will be required of a basic amplifier with negative feedback if load changes result in the basic amplifier voltage gain changing by as much as 20%?

With negative feedback the gain of the required amplifier is given by equation [16] as:

$$\text{overall gain} = 100 = \frac{A_V}{1+\beta A_V}$$

$$A_V(1 - 100\beta) = 100$$

However, when the basic amplifier gain changes to $0.8A_V$ the overall gain must only change to 99. Thus:

$$99 = \frac{0.8A_V}{1+\beta 0.8A_V}$$

$$A_V(0.8 - 79.2\beta) = 99$$

Eliminating A_V between the two equations gives:

$$99(1 - 100\beta) = 99(0.8 - 79.2\beta)$$

Hence $\beta = 0.0096$. Substituting this value in an equation gives A_V = 2500.

Revision

6 An amplifier has an open-circuit voltage gain of 100 and an output resistance of 100 Ω. It is connected across a load of resistance 900 Ω. Negative feedback is provided by connection of a voltage divider across the output so that one twentieth of the output voltage is fed back to be in series with the input voltage and subtract from it. Determine the overall voltage gain of the system.

11.6.2 Oscillators

The function of an oscillator is to produce a constant frequency, constant amplitude sinusoidal signal. Oscillators can be considered to be basically amplifiers with positive feedback (Figure 11.44). Equation [16] thus becomes:

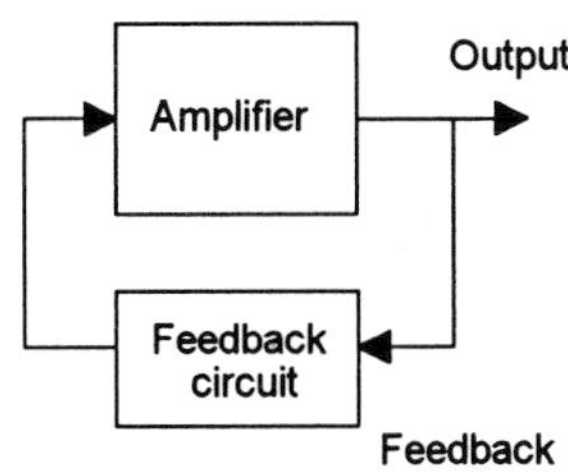

Figure 11.44 *Basic principle of an oscillator*

$$\text{overall gain} = \frac{V_{out}}{V_{in}} = \frac{A_V}{1-\beta A_V} \quad [17]$$

When $\beta A_V = 1$ then the overall gain becomes infinite; the system is thus unstable. This is the condition for oscillation.

If we think of the amplifier as initially having an input of a sinusoidal signal, then it gives an output of a larger sinusoidal signal. Some of this larger signal is then fed back to provide to the input. Provided it is in phase with the original input signal, i.e. the feedback is positive rather than negative, and the overall loop gain is 1 then the result is that the input has been 'replenished'. The circle can then repeat itself and so the system becomes self-perpetuating.

> The conditions for sustained oscillations are:
> 1 The phase shift around the feedback loop must be 0°.
> 2 The voltage gain around the feedback loop must be 1.

The voltage gain around the closed feedback loop is the product of the amplifier gain A_V and the attenuation β of the feedback circuit, i.e. $A_v\beta$. The term *attenuation* is used for a network which has a negative gain, i.e. the output is smaller than the input. Thus since we must have $A_v\beta = 1$, if the amplifier has a voltage gain of 100 then the feedback circuit must have an attenuation of 0.01 to make the loop gain equal to 1.

Sinewave oscillators include some sort of bandpass filter, i.e. a circuit which effectively only passes a narrow band of frequencies, in the feedback loop so that oscillation takes place at primarily one frequency. *LC* oscillator feedback circuits, with a high *Q*-factor and narrow bandwidth, are generally used for the higher frequency ranges above about 10^5 Hz and *RC* oscillator feedback circuits from very low frequencies up to about 10^7 Hz. The use of *RC* at low frequencies, rather than *LC*, is to avoid the requirement for bulky inductors since high values of *L* are required for resonance at low frequencies.

Figure 11.45 shows the basic circuit of a *Wien-bridge oscillator*. This is for use at low frequencies and has a *RC* feedback circuit. The feedback fraction $\beta = Z_2/(Z_1 + Z_2)$. It also has a second feedback circuit round the operational amplifier to set its voltage gain *G*, i.e. $G = R_2/(R_1 + R_2)$.

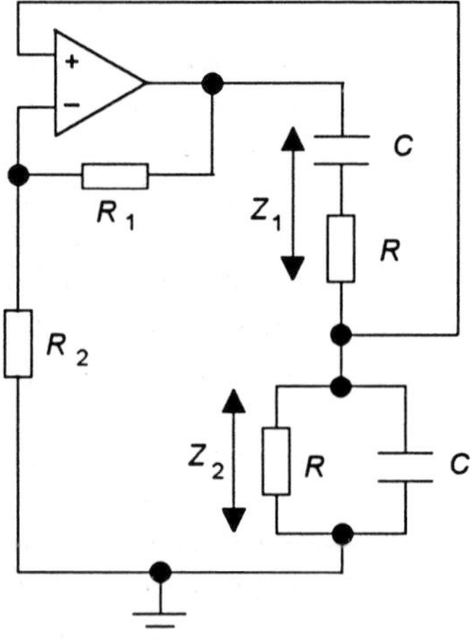

Figure 11.45 *Wien-bridge oscillator*

Figure 11.46(a) shows the basic circuit of the *Colpitts oscillator* and Figure 11.46(b) that of the *Hartley oscillator*; in both cases the circuit components required for biasing the transistor have not been included. Both use a *LC* tuned circuit for the frequency selective part of the feedback loop. With the Colpitts oscillator, the capacitance comprises C_1 and C_2 in series. These form a potential divider circuit so that the fraction of the voltage across C_1 appears between the collector and emitter of the transistor, the fraction across C_2 appearing between the base and the emitter. When the ratio of the capacitors is chosen so that the fraction of the collector voltage fed back to the base is sufficient, the loop gain exceeds 1 at the resonant frequency of the *LC* circuit and so oscillations occur. A similar arrangement is used with the Hartley oscillator but here the voltage divided circuit is provided by subdivision of the inductor.

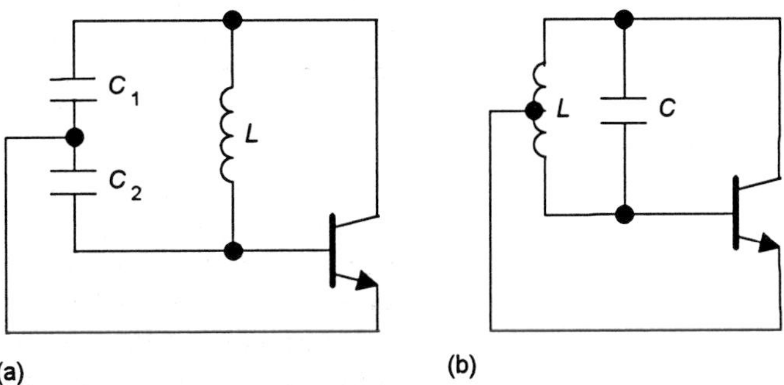

Figure 11.46 *(a) Colpitts oscillator, (b) Hartley oscillator*

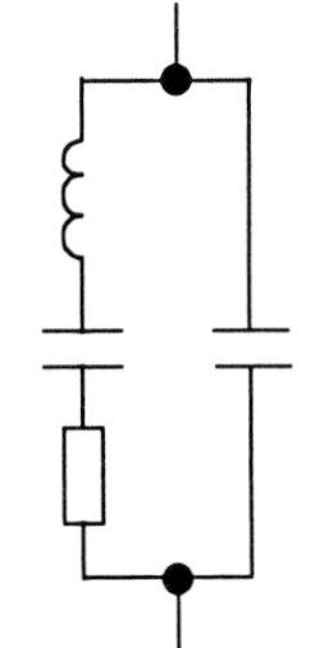

Figure 11.47 *Equivalent circuit of a quartz crystal*

A problem with the above oscillators is that the frequency of the oscillator tends to drift as the parameters of the circuits change with temperature and time. While the use of good quality components can mitigate this, an alternative when a fixed-frequency oscillator is required is to use a quartz crystal. Quartz crystals are piezoelectric materials and when mechanically stressed set up opposite charges on opposing faces. When such a crystal has a sinusoidal alternating voltage applied between electrodes connected to gold plating on opposite crystal faces, the quartz vibrates at the same frequency. This vibration will be maximum amplitude when it coincides with the natural frequency of oscillation of the crystal. The result is effectively a resonant circuit with a very high *Q*-factor, of the order of 10 000. Figure 11.47 shows the equivalent circuit of a quartz crystal. Figure 11.48 shows a basic oscillator circuit, the *Pierce oscillator*, using a quartz crystal. The crystal is replacing the inductor of the Colpitts oscillator.

In considering an oscillator the parameters that have to be taken into account are its frequency range, its frequency stability, its amplitude stability and its distortion level.

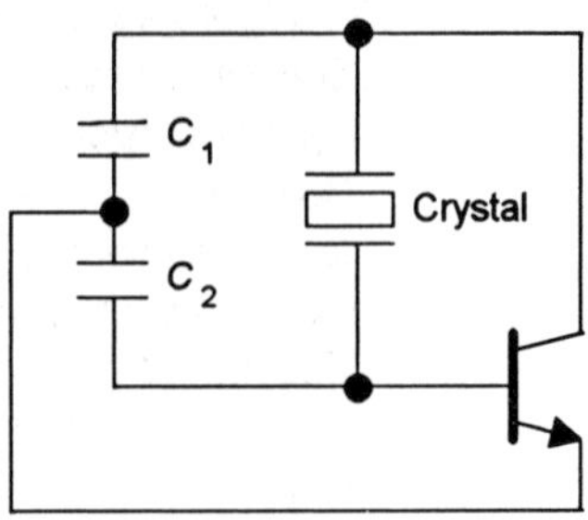

Figure 11.48 *Crystal-controlled oscillator*

11.7 Noise

If the volume control of a hi-fi system is turned right up and there is no deliberate input signal to the system, it is often possible to hear a hiss from the loudspeakers. This arises from what is termed internal *noise*. Noise is the term used with electronic systems for unwanted signals that occur; they can be internally generated within a system or arise from external sources.

For internal noise, we can think of the electrons in a conductor as being like molecules in a gas and in random motion resulting from the effect of being at some temperature above absolute zero. This random motion leads to a randomly varying voltage appearing across a conductor; this is one source of internal noise. Another source of internal noise arises from a direct current being the average rate of flow of charge carriers with time and because the actual number may vary from second to second, a fluctuating signal is produced. Thus such noise can be generated by the electrons moving through a transistor. Internal noise can also arise from changes in the conductivity of semiconductor material. The contribution of each resistor and transistor to the noise output of an amplifier depends on its position in the circuit. Noise generated in the internal resistance of the signal source or the input of the amplifier is amplified by the amplifier and can become very significant and completely exceed any noise generated in a component at later stages in the amplifier. In addition to internally generated noise we can also have externally generated noise which perhaps can arise from electrostatic and electromagnetic interference from the alternating current mains supply.

For a particular signal input power over a defined frequency range, the signal-to-noise ratio for an amplifier is defined as:

$$\text{S/N ratio} = \frac{\text{wanted signal power}}{\text{noise signal power}} \qquad [18]$$

The ratio is usually expressed in decibels (dB) as:

$$\text{S/N ratio} = 10\,\text{lg}\left(\frac{\text{wanted signal power}}{\text{noise signal power}}\right) \qquad [19]$$

The S/N ratio for a communication system determines the quality of the information received. An S/N ratio of about 70 dB gives a signal which appears to be essentially free from noise and is typical of the sound produced by a good quality hi-fi system. With an S/N ratio of 20 dB the noise becomes noticeable. This would be acceptable in a telephone system where speech is being transmitted but unacceptable with music. With an S/N ratio of 6 dB the signal is very badly degraded by noise and when the S/N ratio drops to –10 dB the signal becomes completely lost in noise.

With a digital system, the information processed is binary and the systems are only testing whether the signal received is high or low. The actual size of the high and low signals may vary quite significantly, the precise values not being important as long as the system can distinguish between high and low levels. Thus noise is no problem provided it is not large enough to induce a level change. Thus the transmission of data in digital form offers advantages over transmission in analogue form since an analogue signal will be degraded in transmission as a result of noise being picked up while the noise, provided it is not so large as to permit discrimination between high and low levels, will not affect the quality of transmission of a digital signal.

Example

If the S/N ratio for each of two cascaded amplifiers is 46 dB, what is the overall S/N ratio?

If the wanted signal power is taken as 1 W then the noise signal power for an amplifier is $10^{-4.6} = 2.5 \times 10^{-5}$ W. The total noise signal power is thus $2 \times 2.5 \times 10^{-5}$ W. Thus the total S/N ratio is 10 $\lg(1/5 \times 10^{-5}) = 43$ dB.

Revision

7 The S/N ratio for an amplifier is 40 dB. What is the noise power when there is a wanted signal power of 1 W?

Problems

1 An analogue-to-digital converter has a word length of 12 bits and the analogue signal input varies between 0 and 5 V, what is the resolution?

2 An electronic gauge pressure measurement system employs a sensor which gives an output which is converted to a digital signal by an analogue-to-digital converter and the digital signal then displayed. If the range of gauge pressure to be measured is 0 to 100 kPa and a resolution of 0.5 kPa is required, what should be the word length of the ADC?

3 A microprocessor gives an output of a 4-bit word which is fed through an 4-bit digital-to-analogue converter to control the speed of

a motor. The motor requires the voltage to change from 0 to 10 V to go from stopped to full speed. If the stopped state is given by 0000 and the full speed state by 1111 what will be the change in output to the motor when there is a change of 1 bit in the output from the microprocessor?

4 A sinusoidal voltage supply of 2 mV has an internal resistance of 1 kΩ and is connected to an amplifier of open-circuit voltage gain 10, input resistance 99 kΩ and output resistance 100 Ω. Determine the voltage appearing across a load of 1 kΩ connected across the output terminals of the amplifier.

5 An amplifier is specified as having a voltage gain of 26 dB. Express this as a voltage ratio.

6 Two amplifiers with voltage gains of 10 dB and 5 dB are connected in cascade with the output of one feeding the input of the other. What is the overall voltage gain if it can be assumed that the second amplifier does not affect the gain of the first and vice versa?

7 Two amplifiers with voltage gains of 12 dB and 10 dB are connected in cascade by a cable of attenuation 3 dB with the output of one feeding the input of the other. What is the overall voltage gain if it can be assumed that the second amplifier does not affect the gain of the first and vice versa?

8 A basic amplifier has voltage gain of 1000 and is to be used with negative feedback with an attenuation of 0.001. Determine the overall voltage gain of the feedback system.

9 An amplifier is required which will have an overall voltage gain of 100 and is to be constructed from a basic amplifier with a gain of 500 and negative feedback. Determine the attenuation required of the feedback.

10 The S/N ratio for each of three cascaded amplifiers is 46 dB, what is the overall S/N ratio?

11 The following are specification details for a range of bead thermistors. Select one and devise the required signal processing circuit so that over the temperature range 0°C to 100°C the output voltages are in the range 1 to 6 V.

A range of negative temperature coefficient thermistors with the resistance R_{T1} at temperature T_1 K being given in terms of the resistance R_{T2} at temperature T_2 K by:

$$R_{T1} = R_{T2} \exp\left(\frac{B}{T_1} - \frac{B}{T_2}\right)$$

where B is a constant for a particular thermistor.

R in kΩ at 25°C	B in K	Approx. R in kΩ at 100°C	Approx. R in kΩ at 0°C
4.7	3977	0.318	15.28
15	3740	1.19	45.13
47	4090	2.94	155.6
150	4370	7.73	534.9

Tolerance ±5%
Power 250 mW max
Response time 1.2s
Temperature range –40°C to +125°C,
reducing to 0°C to 55°C at max. power

12 Devise a temperature measurement system, using the LM35 thermotransistor, which will give a binary output of 1 bit for each 1°C change in temperature in the range 0 to 100°C. The LM35 gives an analogue output of 10 mV/°C when it has a supply voltage of 5 V.

12 Control systems

12.1 Introduction

Control systems are everywhere. For example, driving a car along a road involves the brain of the driver as a controller comparing the actual position of the car on the road with the desired position and making adjustments to correct any error between the desired and actual position. Control systems are used to control the speed of rotation of a d.c. or a.c. motor. They are used with the domestic central heating system so that the temperature in the house adjusts to maintain a set value; a room thermostat is used to control the heating system to give the required room temperature by switching the heater on or off to reduce the error between the actual temperature and the required temperature. Control systems are used with an automatic camera to adjust the lens position so that the object being photographed is always in focus and to ensure the exposure is correct. They are used with the domestic washing machine to ensure that each of the steps in the washing sequence are carried out in the right order. Control systems are involved with automation in industry. For example, with an automatic drilling machine, the control system might be required to start lowering the drill when the workpiece is in place, start drilling when the drill reaches the workpiece, stop drilling when the drill has produced the required depth hole, retract the drill and then wait for the next workpiece to be put into place so that the sequence can be repeated. These are just a few examples of control systems. This chapter is a consideration of the basic principles of control systems and the elements involved in such systems.

> *Control systems are systems that are used to maintain a desired result or value.*

12.2 Basic principles

There are two basic types of control systems:

1 *Open-loop*
   In an open-loop control system the output from the system has no effect on the input signal to the plant or process. The output is determined solely by the initial setting. Open-loop systems have the advantage of being relatively simple and consequently cheap with generally good reliability. However, they are often inaccurate since there is no correction for errors in the output which might result from extraneous disturbances.

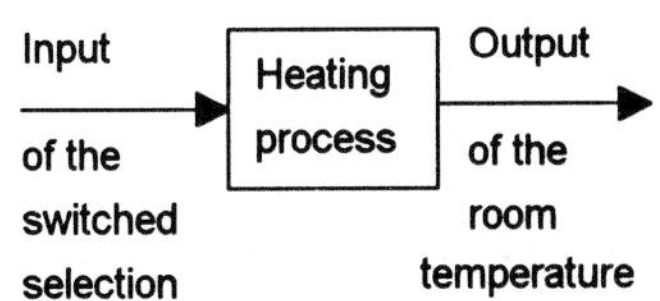

Figure 12.1 *Open-loop system with no feedback of output to modify the input if there are any extraneous disturbances*

As an illustration of an open-loop system, consider the heating of a room to some required temperature using an electric fire which has a selection switch which allows a 1 kW or a 2 kW heating element to be selected. The decision might be made, as a result of experience, that to obtain the required temperature it is only necessary to switch on the 1 kW element. The room will heat up and reach a temperature which is determined by the fact the 1 kW element is switched on. The temperature of the room is thus controlled by an initial decision and no further adjustments are made. Figure 12.1 illustrates this. If there are changes in the conditions, perhaps someone opening a window, no adjustments are made to the heat output from the fire to compensate for the change. There is no information *fed back* to the fire to adjust it and maintain a constant temperature.

Open-loop control is often used with processes that require the sequencing of events by on–off signals with processes being switched on for a prescribed amount to time and then off, e.g. washing machines which require the water to be switched on and then, after a suitable time, switched off followed by the heater being switched on and then, after a suitable time, switched off.

2 *Closed-loop*

In a closed-loop control system a signal indicating the state of the output of the system is *fed back* to the input where it is compared with what was required and the difference used to modify the output of the system so that it maintains the output at the required value (Figure 12.2). The term *closed-loop* refers to the loop created by the feedback path. Closed-loop systems have the advantage of being relatively accurate in matching the actual to the required values. They are, however, more complex and so more costly with a greater chance of breakdown as a consequence of the greater number of components.

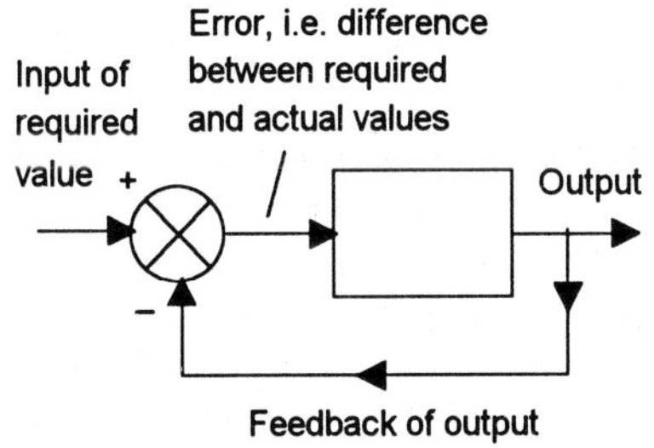

Figure 12.2 *Closed-loop system with feedback of output to modify the input and so adjust for any extraneous disturbances*

As an illustration, consider modifications of the open-loop heating system described above to give a closed-loop system. To obtain the required temperature, a person stands in the room with a thermometer and switches the 1 kW and 2 kW elements on or off, according to the difference between the actual room temperature and the required temperature in order to maintain the temperature of the room at the required temperature. In this situation there is *feedback*, information being fed back from the output to modify the input to the system. Thus if a window is opened and there is a sudden cold blast of air, the feedback signal changes because the room temperature changes and is fed back to modify the input to the system. The input to the heating process depends on the deviation of the actual temperature fed back from the output of the system from the required temperature initially set. Figure 12.3 illustrates this system with the comparison element represented by the summing symbol with a + opposite the set value input and a – opposite the feedback signal to give the sum as + set value – feedback value = error. This error signal is then used to control the process. Because the feedback signal is subtracted from the set value signal, the system is said to have *negative feedback.*

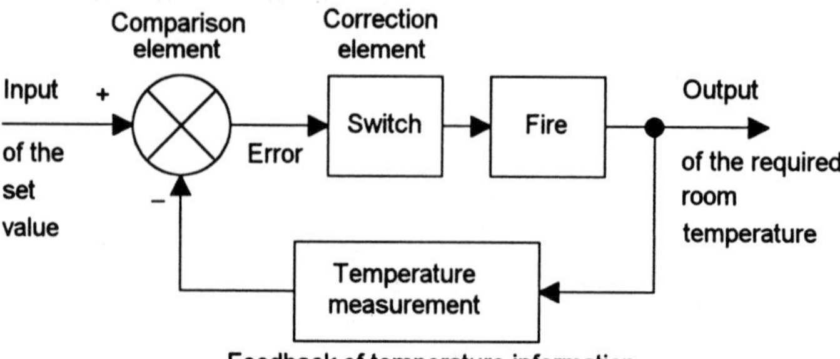

Figure 12.3 *Closed-loop system with feedback being used to modify the input to the controller and so enable the control system to adjust when there are extraneous disturbances*

The system used to control the speed of rotation of a motor might be open-loop or closed-loop. With an *open-loop system*, the motor speed is set by selecting the position of some control knob, the motor then runs to a speed indicated by the knob. However, if the load on the motor changes, there is no mechanism to adjust the motor to compensate for the load change and the speed drops. With a *closed-loop system*, the motor speed is set by the control knob and now, when the load changes there is a feedback signal from some output sensor back to the input to the motor to indicate that the change has occurred and, as a consequence, the speed is adjusted. The closed-loop system can thus take account of load changes, the open-loop system cannot. The primary advantage of open-loop control is that it is less expensive than closed-loop control. The disadvantage is that errors caused by disturbances or changes in loading are not corrected.

An *open-loop control system* does not compare the actual output with the required output to determine the control action but uses a calibrated input setting to obtain the required output. A *closed-loop system* uses feedback to compare the actual output with the required output and so modify its control action in the event of any difference to obtain the required output.

12.2.1 Elements in control systems

The elements of a basic open-loop and a basic closed-loop system are shown in Figure 12.4. The elements of the open-loop system (Figure 12.4(a)) are:

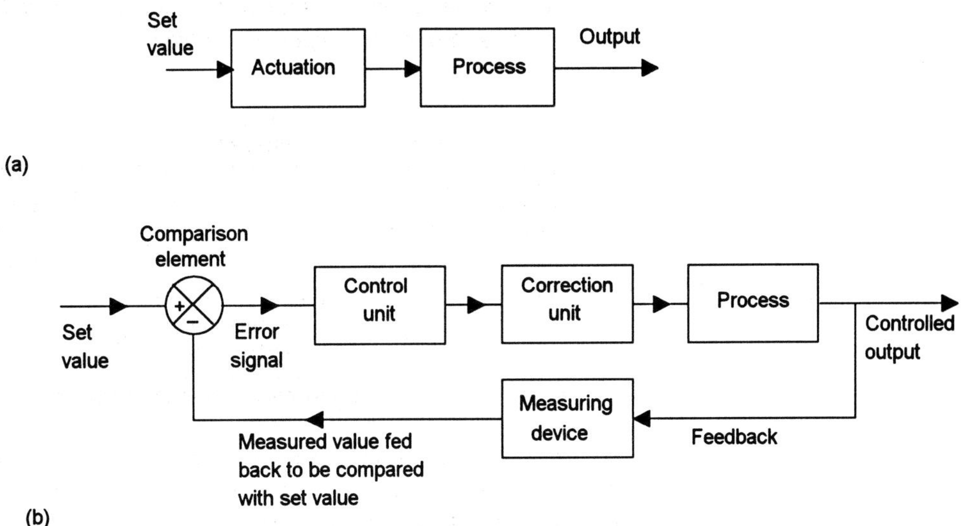

Figure 12.4 *Basic elements of control system: (a) open-loop, (b) closed-loop*

1 *Actuation*
The set value is used to actuate some element, e.g. a switch, which then adjusts the variable in the process to give the required output.

2 *Process*
The process is the system in which there is a variable that is being controlled. Thus the process might be a motor with its shaft rotational speed being controlled.

Note that once the set value has been inputted into the open-loop system there is no further control of the output, no signal being fed back from the input to modify the actuation. With a closed-loop system, however, this is not the case as Figure 12.4(b) illustrates. The basic elements of a closed-loop system are:

1 *Comparison element*
This compares the required, i.e. set, value of the variable condition being controlled with a signal which is representative of the value that is actually occurring and produces an error signal. It can be regarded as adding the reference signal, which is positive, to the measured value signal, which is negative:

Error signal = reference value signal – measured value signal

The feedback of the measured value signal is *negative feedback* because the signal which is fed back subtracts from the input value to give the error signal.

2 *Control element*
This determines the action to be taken when an error signal is received. The control action being used by the element may be just to supply a signal which gives an output which is an on or off signal when there is an error, as in a room thermostat, or perhaps a signal which proportionally opens or closes a valve according to the size of the error.

3 *Correction element*
The correction element acts on the input from the controller to produce a change in the process to correct or change the controlled condition. Thus it might be a switch which switches on a heater and so increases the temperature of the process. The term *actuator* is used for the element of a correction unit that provides the power to carry out the control action.

4 *Process element*
The process is the system in which there is a variable that is being controlled. It could be a room in a house with its temperature being controlled.

5 *Measurement element*
The measurement element produces a signal related to the variable condition of the process that is being controlled. It might be, for example, in a temperature control system a thermocouple which gives an e.m.f. related to the temperature. This signal is fed back to modify the input to the control element.

12.2.2 Types of closed-loop control systems

Closed-loop control systems can be considered to fall into three main groups:

1 *Systems with continuous control*
The control is continuously exercised with the feedback signal being continuously monitored and compared with the set value. As an illustration of a continuous control system consider the basic elements involved in the control of the speed of a motor by the system shown in Figure 12.5. The position of the slider on a potentiometer is used to set the required speed value by giving a set value voltage. The differential amplifier is used to amplify the difference between this set value and a voltage representing the actual speed and provide an error signal which is used to control the speed of the motor. Basically this might be just a signal which is proportional to the error signal. The speed of the rotating shaft is monitored by a sensor, possibly a tachogenerator, which provides a voltage signal related to the speed of the rotating motor shaft. This signal is then fed back for comparison with the set value signal.

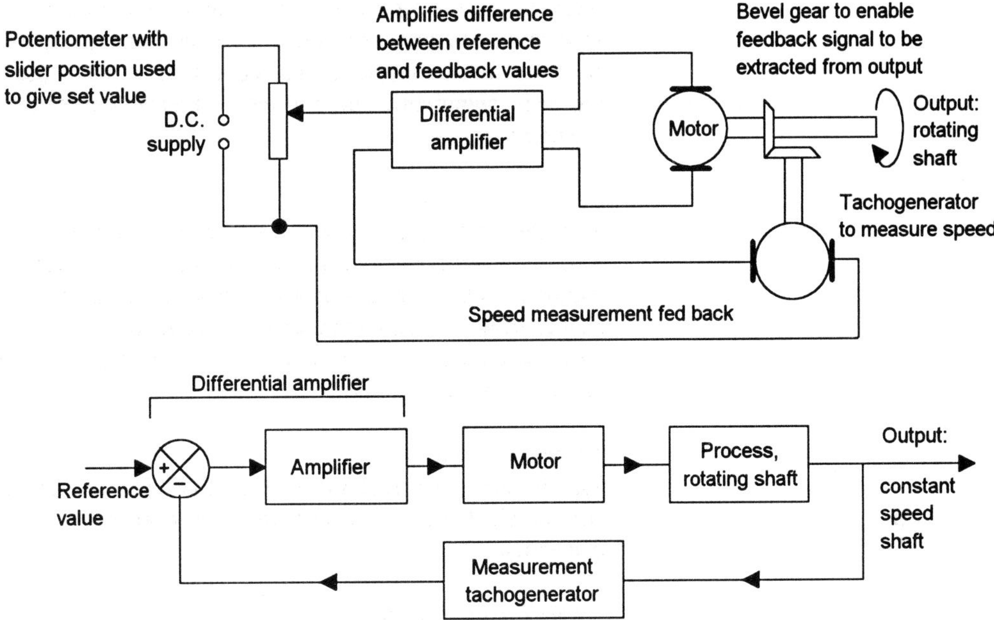

Figure 12.5 *Motor speed control*

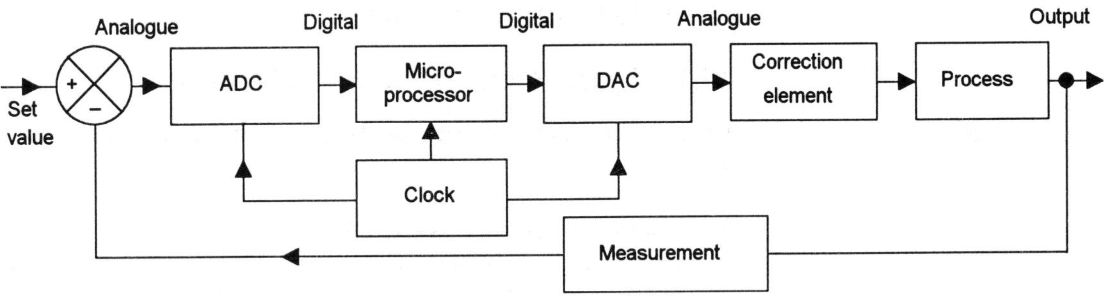

Figure 12.6 *Digital control system*

2 *Systems using digital control which involves sampling*
Digital control, with a system where the error signal is an analogue signal, involves the error being sampled at regular intervals and converted to a digital signal by an analogue-to-digital converter. The control action is then exercised by a microprocessor. Figure 12.6 shows the basic elements involved in such a system. The error between the set value and the actual value, both assumed to be analogue quantities, is converted to a digital signal by an analogue-to-digital converter which samples at regular intervals the analogue error signal. The clock is used to provide the timing signals used to

determine when the sampling is to occur. The resulting digital signal is then processed by a microprocessor. This processing involves the microprocessor carrying out calculations based on the error value and possibly stored values of previous inputs and outputs to generate its output signal. Its output is then converted to an analogue signal by a digital-to-analogue converter and used with some correction element to modify the variable in the process system.

3 *Systems using discrete-event control*
This is often termed *sequential control* since the controller is used to sequence a number of discrete events. This involves the control actions being determined by observed sequential conditions or combinations of a set of conditions. For example, a car washer might be controlled so that it comes on when a car is detected as being in the required position under the washing machine and when the correct coins have been inserted in the car wash machine. An automatic machine tool might be controlled to carry out a machining process when sensors detect that the workpiece is in position and then carry out a prescribed sequence of events.

12.2.3 Examples of control systems

The following are examples of control systems to show how control is exercised in a range of situations

As an illustration of a system using both continuous, or digital, and discrete event control, consider a domestic washing machine (Figure 12.7). Discrete event control is used to switch the various valves and pumps on and off in the required sequence to achieve the selected wash program. However, continuous or digital control is used to control the wash drum rotational speed with the actual drum speed being compared with that required. The following is a typical control program.

1 When the start switch is pressed, the valves open to allow water into the drum.

2 When the full water level is reached, the valves are closed.

3 Then the heater is switched on.

4 The water temperature is controlled and when the correct temperature is reached the washer motor is turned on.

5 The washer motor is run for a set time.

6 The pump is switched on to empty the drum.

7 When empty, the pump is switched off.

8 The valves are opened to fill with water.

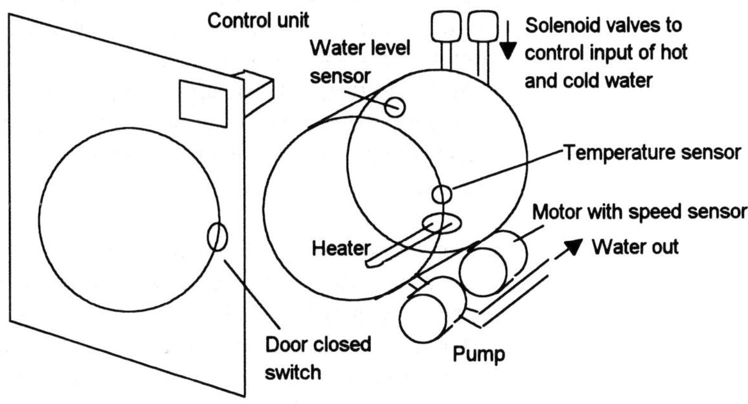

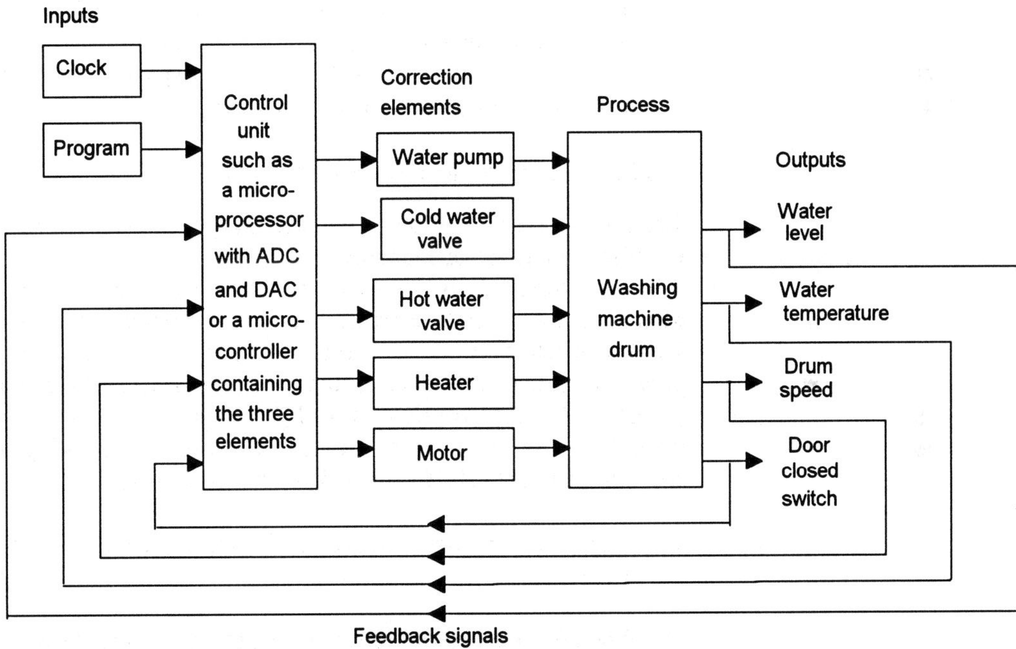

Figure 12.7 *Washing machine system*

9 The water level is controlled and when the full level is reached the valves are closed.

10 The rinse action of rotating the drum first one way and then the other is then started.

11 Operations 6, 7, 8, 9 and 10 are repeated a number of times.

12 When the drum is empty after rinsing, the pump is turned off and the drum spun for a set time.

13 End of program.

Modern cars contain many control systems, e.g. the power steering system used with a car. This comes into operation whenever the resistance to turning the steering wheel exceeds a predetermined amount and enables the movement of the wheels to follow the dictates of the angular motion of the steering wheel. The input to the system is the angular position of the steering wheel. This mechanical signal is scaled down by gearing and has subtracted from it a feedback signal representing the actual position of the wheels. This feedback is via a mechanical linkage. Thus when the steering wheel is rotated and there is a difference between its position and the required position of the wheels, there is an error signal. The error signal is used to operate a hydraulic valve and so provide a hydraulic signal to operate a cylinder. The output from the cylinder is then used, via a linkage, to change the position of the wheels. Figure 12.8 shows a block diagram of the system.

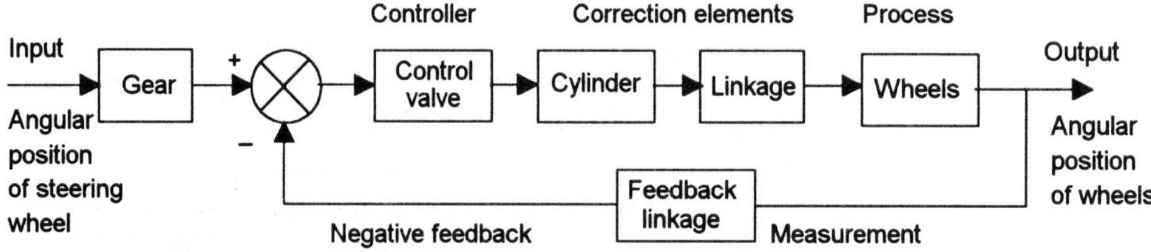

Figure 12.8 *Power assisted steering*

The modern car involves many control systems. For example, there is the *engine management system* aimed at controlling the amount of fuel injected into each cylinder and the time at which to fire the spark for ignition. Part of such a system is concerned with delivering a constant pressure of fuel to the ignition system. Figure 12.9(a) shows the elements involved in such a system. The fuel from the fuel tank is pumped through a filter to the injectors, the pressure in the fuel line being controlled to be 2.5 bar (2.5×0.1 MPa) above the manifold pressure by a regulator valve. Figure 12.9(b) shows the principles of such a valve. It consists of a diaphragm which presses a ball plug into the flow path of the fuel. The diaphragm has the fuel pressure acting on one side of it and on the other side is the manifold pressure and a spring. If the pressure is too high, the diaphragm moves and opens up the return path to the fuel tank for the excess fuel, so adjusting the fuel pressure to bring it back to the required value. The pressure control system can be considered to be represented by the closed loop system shown in Figure 12.10. The set value for the pressure is determined by the spring tension. The comparator and control law is given by the diaphragm and spring. The correction element is the ball in its seating and the measurement is given by the diaphragm.

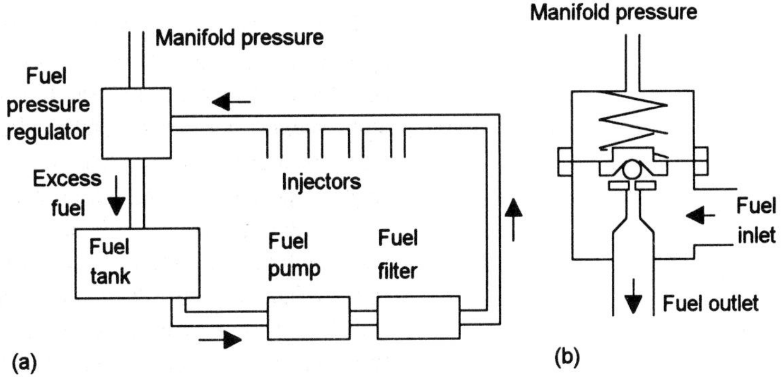

Figure 12.9 *(a) Fuel supply system, (b) fuel pressure regulator*

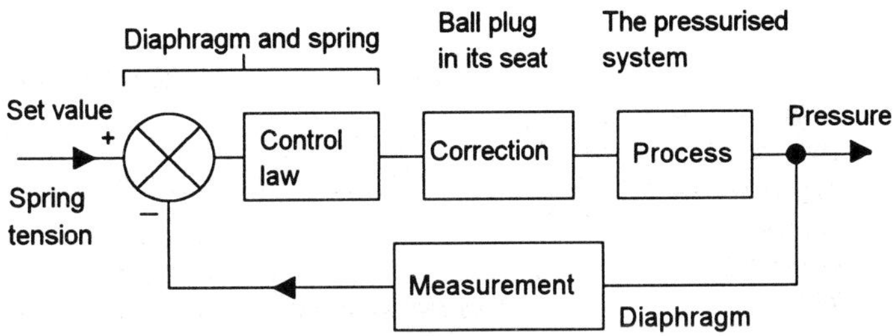

Figure 12.10 *Fuel supply control system*

Another example of a control system used with a car is the *antilock brake system (ABS)*. If one or more of the vehicle's wheels lock, i.e. begins to skid, during braking, then braking distance increases, steering control is lost and tyre wear increases. Antilock brakes are designed to eliminate such locking. The system is essentially a control system which adjusts the pressure applied to the brakes so that locking does not occur. This requires continuous monitoring of the wheels and adjustments to the pressure to ensure that, under the conditions prevailing, locking does not occur. Figure 12.11 shows the principles of such a system.

The two valves used to control the pressure are solenoid-operated valves, generally both valves being combined in a component termed the modulator. When the driver presses the brake pedal, a piston moves in a master cylinder and pressurises the hydraulic fluid. This pressure causes the brake calliper to operate and the brakes to be applied. The speed of the wheel is monitored by means of a sensor. When the wheel locks, its speed changes abruptly and so the feedback signal from the sensor changes. This feedback signal is fed into the controller where it is compared with what signal might be expected on the basis of data stored in the controller memory. The controller can then supply output signals which operate the valves and so adjust the pressure applied to the brake.

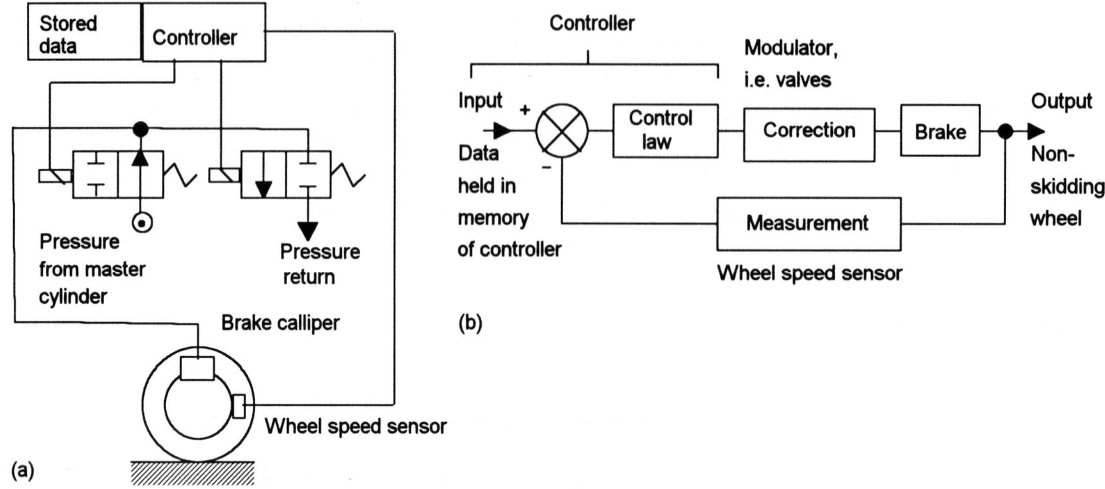

Figure 12.11 *Antilock brakes: (a) schematic diagram, (b) block form of the control system*

As an illustration of a process control system, Figure 12.12 shows the type of system that might be used to control the *thickness of sheet* produced by rollers, Figure 12.13 showing the block diagram description of the system. The thickness of the sheet is monitored by a sensor such as a linear variable differential transformer (LVDT). The position of the LVDT probe is set so that when the required thickness sheet is produced, there is no output from the LVDT. The LVDT produces an alternating current output, the amplitude of which is proportional to the error. This is then converted to a d.c. error signal which is fed to an amplifier. The amplified signal is then used to control the speed of a d.c. motor, generally being used to vary the armature current. The rotation of the shaft of the motor is likely to be geared down and then used to rotate a screw which alters the position of the upper roll, hence changing the thickness of the sheet produced.

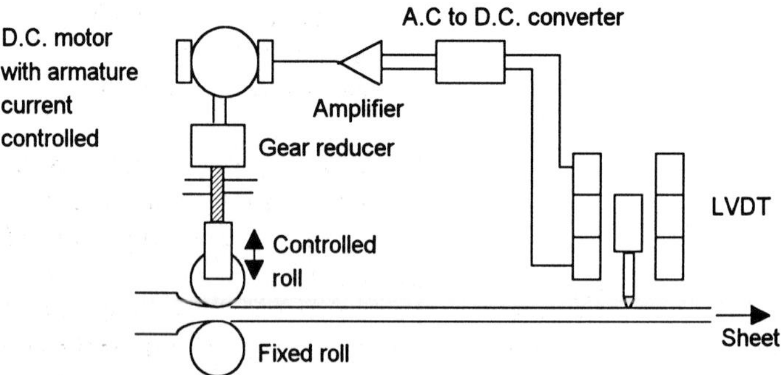

Figure 12.12 *Sheet thickness control system*

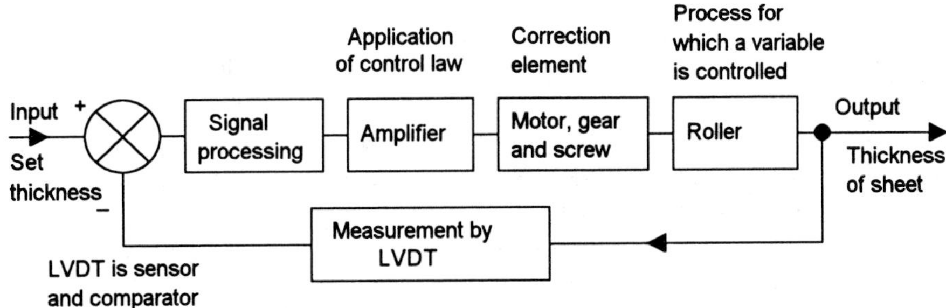

Figure 12.13 *Sheet thickness control system*

12.2.4 The programmable logic controller

For a control system, e.g. an automatic drilling machine, we could wire up electrical circuits in which the closing or opening of switches would result in motors being switched on or valves being actuated. Thus we might have the closing of a switch activating a relay which, in turn, switches on the current to a motor and causes the drill to rotate. Another switch might be used to activate a relay and switch on the current to a pneumatic or hydraulic valve which results in pressure being switched to drive a piston in a cylinder and so results in the workpiece being pushed into the required position. Such electrical circuits would have to be specific to the automatic drilling machine. The controller circuits we devise for different situations would be different. If we change any part of the operation, the wiring has to be changed.

However, instead of hardwiring each control circuit for each control situation we can use the same basic system for all situations if we use a microprocessor-based system and write a program to instruct the microprocessor how to react to each input signal from, say, switches and give the required outputs to, say, motors and valves. Thus we might have a program of the form:

If switch A closes
Output to motor circuit
If switch B closes
Output to valve circuit

By changing the instructions in the program we can use the same microprocessor system to control a wide variety of situations.

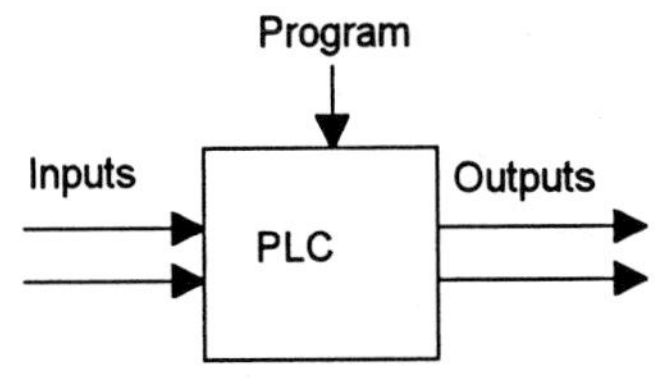

Figure 12.14 *A programmable logic controller*

A *programmable logic controller* (PLC) is a special form of microprocessor-based controller that uses a programmable memory to store instructions and to implement functions such as logic, sequencing, timing, counting and arithmetic in order to control machines and processes (Figure 12.14) and are designed to be operated by engineers with perhaps a limited knowledge of computers and computing languages. They are not designed so that only computer programmers can set up or change the programs. Thus, the designers of the PLC have pre-programmed it so that the control program can be entered using a simple, rather intuitive, form of language. The term *logic* is used because programming is primarily concerned with implementing logic and switching operations, e.g. if A or B occurs switch on C, if A and B occurs switch on D. Input devices, e.g. sensors such as switches, and output devices in the system being controlled, e.g. motors, valves, etc., are connected to the PLC. The operator then enters a sequence of instructions, i.e. a program, into the memory of the PLC. The controller then monitors the inputs and outputs according to this program and carries out the control rules for which it has been programmed.

PLCs have the great advantage that the same basic controller can be used with a wide range of control systems. To modify a control system and the rules that are to be used, all that is necessary is for an operator to key in a different set of instructions. There is no need to rewire. The result is a flexible, cost effective, system which can be used with control systems which vary quite widely in their nature and complexity.

12.3 Electrical switching

Electrical actuators are the elements which are responsible for transforming an electrical output of a controller such as a microprocessor into a controlling action in the process concerned. For example, the electrical output might be transformed into a rotatory or linear motion. Generally there need to be intermediate elements which take the output from the controller, e.g. a microprocessor, and uses it to switch or control a larger current or voltage change which is able to operate the actuation element. The following are examples of such switching elements:

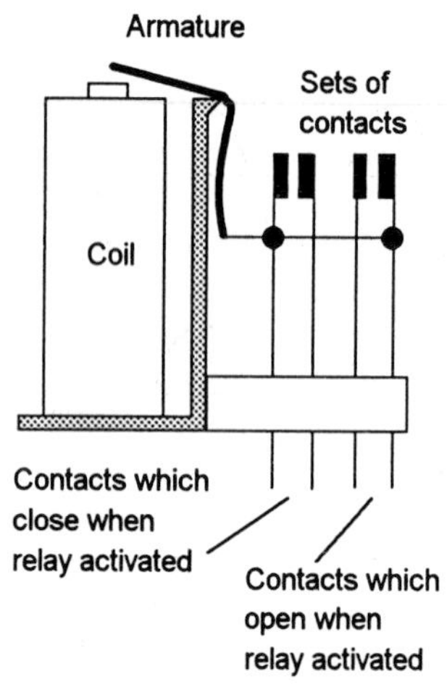

Figure 12.15 *Relay*

12.3.1 Relays

The electrical relay (Figure 12.15) offers a simple on–off switching action in response to a control signal. When a current flows through the coil of wire a magnetic field is produced. This pulls a movable arm, the armature, that forces the contacts to open or close; usually there are two sets of contacts with one being opened and the other closed by the action. These contacts can then be used to switch on or off a much larger current, e.g. the current through a heater or the current used to operate a motor.

12.3.2 Thyristors

A junction diode allows a significant current in one direction through it and barely any current in the reverse direction, i.e. a low resistance in one direction and a high one in the reverse. The *thyristor* or *silicon controlled rectifier* (SCR) can be considered to be a diode which can be switched on to conducting, i.e. switched to a low resistance from a high resistance, at a particular forward direction voltage. The thyristor passes negligible current when reverse biased and when forward biased the current is also negligible until the forward breakdown voltage, e.g.

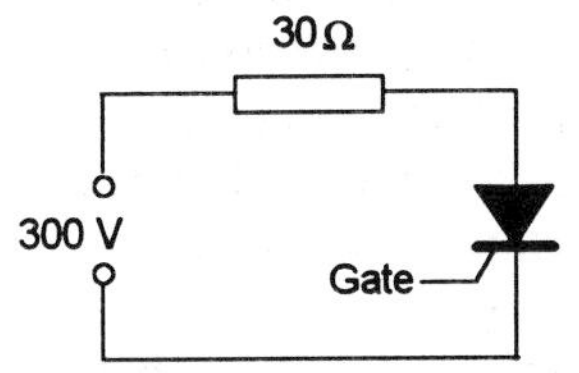

Figure 12.16 *Thyristor circuit*

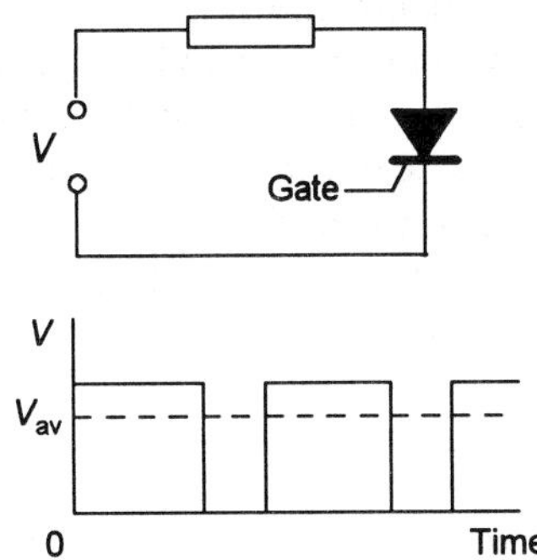

Figure 12.17 *Thyristor d.c. control*

300 V, is exceeded. Thus if such a thyristor is used in a circuit in series with a resistance of 30 Ω (Figure 12.16) then before breakdown we have a very high resistance in series with the 30 Ω and so virtually all the 300 V is across the thyristor with its high resistance and there is negligible current. When forward breakdown occurs, the resistance of the thyristor drops to a low value and now, of the 300 V, only about 2 V might be dropped across the thyristor. Thus there is now 300 – 2 = 298 V across the 30 Ω resistor and so the current rises from its negligible value to 298/30 = 9.9 A. When once switched on the thyristor remains on until the forward current is reduced to below a level of a few milliamps. The voltage at which forward breakdown occurs is controlled by a gate input, being determined by the current entering the gate, the higher the current the lower the breakdown voltage. Thus by controlling the gate current we can determine when the thyristor will switch from a high to low resistance.

As an illustration of the use of a thyristor, Figure 12.17 shows how it can be used to control the power supplied to a resistive load by chopping a d.c. voltage V. An alternating current signal is applied to the gate so that periodically the voltage V becomes high enough to switch the thyristor off and so the voltage V off. The supply voltage can be chopped and an intermittent voltage produced with an average value which is varied and controlled by the alternating signal to the gate.

Application

An example of control using a thyristor is that of a.c. for electric heaters, electric motors or lamp dimmers. Figure 12.18 shows a circuit that can be used. The alternating current is applied across the load, e.g. the lamp for a lamp dimming circuit, in series with a thyristor. R_1 is a current-limiting resistor and R_2 is a potentiometer which sets the level at which the thyristor is triggered. The diode in the gate input is to prevent the negative part of the alternating voltage cycle being applied to the gate. By moving the potentiometer slider the gate current can be varied and so the thyristor can be made to trigger at any point between 0° and 90° in the positive half-cycle of the applied alternating voltage. When the thyristor is triggered near the beginning of the cycle it conducts for the entire positive half-cycle and the maximum power is delivered to the load. When triggering is delayed to later in the cycle it conducts for less time and so the power delivered to the load is reduced. Hence the position of the potentiometer slider controls the power delivered to the load; with the light dimming circuit the slider position controls the power delivered to the lamp and so its brightness.

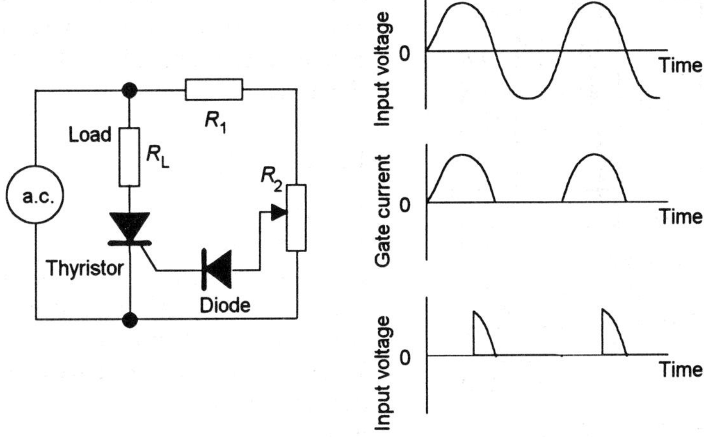

Figure 12.18 *Thyristor control for a.c. power to a load*

12.3.3 Transistors

For the junction transistor in the circuit shown in Figure 12.19(a), when the base current I_B is zero both the base-emitter and the base-collector junctions are reverse biased. When the base current I_B is increased to a high enough value the base-collector junction becomes forward biased. By switching the base current between 0 and such a value, bipolar transistors can be used as switches. When there is no input voltage V_{in}

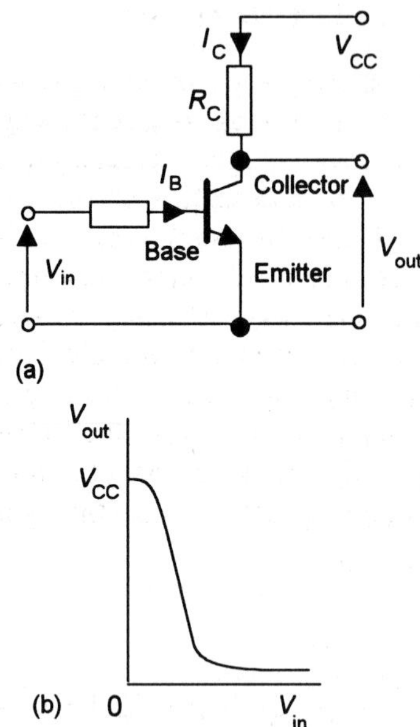

Figure 12.19 *Transistor switch*

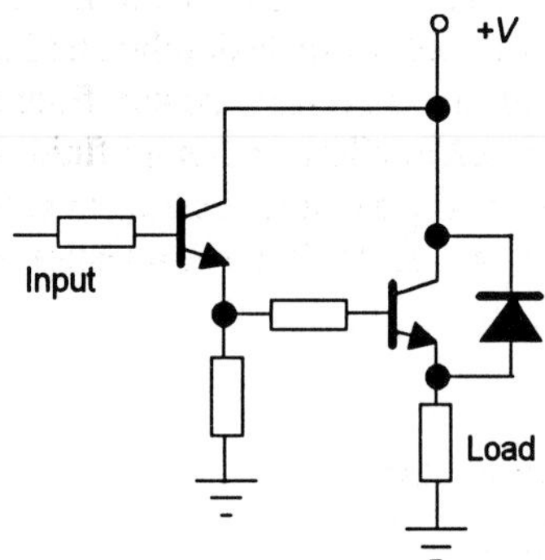

Figure 12.20 *Switching a load*

then virtually the entire V_{CC} voltage appears at the output as the resistance between the collector and emitter is high. When the input voltage is made sufficiently high so that the resistance between the collector and emitter drops to a low value, the transistor switches so that very little of the V_{CC} voltage appears at the output (Fig. 12.19(b)). We thus have an electronic switch.

Because the base current needed to drive a bipolar power transistor is fairly large, a second transistor is often needed to take the small current and produce a large enough current to the base of the transistor used for the switching and so enable switching to be obtained with the relatively small currents supplied, for example, by a microprocessor. Such a pair of transistors (Figure 12.20) is termed a *Darlington pair* and they are available as single-chip devices. Since such a circuit is often used with inductive loads and large transient voltages can occur when switching occurs, a protection diode is generally connected in parallel with the switching transistor to prevent damage to it when it is switched off. As an indication of what is available, the integrated circuit ULN2001N contains seven separate Darlington pairs, each pair being provided with a protection diode.

12.3.4 Limit switches

The term *limit switch* is used for a mechanical switch which is used to detect the presence or passage of a moving part. It can be actuated by a cam, roller or lever. Figure 12.21 shows some examples. The cam (Figure 12.21(c)) can be rotated at a constant rate and so switch the switch on and off for particular time intervals. Limit switches find use in sensing the presence or otherwise of moving mechanical items in control systems. An example is given later in this chapter.

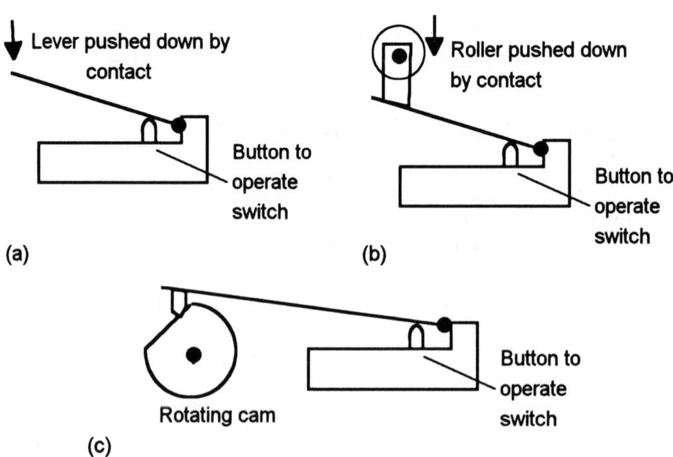

Figure 12.21 *Limit switches*

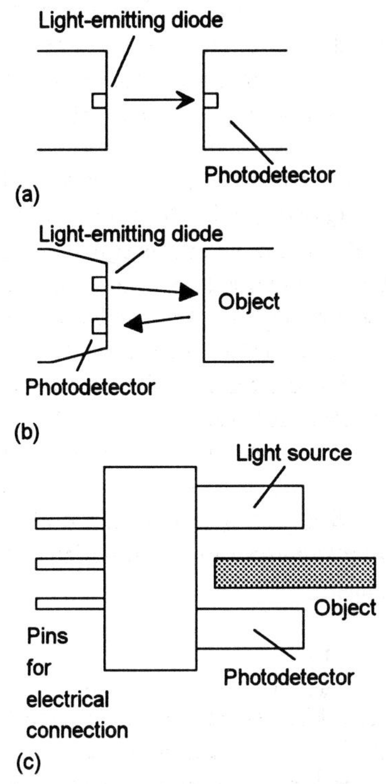

Figure 12.22 *Photoelectric sensors*

12.3.5 Photoelectric switches

Photoelectric switch devices can either operate as *transmissive types* where the object being detected breaks a beam of light, usually infrared radiation, and stops it reaching the detector (Figure 12.22(a)) or *reflective types* where the object being detected reflects a beam of light onto the detector (Figure 12.22(b)). In both types the radiation emitter is generally a *light-emitting diode (LED)*. The radiation detector might be a *phototransistor*, often a pair of transistors, known as a *Darlington pair*. The Darlington pair increases the sensitivity. Depending on the circuit used, the output can be made to switch to either high or low when light strikes the transistor. Such sensors are supplied as packages for sensing the presence of objects at close range, typically at less than about 5 mm. Figure 12.22(c) shows a U-shaped form where the object breaks the light beam.

12.3.6 Directional control valves

A *solenoid operated valve* is often used, by the presence or not of a current through a solenoid, to control the directions of flow of pressurised air or oil and so operate other devices such as a piston moving in a cylinder.

Figure 12.23 shows one such form, a spool valve, used to control the movement of a piston in a cylinder. Pressurised air or hydraulic fluid is inputted from port P, this being connected to the pressure supply from a pump or compressor and port T is connected to allow hydraulic fluid to return to the supply tank or, in the case of a pneumatic system, to vent the air to the atmosphere. The term port is used to describe an inlet or outlet from the valve.

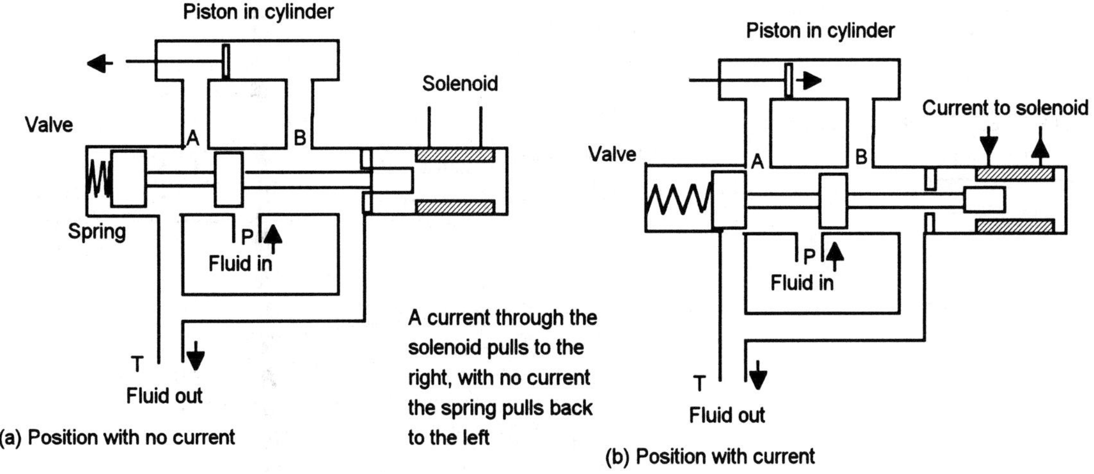

Figure 12.23 *An example of a solenoid operated valve*

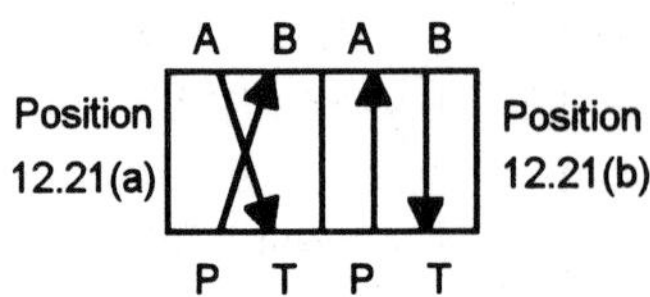

Figure 12.24 *The 4/2 valve*

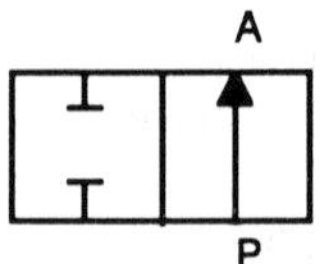

2/2 valve: flow from P to A switched to no flow

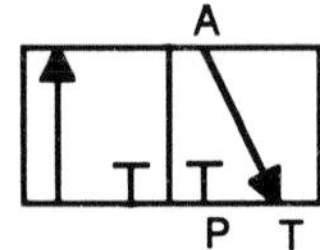

3/2 valve: no flow from P to A and flow from A to T switched to T being closed and flow from P to A

Figure 12.25 *Directional control valves*

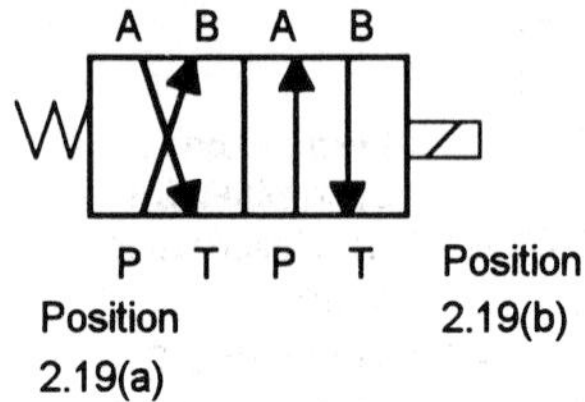

Figure 12.26 *The 4/2 valve*

With no current through the solenoid (Figure 12.23(a)) the hydraulic fluid of pressurised air is fed to the right of the piston and exhausted from the left, the result then being the movement of the piston to the left. When a current is passed through the solenoid, the spool valve switches the hydraulic fluid or pressurised air to the left of the piston and exhausted from the right. The piston then moves to the right (Figure 12.23(b)).

With the above valve there are the two control positions shown in Figure 12.23(a) and (b). Directional control valves are described by the number of ports they have and the number of control positions. The valve shown in Figure 12.24 has four ports, i.e. A, B, P and T, and two control positions. It is thus referred to as a 4/2 valve. The basic symbol used on drawings for valves is a square, with one square being used to describe each of the control positions. Thus the symbol for the valve in Figure 12.23 consists of two squares. Within each square the switching positions are then described by arrows to indicate a flow direction or a terminated line to indicate no flow path. Figure 12.24 shows this for the valve shown in Figure 12.23. Figure 12.25 shows some more examples of direction valves and their switching positions.

The actuation methods used with valves are added to the diagram symbol; Figure 12.26 shows the valve of Figure 12.24 with a spring to give one position and a solenoid to give the other.

Direction valves can be used to control the direction of motion of pistons in cylinders, the displacement of the pistons being used to implement the required actions. Figure 12.27 shows how a valve can be used to control the direction of motion of a piston in a single-acting cylinder, the term *single acting cylinder* is used for one which is powered by the pressurised fluid being applied to one side of the piston to give motion in one direction, it being returned in the other direction by possibly an internal spring.

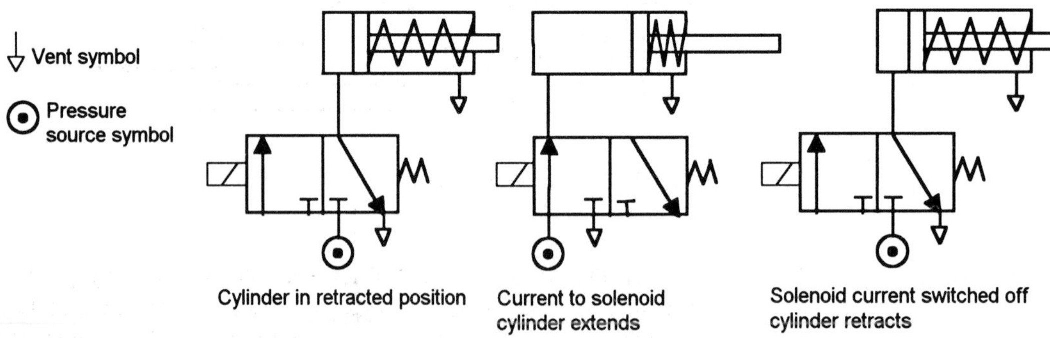

Figure 12.27 *Control of a single-acting cylinder*

Application

Figure 12.28 shows a possible system for an automatic door which opens when a person approaches it, remains open for say 5 s, and then closes. Heat sensitive semiconductor elements can be used to give voltage signals when infrared radiation from body heat falls on them and so detects the presence of a person. There will also be inputs to the controller probably from limit switches to indicate when the door is fully open and a timer to keep the door open for the required time. The output from the controller can be to solenoid operated pneumatic valves which use the movement of pistons in a cylinder to open and close the door. When a person approaches from outside, there is an output to the solenoid to open the door inwards, the air pressure is applied to the unvented side of the piston and causes it to move. When this solenoid is no longer energised, the spring returns the piston back by connecting the unvented side to a vent to the atmosphere. A similar arrangement is used for opening the door outwards.

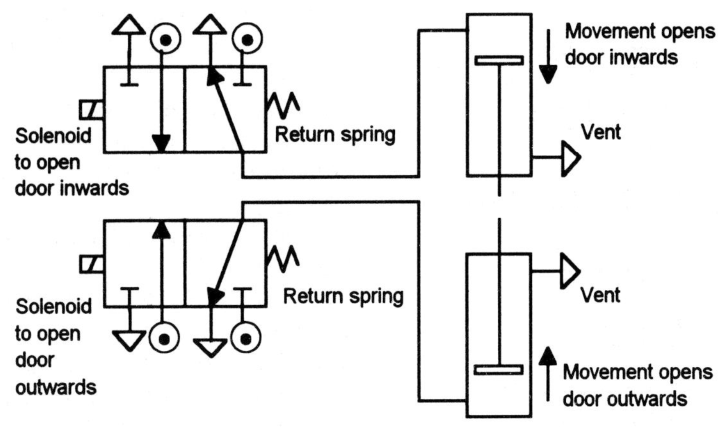

Figure 12.28 *Pneumatic valve door opening system*

Figure 12.29 shows a system that might be used for an automatic drill. Valves are used to actuate the movement of pistons in cylinders with limit switches to provide information about the location of the drill and the workpiece.

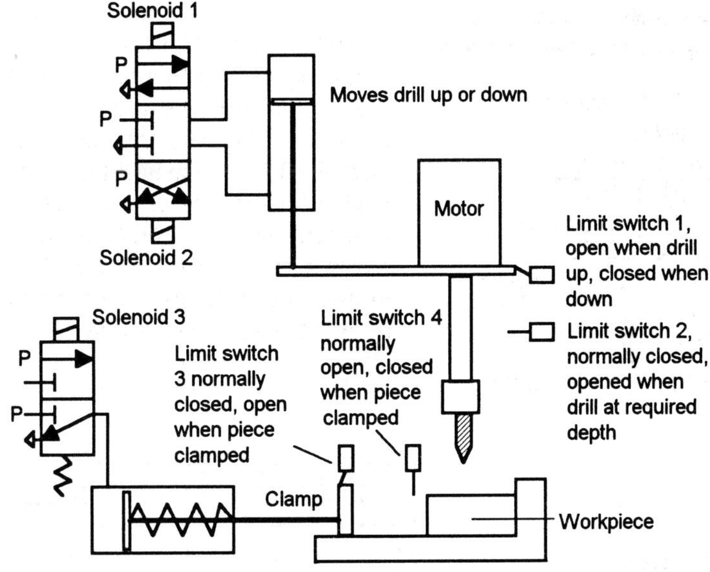

Figure 12.29 *Automatic drill*

12.4 Speed control of motors

Consider the open-loop control of d.c. motor speed by *pulse-width modulation* (PWM). This technique involves the switching on and off of a d.c. voltage to control its average value (Figure 12.30). The greater the fraction of a cycle that the d.c. voltage is switched on the closer its average value is to the input voltage.

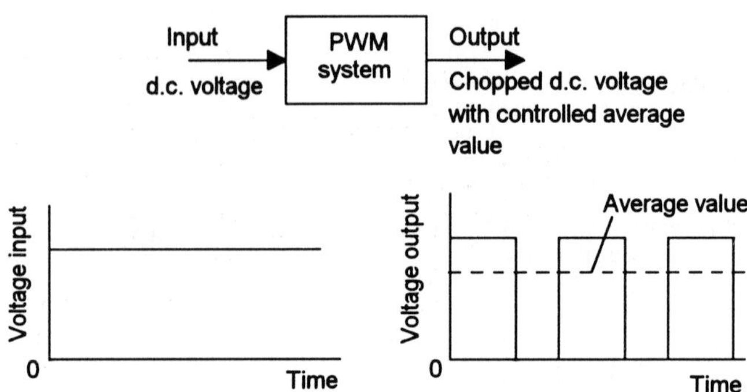

Figure 12.30 *Pulse-width modulation*

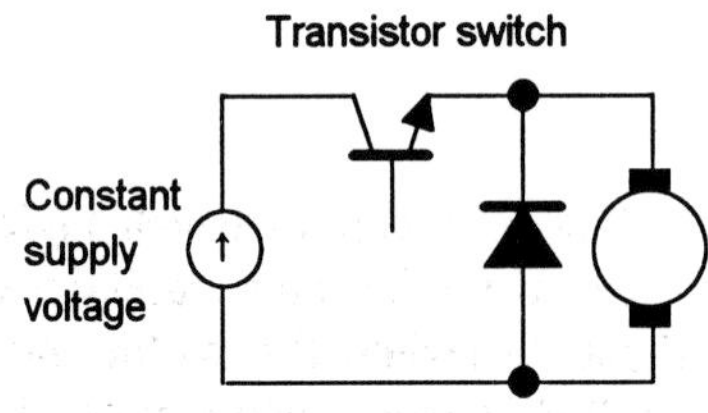

Figure 12.31 *PWM control circuit*

Figure 12.31 shows how pulse width modulation can be achieved by means of a basic transistor circuit. The transistor is switched on and off by means of a signal applied to its base, e.g. the signal from a microprocessor as a sequence of pulses. By varying the time for which the transistor is switched on so the average voltage applied to the motor can be varied and its speed controlled. Because the motor when rotating acts as a generator, the diode is used to provide a path for current which arises when the transistor is off.

Such a basic circuit can only drive the motor in one direction. A circuit (Figure 12.32) involving four transistors, in what is termed an H-circuit, can be used to control both the direction of rotation of the motor and its speed. The motor direction is controlled by which input receives the PWM voltage. In the forward speed motor mode, transistors 1 and 4 are on and current flow is then from left-to-right through the motor. Thus input B is kept low and the PWM signal is applied to input A. For reverse speed, transistors 2 and 3 are on and the current flow is from right-to-left through the motor. Thus input A is kept low and the PWM signal is applied to input B.

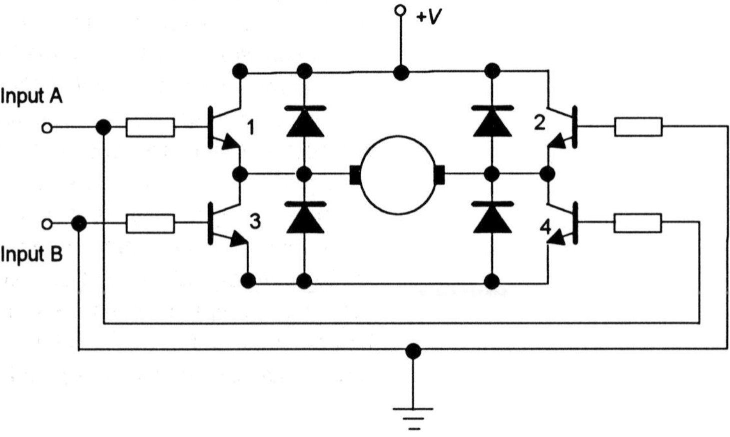

Figure 12.32 *H-circuit*

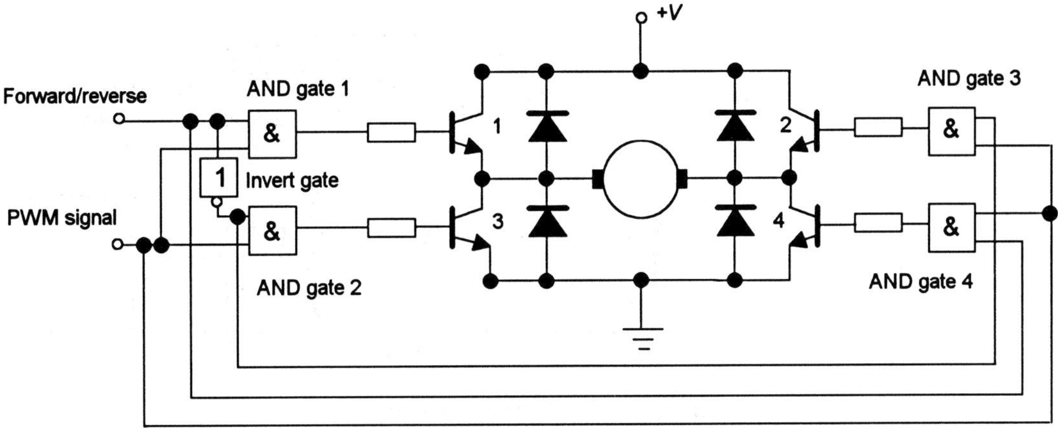

Figure 12.33 *Circuit for microprocessor control of a motor*

Figure 12.33 shows a better version of the H-circuit in which logic gates are used to control inputs A and B to achieve the above conditions with now one input supplied with a signal to switch the motor into forward or reverse and the other input the PWM signal. Such a circuit is better suited to microprocessor control for d.c. motors. A high input to the forward/reverse input means that when there is a high PWM signal the AND gate 1 puts transistor 1 on because the two inputs to it are high and so its output is high. The inverter means that AND gate 2 receives a low pulse when the forward/reverse input is high. As a result, transistor 3 is switched off. Because the AND gates 3 and 4 receive the same inputs, transistor 4 is on and transistor 2 is off. The situations are reversed when the signal to the forward/reverse input goes low.

The above methods of speed control using PWM have been open-loop systems with the speed being determined by the input to the system and no feedback to modify the input in view of changing load conditions. For a higher grade of speed control than is achieved by the open-loop system, feedback is required. This might be provided by coupling a tachogenerator to the drive shaft (see Figure 12.5). A tachogenerator gives a voltage proportional to the rotational speed. This voltage can be compared with the input voltage used to set the required speed and, after amplification, the error signal used to control the speed of the motor.

Figure 12.34 shows how such a closed-loop system might appear when a microprocessor is used as the controller. The analogue output from the tachogenerator is converted to a digital signal by an analogue-to-digital converter. The microprocessor is programmed to compare the digital feedback signal with the set value and give an output based on the error. This output can then be used to control a PWM circuit and so supply a d.c. signal to the motor to control its speed.

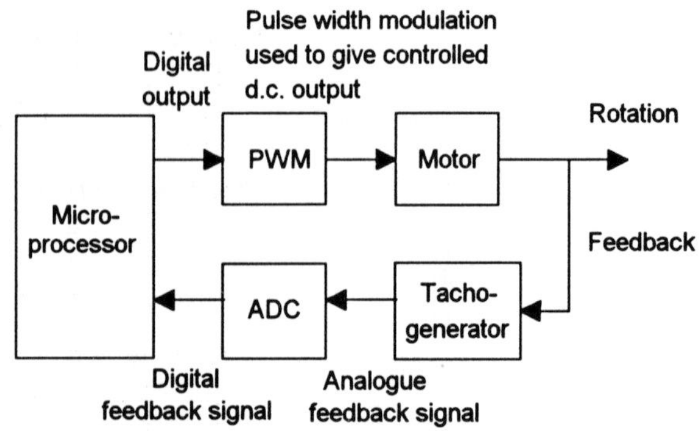

Figure 12.34 *Microprocessor controller with feedback*

12.4.1 Speed control of a.c. motors

The speed of rotation of an a.c. motor depends on the speed of a rotating magnetic field which is determined by the frequency of the a.c. supply to the motor, thus speed control can be achieved by controlling the frequency of this supply. One method of doing this involves first rectifying the a.c. to give d.c. by means of a converter. Then the d.c. voltage is inverted back to a.c. again but at a frequency that can be selected (Figure 12.35).

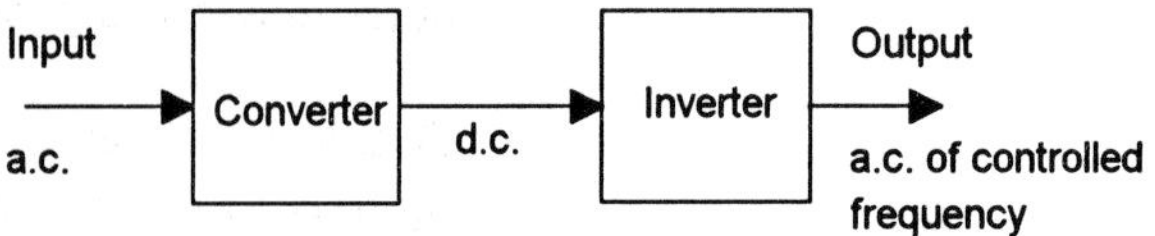

Figure 12.35 *Controlling the speed of an a.c. motor*

A common form of inverter is the H circuit described earlier for pulse width modulation. The supply to it is a d.c. voltage which is chopped to give an on–off voltage output. The output is thus a square wave voltage, the frequency of which can be varied by controlling the frequency with which the input pulses are applied to switch the transistors on or off. A near-sinusoidal voltage of variable frequency can, however, be produced by varying the duration of each voltage pulse in the way shown in Figure 12.36. This technique is termed *sinusoidal pulse-width modulation.*

Problems

1 Explain the difference between open-loop and closed-loop control.

2 Give an example of (a) continuous closed-loop control, (b) open-loop control, (c) sequential control.

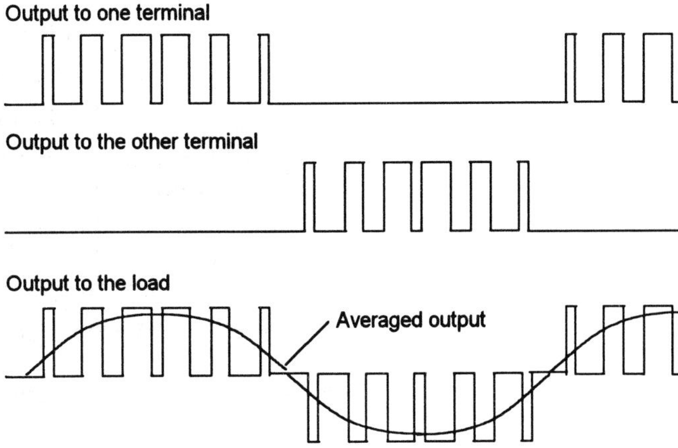

Figure 12.36 *Sinusoidal pulse-width modulation*

3 Explain how (a) relays, (b) thyristors and (c) transistors can be used as switches for actuators in control systems.

4 Figure 12.37 shows a block diagram of a domestic central heating system. Explain its operation.

5 Describe a circuit which uses transistor switching to control the speed and direction of rotation of a d.c. motor.

6 Describe a circuit which could be used as a light dimmer when a light source is connected to the mains power supply.

7 Identify the basic functional elements that might be used in the closed-loop control systems involved in:
(a) A temperature-controlled water bath.
(b) A speed-controlled electric motor.
(c) Rollers in a steel strip mill being used to maintain a constant thickness of strip steel.

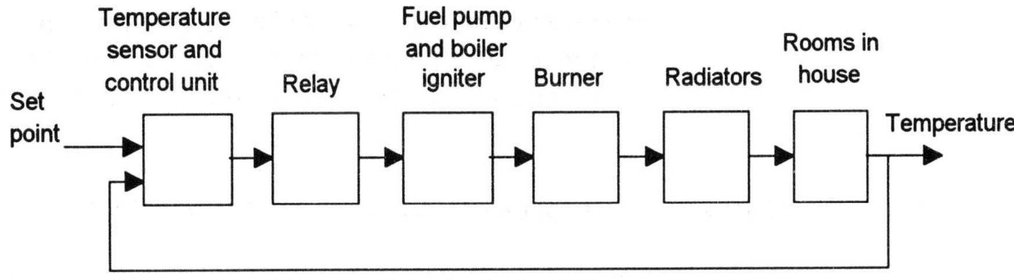

Figure 12.37 *Problem 4*

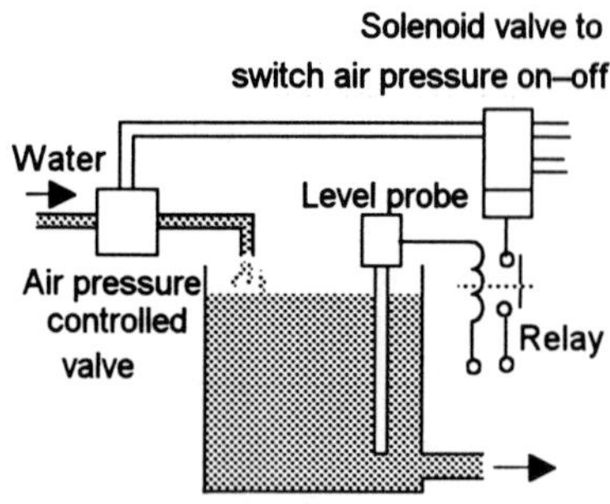

Figure 12.38 *Problem 8*

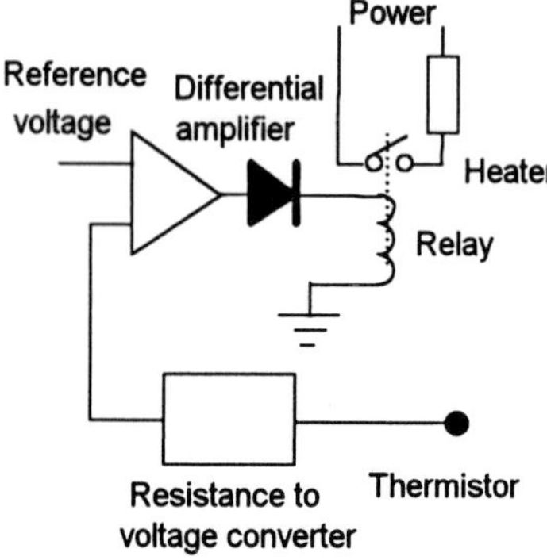

Figure 12.39 *Problem 8*

8 Figure 12.38 shows a temperature control system and Figure 12.39 a water level control system. Identify the basic functional elements of the systems.

9 Suggest sensors that could be used with control systems to give measures of (a) the temperature of a liquid, (b) whether a workpiece is on the work table, (c) the varying thickness of a sheet of metal, (d) the rotational speed of a motor shaft.

10 For each of the valve symbols in Figure 12.40, state the method and outcomes of actuation.

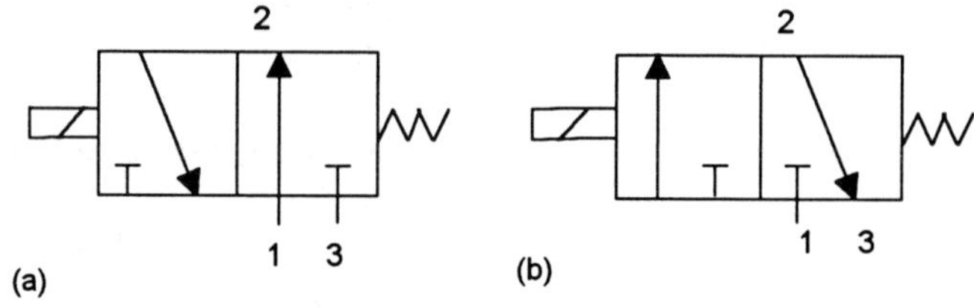

Figure 12.40 *Problem 10*

11 State the outcomes of the switching on and then off of currents through the solenoids with regard to valves shown in Figure 12.41.

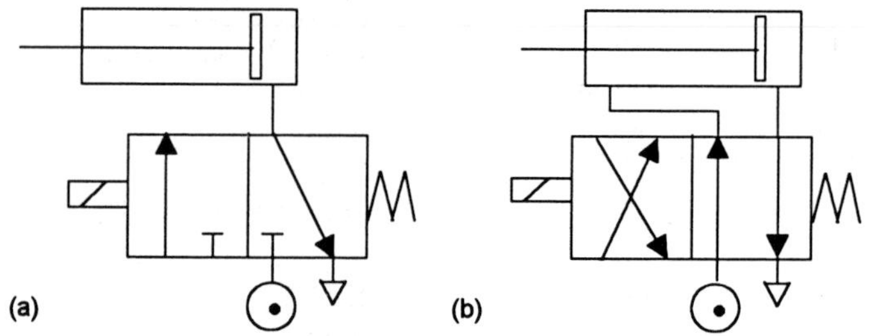

Figure 12.41 *Problem 11*

12 Draw a block diagram for a negative-feedback system that might be used to control the level of light in a room to a constant value.

13 Draw block diagrams which can be used to present the operation of a toaster when it is (a) an open-loop system, (b) a closed-loop system.

14 Explain how a ball valve is used to control the level of water in a cistern.

Answers

Chapter 1

Revision

1 31.8 MPa
2 –8.9 MPa
3 -5×10^{-4}
4 0.56 mm
5 0.30 mm
6 3.6 MPa, 51.7 MPa
7 61.1 MPa, 27.5 MPa, 0.153 mm
8 283.5 kN
9 36.0 MPa, 75.8 MPa
10 68.4 kN
11 6.93 mm
12 0.012 mm, 0.004 mm

Problems

1 101.9 MPa, 500×10^{-6}
2 163 MPa, 12.2 mm
3 32.4 MPa, 0.24 mm
4 3.0 mm
5 20 mm
6 0.20 mm
7 64.6 MPa, 38.8 MPa
8 93.7 MPa, 6.6 MPa
9 234 kN
10 76 mm
11 1.80 mm, -5.41×10^{-3} mm

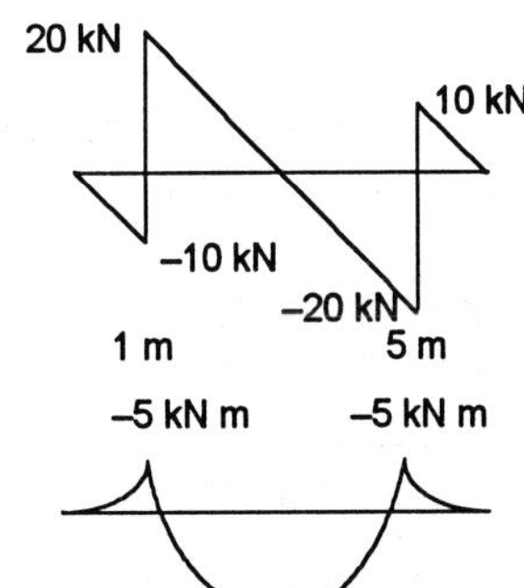

Figure A.1 *Revision problem 7, Chapter 2*

Chapter 2

Revision

1 Vertically 3 kN and 3 kN, horizontally 0
2 Vertically 4.1 kN and 3.9 kN, horizontally 0
3 Vertically 40 kN and 40 kN, horizontally 0
4 (a) –250 N, + 250 N m, (b) –250 N, + 375 N m
5 (a) +8 kN, –12 kN m, (b) +8 kN, –8 kN m
6 +20 kN, – 20 kN m
7 See Figure A.1, 15 kN m at midpoint
8 See Figure A.2, 27.5 kN m at midpoint
9 See Figure A.3, 24 kN m at midpoint
10 See Figure A.4, 54 kN/m at midpoint
11 0.96 mm

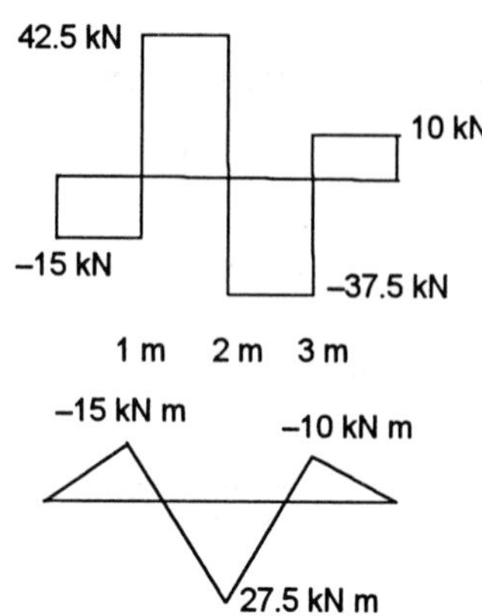

Figure A.2 *Revision problem 8, Chapter 2*

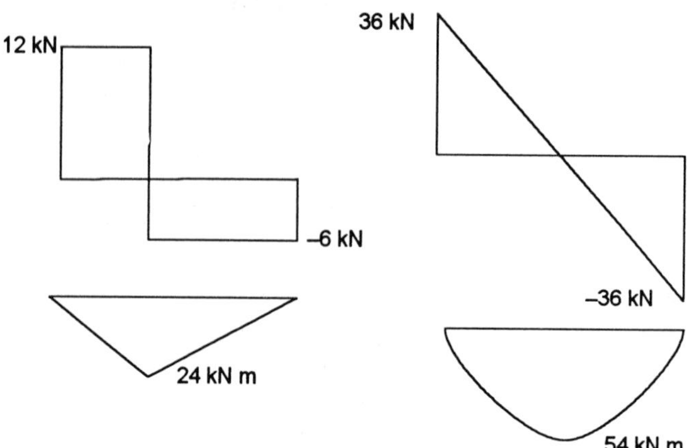

Figure A.3 *Revision problem 9, Chapter 2*

Figure A.4 *Revision problem 10, Chapter 2*

12 97 mm from base
13 317.7×10^6 mm^4
14 339 kN m
15 720 N m
16 4.8 kN m
17 2.95×10^{-4} m^3
18 80
19 77, about 187 MPa
20 (a) 4 MPa compressive, (b) 0.8 MPa to 7.2 MPa compressive
21 40 mm

Problems

1 (a) –10 kN, + 5kN m, (b) –10 kN, +10 kN m
2 (a) +15 kN, +8.75 kN m, (b) +10 kN, +17.5 kN m
3 As in Figure 2.14, maximum bending moment 60 kN m at midpoint
4 See Figure A.5, 80 kN m at 2 m
5 20 kN, –20 kN m
6 See Figure A.6, maximum shear force 60 kN at fixed end, maximum bending moment 130 kN m at fixed end
7 See Figure A.7, maximum shear force 78 kN, 8 m from supported end, maximum bending moment –112 kN m, 8 m from supported end
8 420 MPa
9 89.5 mm from base
10 4.17×10^6 mm^4
11 76.3 mm from base, 2.03×10^6 mm^4
12 15 kN m
13 ±130 MPa
14 3.57×10^{-3} m^3
15 (a) 55 mm from bottom, 5.89×10^6 mm^4, 1.07×10^5 mm^3, (b) 30 mm from bottom, 5.54×10^5 mm^4, 1.85×10^4 mm^3, (c) 15 mm

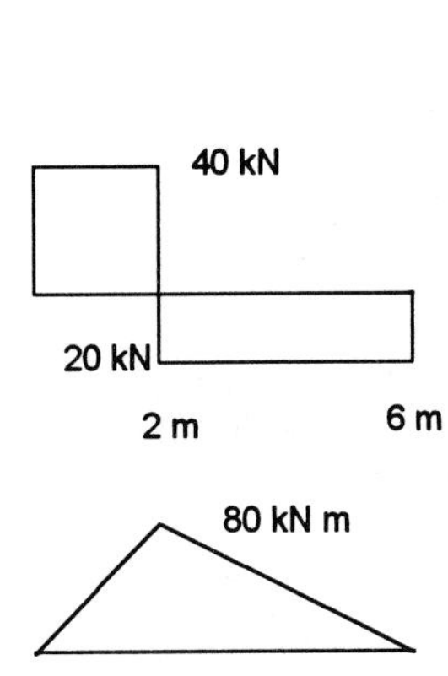

Figure A.5 *Problem 4, Chapter 2*

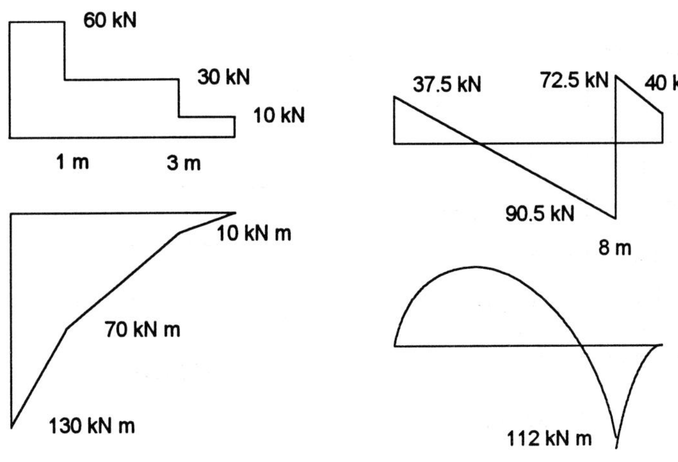

Figure A.6 *Problem 6, Chapter 2*

Figure A.7 *Problem 7, Chapter 2*

from bottom, 2.08×10^5 mm^4, 5.95×10^3 mm^3, (d) 22.5 mm from bottom, 8.61×10^5 mm^4, 2.30×10^4 mm^3

16 26 MPa
17 (a) 55.6, (b) 79.4
18 11.6 m
19 (a) 74.4, (b) about 187 MPa
20 Eg. $203 \times 203 \times 71$ kg/m
21 (a) 10.5 MPa uniform across section, (b) 11.6 MPa tension on one face to 35.6 MPa compressive on the opposite
22 40 mm
23 1.1 MPa compressive on one face to 6.9 MPa compressive on the opposite face, 16.7 mm
24 5.3 MPa tension on one face to 8.3 MPa compressive on the opposite face
25 2.2 MPa tension on one face, 45.8 MPa compressive on the opposite face

Chapter 3

Revision

1 14.6 mm
2 12.6 mm
3 141 kN
4 0.01 mm
5 2.5×10^{-4}
6 1.7 kN m
7 1.8 kN m
8 0.127 m
9 150 mm
10 33.3 MPa
11 15.8 MPa
12 501 kW

Problems

1 2.5 kN
2 0.006 mm
3 22 kN
4 19.9 MPa
5 49.6 mm
6 1.26 kN m
7 (a) 65.2 MPa, (b) 149.4 GPa, (c) 4.36×10^{-4}
8 (a) Tubular shaft stress is 1.15 that in solid, (b) tubular shaft angle is 1.15 that in solid, (c) tubular shaft mass is 0.64 that of solid
9 53.1 MPa
10 Tubular shaft torque is 1.44 that of solid shaft
11 33.3 MPa, 46.9 MPa
12 165 MPa
13 10 kN m
14 17.7 Hz
15 4.4 mm
16 (a) 45.0 MPa, (b) 4.3°
17 257.8 N m, 81.0 kW
18 290 kW

Chapter 4

Revision

1 128 m
2 40 m
3 7.5 s, 112.5 m
4 (a) -2 m/s^2, (b) 2 m
5 2.0 s
6 11.5 m
7 27.6 m
8 23.5 km/h 70° east of north
9 1.37 m/s east, 3.76 m/s north
10 1.27 m, 8.83 m
11 31.0°, 59.0°
12 13.2 m, 3.28 s, 62.7 m
13 8.64 rad/s^2, 25
14 0.21 rad/s^2, 5
15 21.2
16 72.7 rad/s
17 20 rad/s^2
18 21.2 rad/s, 2.67 rad/s^2
19 400 mm
20 150 N
21 9860 N
22 5.75×10^{-3} m/s
23 23.5 N, 1.96 m/s^2
24 5.12 m/s^2
25 0.58
26 2.19 m/s^2
27 16.2 N

28 (a) 520 N, (b) 980 N
29 0.15
30 (a) 62 kg, (b) 6.0 kg
31 0.16 kg m^2
32 4.29 kg m^2
33 1.196 kg m^2
34 11.8 kg m^2
35 $(128mL^4/3) - \pi mr^2(r^2/4 + L^2)$
36 0.375 N m
37 102 N
38 3.27 m/s^2
39 89 rad/s

Problems
1 15 m/s
2 16 m
3 5 s
4 (a) 16 m, (b) 7 m
5 16 m, 4 s
6 40 m/s
7 4 ms, 500 m/s^2
8 61.2 s
9 50 s
10 2 m/s^2
11 (a) 26 m/s, (b) 8 m/s^2
12 22.6 m
13 (a) 9.1 m, (b) 8.4 m
14 25 m/s 16° east of north
15 5 m/s at 36.9° to the horizon
16 1.4 m/s north, 1.4 m/s east
17 40.8 m
18 16.3 km
19 20.7 m/s
20 17.1 m
21 28.1 km
22 0.84 rad/s^2, 19.2 revs
23 0.31 rad/s^2, 7.5 revs
24 1.26 rad/s^2
25 1.05 m/s
26 (a) 0.57 rad/s^2, (b) 0.10 m/s^2
27 8.3 rad/s^2
28 300 mm
29 7.5 rad/s^2
30 100 rev/min
31 2.39 m/s^2, 24.4 N
32 3.68 m/s^2, 18.4 N
33 27.2 kg
34 As given in the problem
35 1.67 m/s^2, 1300 N

36 $\sqrt{(2d\{M+m\}/Mg)}$
37 0.052
38 0.41
39 0.24, 22.9 N
40 As given in the problem
41 0.02 kg m^2
42 1.35 kg m^2
43 0.5 m
44 22.8 kg m^2
45 45.9 kg m^2
46 14.8 rad/s
47 24.5 kg m^2
48 20 rad/s^2
49 135.7 N m
50 0.029 N m, 0.73 kg m^2
51 4.7 s
52 4.2 N m
53 2560 kg m^2, 1690 N m

Chapter 5

Revision
1 (a) 300 J, (b) 260 J
2 52.2 J
3 1150 J
4 1.6 J
5 754 J
6 20.9 J
7 188 J
8 15 MW
9 50 W
10 88 W
11 18.8 kW
12 98 J
13 0.2 J
14 1.6 J
15 45 kN
16 45 J
17 444 kJ
18 0.8 J
19 157 J
20 14 m/s
21 20 kW
22 14.0 rad/s
23 3.1 rad/s
24 137 N
25 (a) 8 rev/s, (b) 1500 N m
26 1056 W

Problems

1 32 J
2 735 J
3 1970 J
4 6370 J
5 20 J
6 45 J
7 754 J
8 377 J
9 1.1 kJ
10 1500 kW
11 24 kW
12 1286 N
13 (a) 225 kW, (b) 409 kW
14 (a) 1 J, (b) 1.44 J
15 1.2 J
16 600 J
17 3 J
18 294 J
19 9.4 m/s
20 2.5 m
21 20 m
22 1.8 m
23 318 W
24 7.9 kW
25 1740 s
26 16.1 m/s
27 58.8 kN
28 2.8 m/s
29 8.1 m/s
30 4.9 rad/s
31 31.8 N m, 159.2 N m
32 189 N m
33 83.3%
34 18.8 m/s
35 18.9 kW
36 (a) 800 N m, (b) 25.1 kW

Chapter 6

Revision

1 0, 10.1 m/s, 2.53×10^3 m/s^2, 0
2 5 m/s
3 26.6 N
4 50 mm/s, 100 m/s^2
5 2.22 J
6 2.23 Hz
7 2.25 Hz
8 $\frac{1}{2\pi}\sqrt{\frac{9}{\left(\frac{4}{k_1}+\frac{1}{k_2}\right)M}}$

9 11.3 Hz
10 1.23 Hz
11 Period increases to 1.001 s
12 0.59 Hz

Problems

1 3.77 m/s
2 2.25 Hz
3 3.0 m/s, 56.8 m/s^2
4 4.44 s
5 3.1 m/s, 2.7 m/s
6 3.87 J
7 790 N/m
8 1.125 Hz
9 2.25 Hz
10 27.6 Hz
11 1.29 Hz
12 Period increases to 1.002 s
13 1.59 Hz
14 3.53 Hz
15 As given in the problem
16 As given in the problem
17 0.040 m
18 $\frac{1}{2\pi}\sqrt{\frac{2k}{m}}$
19 (a) 2.5 s, (b) 12.1 s
20 Reduced by factor of $1/\sqrt{2}$
21 24 mm, 19 mm, 16 mm

Chapter 7

Revision

1 (a) 20 V, (b) 2 Ω
2 35 mA
3 (a) 12.5 mA, (b) 1.25 V, 2.5 V, 6.25 V
4 750 Ω
5 (a) 1.2 A, (b) 0.4 A
6 28.6 Ω
7 12 V
8 3 mA
9 (a) 8 V, (b) 20 V
10 0.505 Ω
11 (a) 9 V, (b) 20 V, (c) 8 V
12 (a) 1.144 A, 0.429 A, 0.714 A, (b) 182 mA, 273 mA, 91 mA
13 As problem 12
14 (a) 10 μF, (b) 110 μF
15 500 pC, 100 pC, 200 pC
16 8 s
17 (a) 0.2 mA, (b) 0.121 mA, (c) 0.0736 mA
18 (a) 1 s, (b) 5 V, 5 V
19 (a) 0.3 ms, (b) 0.446 mA, (c) 11.08 V, 8.92 V

20 1000 V/s
21 200 s
22 (a) 0.2 s, (b) 1 mA, (c) 0.61 mA
23 0.8 μA, 8.2 V
24 (a) 6.32 V, (b) 3.83 V
25 20 V
26 50 A/s
27 (a) 0.5 ms, (b) 480 A/s, (c) 10.8 V, (d) 13.2 V, (e) 0.24 A
28 (a) 12 V, (b) 0.6 V
29 0.149 A
30 (a) 100 mA, (b) 77.9 mA
31 2 A, 100 A/s

Problems

1 50 mA
2 11.6 V
3 >1.6 W
4 0.20 A, 4.0 V, 6.0 V, 10.0 V
5 750 Ω
6 500 Ω and 1900 Ω in series
7 (a) 0.41 W, 1.02 W, 1.43 W, (b) 5 W, 2 W, 7 W
8 (a) 30 Ω, (b) 2.73 Ω
9 (a) 6.32 kΩ, (b) 0.0911 W
10 6 Ω
11 (a) 0.25 A, (b) 2 V, 3 V
12 (a) 1 A, 0.33 A, (b) 5.33 W
13 6.9 mA
14 2.03 V, 9.97 V
15 0.57, 0.29, 0.14
16 (a) 22 Ω, 0.55 A, (b) 6 Ω, 2 A, (c) 12.8 Ω, 0.94 A
17 (a) 30 Ω, 0.5 A, (b) 10.8 Ω, 1.11 A, (c) 24 Ω, 0.5 A, (d) 20 Ω, 0.5 A, (e) 4.25 Ω, 2.8 A, (f) 1.2 Ω, 10 A
18 (a) 1 A, 5 V; 0.5 A, 5 V; 0.5 A, 5 V, (b) 1 mA, 8 V; 0.67 mA, 4 V; 0.33 mA, 1.3 V; 0.33 A, 2.7 V, (c) 0.2 A, 12 V; 0.2 A, 12 V; 0.1 A, 10 V; 0.1 A, 10 V; 0.2 A, 10 V
19 2.5 V
20 (a) 0.22 A, (b) –0.09 A, (c) 0.22 A
21 4 V
22 (a) +16 μC, –16 μC, (b) 8 V, 4 V
23 (a) 1.1 μF, (b) 14 μF
24 (a) 6 μF, (b) +10 μC, –10 μC, +20 μC, –20 μC, +30 μC, –30 μC, (c) 10 V
25 (a) 10 μF, (b) 360 μC, (c) 18 V, 12 V, 6 V
26 (a) 370 pF, (b) 3700 pC, (c) 10 V
27 20 000 V/s
28 30 V/s
29 (a) 0.4 mA, (b) 0.147 mA
30 (a) 0.3 ms, (b) 0.736 mA, (c) 0.446 mA, (d) 0
31 (a) 30 V, (b) 11.0 V, (c) 1.07 V

32 (a) 0.25 s, (b) 6.3 V, (c) 8.6 V
33 (a) 1.29 s, (b) 1.07 mA
34 (a) 5 μA, (b) 4.6 s
35 (a) 0.767 V, (b) –0.767 V
36 2 H
37 0.4 V
38 (a) 0.05 s, (b) 40 A/s, (c) 1.73 A, (d) 2 A
39 (a) 0.1 s, (b) 40 A/s, (c) 3.1 A, (d) 4 A
40 73 mA
41 (a) 5 A, (b) 0.092 A
42 4.6 ms
43 (a) 80 V, (b) 2.75 A/s
44 (a) 0.95 A, (b) 0.69 s
45 0.1 H
46 20 H

Chapter 8 *Revision*

1 (a) 2.55 V, 2.83 V, (b) 0.76 A, 0.85 A
2 5.4 V at 21.8° leading 5 V phasor
3 3.6 A at 56.4° leading 2 A phasor
4 10.8 V at 21.8° lagging from sum phasor
5 2500 Ω, 0.0004 S
6 19.9 Ω
7 1 kΩ, 16 sin 2000t mA
8 32 Ω, 3.2 sin 400t V
9 500 Ω, 53.1°
10 134 mA, –26.6°
11 3805 mA, –29.9°
12 2.88 mA, 87.3°
13 103.1 Ω, –14.0°
14 5033 Hz, 0.25 A
15 91.9 Hz, 2.5 A
16 205 Hz, 129 V at –90°
17 21.1 Ω
18 0.400 A, 72.6° lagging, 30 Ω
19 14.2 mA, 32.1° leading, 845 Ω
20 4490 Hz, 2000 Ω, 5 mA
21 59.3 Hz
22 495 Hz
23 316
24 576 Ω
25 1.14 W
26 0.124 W
27 82.99 W, 154.5 V A
28 0.6 lagging, 4.61 kV A, 3.69 V Ar, 2.77 kW
29 244 kV Ar
30 35.4 μF

Problems

1 (a) $v = 10 \sin 2\pi 50t$ V, (b) (i) 5.88 V, (ii) 9.51 V, (iii) –5.88 V
2 (a) $i = 50 \sin 2\pi \times 2 \times 10^3 t$ mA, (b) (i) –47.6 mA, (ii) –29.4 mA, (iii) 47.6 mA
3 (a) 0, (b) 10 V, (c) 1.99 V
4 (a) 1592 Ω, (b) 2 Ω, (c) 4167 Ω
5 (a) 2513 Ω, (b) 4000 Ω, (c) 10 Ω
6 0.4 sin 1000t mA
7 26.5 kHz
8 79.6 Hz
9 21.7 Ω, 11.1 A, 46.3° current lagging voltage
10 17 Ω, 28.1° voltage leading current
11 (a) 75 Ω, (b) 1.33 A, (c) 59.9 V, 79.8 V
12 (a) 24.1 Ω, (b) 4.15 A at 51.4° lagging voltage
13 148 μF
14 (a) 75.2 Ω, (b) 3.19 A, 57.9° leading voltage
15 (a) 48.0 Ω, (b) 0.5 A, 33.5° leading voltage
16 (a) 70.7 V, 45° current lagging voltage, (b) 50 V, 100 V, 50 V
17 (a) 0.86 A, 49.9° lagging voltage, (b) 174 V, (c) 86 V
18 1125 Hz, 1.2 A
19 356 Hz
20 15 V at 0°, 300 V at 90°, 300 V at –90°
21 (a) 8.4 μF, (b) 9.4
22 (a) 118.6 Hz, (b) 11.2
23 28.4 A, 32.5° lagging the voltage
24 5 A, 53.1° lagging the voltage
25 200 Ω, 5.3 μF
26 (a) 1.73 A, 1.0 A, (b) 57.7 Ω, 100 Ω
27 7118 Hz
28 2251 Hz, 100 kΩ, 1 mA, 141
29 57.8 Hz, 200 Ω, 0.5 A, 4.6
30 105
31 39.8 mV
32 0.099 leading, 22.8 V A, 22.69 V A r, 2.26 W
33 (a) 38.7 mA, 83.1° lagging, (b) 0.12, (c) 1.11 W, (d) 9.28 V A, (e) 9.21 V A r
34 (a) 0.53, (b) 8.15 W, (c) 15.3 V A, (d) 10.8 V Ar
35 133.3 kV A, 106.7 kV Ar
36 112.5 kV Ar
37 24.9 μF

Chapter 9

Revision

1 Second and fourth (and a d.c. term)
2 (a) See Figure A.8, (b) sum graph of Figure A8 lifted up by 5
3 0.32 + 0.5 cos 100t + 0.21 cos 200t mA
4 0.5 cos(100t – 90°) + 0.21 cos(200t – 90°) A
5 3.2 sin(100t + 90°) + 3.2 sin(200t + 90°) mA
6 15%
7 (a) 72 mV, (b) 2.1%

Problems
1 See Figure A.9
2 See Figure A.10
3 0.2 sin 500t + 0.1 sin 1000t A
4 0.1 sin(100t + 90°) + 0.84 sin200t
5 3.2 sin(100t + 90° + 3.2 sin(200t + 90° mA
6 43%
7 10%

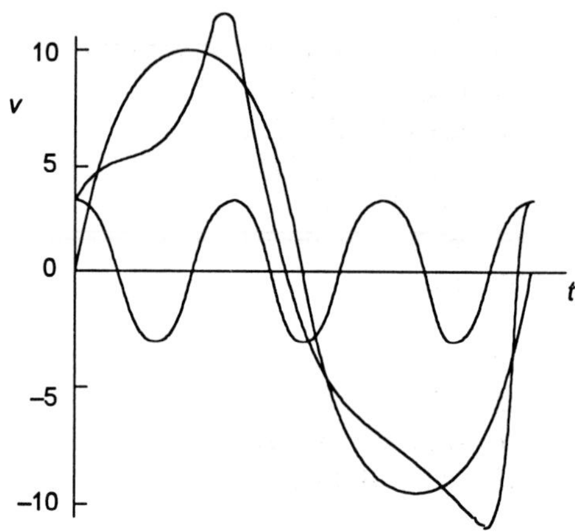

Figure A.8 *Revision problem 2, Chapter 9*

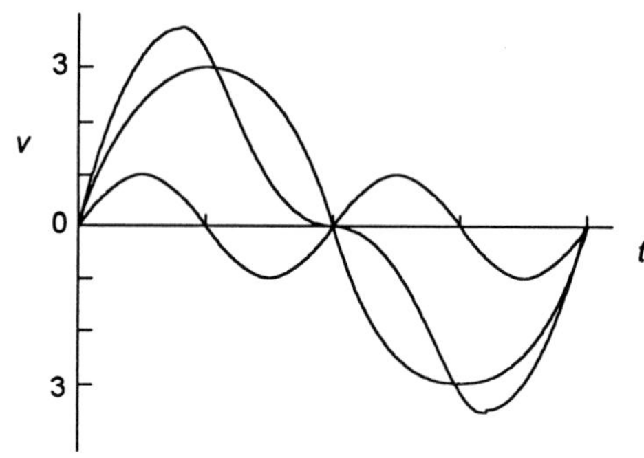

Figure A.9 *Problem 1, Chapter 9*

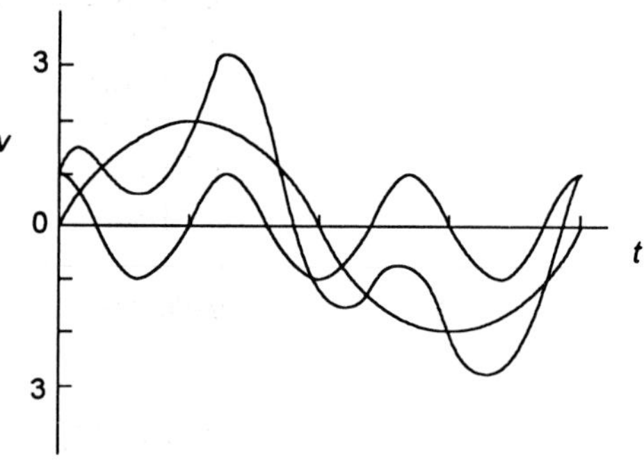

Figure A.*10* *Problem 2, Chapter 9*

Chapter 10

Revision
1 60 V
2 100
3 20, 12.5 A
4 (a) 20 V, (b) 12 A
5 (a) 40, (b) 400 A, (c) 10 A
6 25 A, 62.5 A
7 95.9%

8 (a) 97.8%, (b) 97.4%
9 2.5 W
10 10 Ω, –5 Ω

Problems
1 840 V
2 (a) 1.33, (b) 2.25 A
3 (a) 75 V, (b) 5 A, (c) 1.5 A, (d) 375 W
4 (a) 1.7 A, 20.8 A, (b) 240 V
5 1920 V
6 (a) 25 V, (b) 600 ampere-turns
7 0.42 A, 8.33 A
8 (a) 91%, (b) 90%
9 (a) 98%, (b) 97%
10 287 W, 530 W
11 97%
12 1/4.47
13 1/4.0
14 11
15 250 Ω
16 (a) 0.375 W, (b) 0.5 W, 1/3
17 25 Ω, 10 mW

Chapter 11

Revision
1 9
2 33.3, 221 762
3 7.94
4 30 dB
5 17 dB
6 16.4
7 10^{-4} W

Problems
1 1.22 mV
2 8
3 0.625 V
4 18 mV
5 19.95
6 15 dB
7 19 dB
8 500
9 0.008
10 41.2 dB
11 Using Figure 11.29 with the 4.7 kΩ thermistor, the variable resistor might be 0 to 10 kΩ to enable the sensitivity of the arrangement to be altered. However, if the variable resistor was set to zero resistance then, without a protective resistor, we could possibly have a large current passed through the thermistor. The protective resistor is to

prevent this occurring. With a 6 V supply, the variable resistor set to zero resistance, the protective resistance of R, and the thermistor at 100°C, the current I through the thermistor is given by $V = IR$ as $6 = I(0 + R + 318)$ and so $I = 6(R + 318)$. The maximum power that the thermistor can withstand is specified as 250 mW. The power dissipated by the thermistor is $I^2 \times 318$ and so if we want this to be significantly below the maximum possible, say 100 mW, in order to be able to use at 100°C, then we have: $0.100 = 6^2 \times 318/(R + 318)^2$. Hence R needs to be about 20 Ω. When the temperature of the thermistor is 0°C its resistance is 15.28 kΩ. If we set the variable resistor as, say, 5 kΩ and the protective resistor as 20 Ω then the voltage output when the supply is 6 V is $5.020 \times 6/(15.28 + 5.020) = 1.48$ V. When the temperature rises to 100°C the output voltage becomes $5.020 \times 6/(0.33 + 5.020) = 5.63$ V. Thus, over the required temperature range, the voltage output varies from 1.48 V to 5.63 V.

12 Figure A.11 shows a possible arrangement.

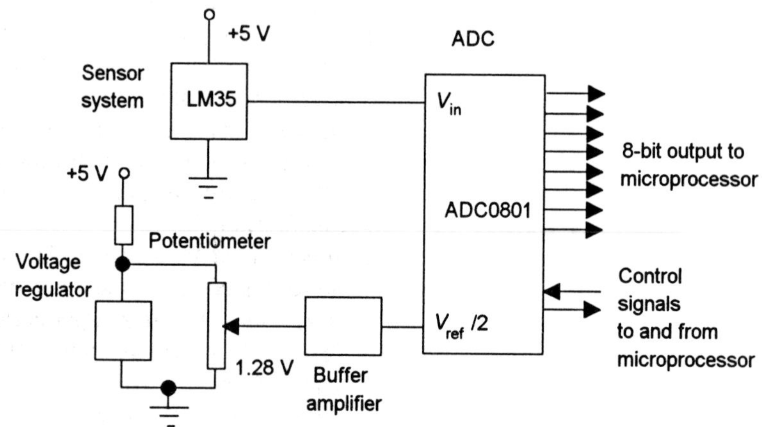

Figure A.11 *Chapter 11, problem 12*

The output from the LM35, which is analogue, is applied to an analogue-to-digital converter (ADC) to give a digital output. An 8-bit analogue-to-digital converter (ADC) is required since changes corresponding to a total range of 100 bits is required and we required $2^n = 100$ and so a 6.6 bit ADC; the nearest ADC is 8-bit. Since 1°C generates 10 mV, we need the resolution of the ADC to be 10 mV so that each step of 10 mV will generate a change in output of 1 bit. Suppose we use an 8-bit successive approximations ADC, e.g. ADC0801. This type of ADC requires an input of a reference voltage which when subdivided into $2^8 = 256$ bits gives 10 mV per bit. Thus a reference voltage of 2.56 V is required. For this to be obtained the reference voltage input to the ADC0801 has to be $V_{ref}/2$ and so an accurate input voltage of 1.28 V is required. Such a voltage can be obtained by using a potentiometer circuit across a 5 V supply with a buffer amplifier (a voltage follower) to avoid loading problems (the buffer amplifier is a non-inverting operational amplifier where all the output voltage is fed back with no attenuation to an inverting input, the result being a total voltage gain of 1). The buffer amplifier has a very high input impedance and very low output impedance and is used for interfacing high impedance sources and low impedance loads. Because the voltage has to remain steady at 1.28 V, even if the 5 V supply voltage fluctuates, a voltage regulator is likely to be used, e.g. a 2.45 V voltage regulator ZN458/B. For an output to a microprocessor, control signals to and from the microprocessor are used to ensure that the ADC supplies signals when the microprocessor is ready.

Chapter 12

Problems

1 Open-loop has no feedback and so output determined solely by set point setting. Closed-loop control has feedback and so can react to changes on the output variable, e.g. loading.

2 For example you might have: (a) thermostatically controlled heating system, (b) an electric fire which is purely controlled by the set point determined by how many bars are switched on, (c) a washing machine where one event such as water at temperature is followed by the washing motor being switched on, this in turn being followed after a set time by another operation.

3 See Section 12.3

4 It is a closed-loop system. The control unit compares the set and actual temperature values and gives a signal to operate a relay. This is able to switch on a larger current and hence the fuel pump and igniter. This gives an input to the boiler of heat and an output of hot water. This is then the input to radiators which give an output of warmed air to the process which are the rooms in the house. The result is a temperature output. The temperature signal is fed back to the control unit.

5 See Figure 12.30

6 See Figure 12.17

7 (a) Measurement – temperature sensor, controller – thermostat, correction – heater, process – water bath, (b) measurement – rotary speed sensor, controller – motor, process – shaft, (c) measurement – sensor of thickness, e.g. LVDT, controller – differential amplifier, correction – rollers, process – steel strip

8 Measurement – thermistor with resistance-to-voltage converter, comparison – differential amplifier, correction - relay and heater, process – the enclosure being controlled; measurement – level probe, controller – relay and solenoid valve, correction – flow control valve, process – water tank, measurement – level probe

9 Eg. (a) resistance temperature detector, (b) limit switch, (c) LVDT, (d) tachogenerator

10 (a) Solenoid activated, spring return: initially output 2 is at pressure, when solenoid activated output 2 is vented, (b) solenoid activated,

spring return: initially output 2 is vented, when solenoid activated output 2 is pressurised

11 (a) Solenoid activation causes piston to move to left, release gives return, (b) solenoid activation causes piston to move to left, release gives movement to right

12 See Figure A.12

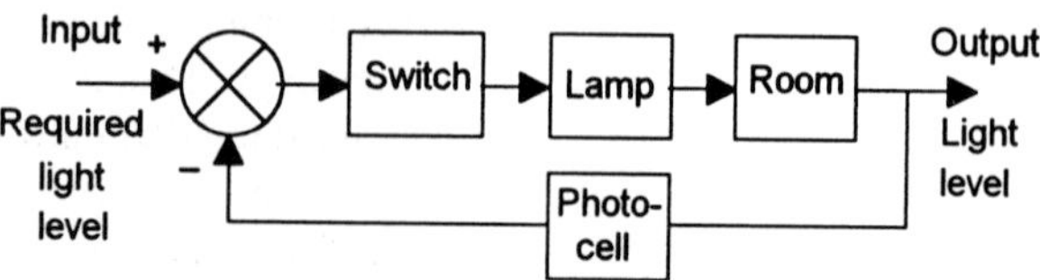

Figure A.12 *Chapter 12, problem 12*

13 (a) See Figure A.13(a), (b) see Figure A.13(b)

14 Ball operates a lever which opens or shuts a valve, depending on the height to which the ball is floating. The comparison element is the position of the lever, the control element is the lever, the correction element the valve, the process the water in the cistern and the measurement is the floating ball.

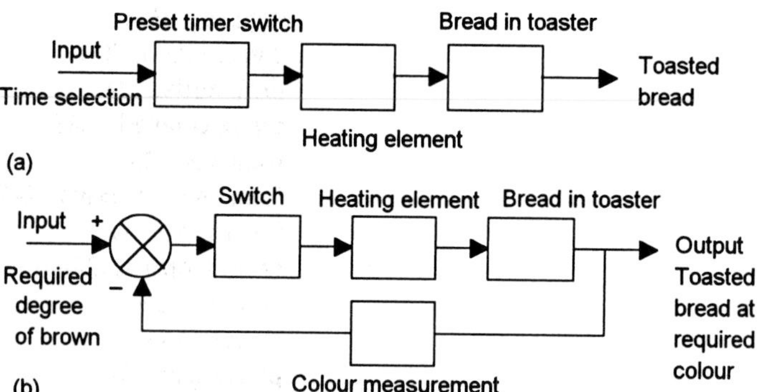

Figure A.13 *Chapter 12, problem 13*

Index

Acceleration,
- average, 58
- constant acceleration motion, 59
- defined, 58
- gravitational, 61
- instantaneous, 58
- uniform, 58

Accuracy, 251
Actuators, 265
Admittance, 184
Alternating current, defined, 177
Ampere-turns, 224
Amplifier,
- buffer, 244
- clipping, 241
- current gain, 241
- differential, 244
- direct coupled, 241
- feedback, 253
- frequency selective, 242
- inverting, 244
- non-inverting, 244
- op-amp, 254
- power, 244
- power gain, 241
- system, 235, 241
- voltage gain, 241

Amplitude modulation, 236
Analogue,
- signals, 236
- to digital conversion, 239

Angular,
- acceleration, 67
- displacement, 66
- velocity, 66

Armature, 167
Attenuation, 256
Average value, 176

Beams,
- built-in, defined, 14
- cantilever, defined, 14
- defined, 13
- loads, 16
- simply supported, defined, 14
- supports, 14
- universal, 17

Bel, 242
Belt drives, 70, 109
Bending moment,
- defined, 18
- diagrams, 20
- hogging, 18
- sagging, 18

Bending stresses, 25
Binary system, 237
Bits, 237
Block diagrams,
- connected systems, 233
- summing junction, 233
- system, 232
- take-off point, 234

Bone, torsional strength, 49
Bridge,
- materials, 4, 32
- stresses, 32
- supports, 14

Buckling, 35, 36

Cantilever,
- bending moment diagram, 21, 24
- defined, 14
- oscillations, 124
- shear force diagram, 21, 24

Capacitance, 149
Capacitor,
- ceramic, 150
- charging, 151
- defined, 149
- discharging, 155

Capacitor *(continued)*
- electrolytic, 150
- paper, 150
- parallel, 150
- phasors, 182
- plastic, 150
- series, 150
- types of, 150

Car,
- antilock brake system, 271
- engine management system, 270
- power steering system, 270
- shock absorbers, 129

Central heating system, 234
Charge, defined, 135
Charpy test, 106
Choke, 160
Circuit,
- conventions, 136
- symbols, 136

Column,
- critical buckling load, 35
- defined, 35
- eccentric loading, 39
- effective length, 35
- slender, 35
- slenderness ratio, 35

Commutator, 168
Complex waveforms, 210
Compound bars, 6
Compression, 1
Conductance, 137
Control systems,
- closed-loop, 263, 266
- continuous control, 266
- digital control, 267
- discrete-event control, 268
- elements, 264
- open-loop, 262
- principles, 262
- sequential control, 268

Current,
- defined, 135
- division, 142

Cycle, 115, 176
Cylinder, 278

Damping, 116, 128
Darlington pair, 276
Data presentation, 248
Decibel, 242, 258
Diaphragm pressure gauge, 250
Digestive biscuit, modulus of elasticity, 4
Digital signals,
- defined, 237
- parallel transmission, 237
- serial transmission, 237
- to analogue conversion, 240

Directional control valves, 277
Displacement, defined, 56
Distance, defined, 56
Door, automatic opening, 279
Drill, automatic, 279
Dynamic resistance, 198

Eddy currents, 225
E.m.f., 136
Energy,
- conservation, 94, 104
- simple harmonic motion, 120
- transfer, 94

Equilibrium conditions, 15

Faraday's laws, 176
Feedback, 253, 263
Field,
- coils, 167
- poles, 167

Filter,
- band-pass, 192
- cut-off frequency, 186
- high-pass, 186
- low-pass, 186

First moment of area, 29
Flywheel, 104
Fourier series, 210
Free body, 17, 74
Frequency,
- defined, 115, 176
- modulation, 236
- natural, 115

Friction,
- angle of, 78
- coefficient of, 78
- defined, 77

Gears, 70, 108
Gravitational potential energy, 101

Gravity,
  acceleration due to, 61
  vertical motion under, 61

Hall effect, 251
Harmonics,
  defined, 211
  non-linear circuit elements, 218
  production of, 217
H-circuit, 280
Hi-fi system, 233
Hooke's law, 4
Humidity, 249
Hydroelectricity, 104
Hysteresis loop, 225

Impact tests, 106
Impedance,
  defined, 184
  matching, 228
Inductance, defined, 160
Inductors,
  current decay, 164
  current growth, 160
  defined, 160
  ferrite core, 160
  ferromagnetic core, 160
  phasors, 182
Information, 235
Izod test, 106

Kinetic energy,
  angular, 103
  linear, 102
Kirchhoff's laws, 141

Lamp dimmer circuit, 275
LED, series resistor, 138
Limit switch, 276
Linear circuits, 210
Linear variable differential transformer, 249, 272
Line spectrum, 213
LM35, 251, 252
LVDT, 249, 272

Maximum power transfer, 228
Mean value, 176
Measurement systems, 247
Mesh analysis, 146
Middle-third rule, 40
Mobile telephone system, 238
Modulus of elasticity, 4
Moment of inertia,
  defined, 81
  table, 82
Momentum, 73
Motor,
  compound, 170
  d.c., 167
  permanent magnet, 169
  separately excited, 170
  series wound, 169
  shaft, 45
  shunt wound, 170
  speed control system, 266, 279

Neutral axis, 25
Newton's laws, 73
Node analysis, 145
Noise, 258
Non-linearity error, 252

Ohm's law, 137
Oscillations,
  damped, 116, 128
  forced, 116, 130
  free, 115
  terms, 115
  torsional, 127
  undamped, 121
Oscillators,
  Colpitts, 257
  Hartley, 257
  Pierce, 257
  principles, 255
  quartz crystal, 257
  Wien-bridge, 256
Output impedance, 252

Parallel axes theorem, 30, 83
Pendulum, simple, 125
Periodic time, 115, 176
Phase angle, 179
Phasors,
  addition, 180
  capacitors, 182
  defined, 179

Phasors *(continued)*
- inductors, 182
- parallel circuits, 193
- pure components, 182
- resistors, 182
- series circuits, 185
- subtraction, 180

Photoelectric switches, 277
Piezoelectric sensor, 251
PLC, 273
Poisson's ratio, 10
Polar second moment of area, 49
Polar section modulus, 51
Potential difference, 136
Potential energy, 100
Power,
- active, 203
- apparent, 203
- defined, 99
- electrical, 135, 199
- factor, 202
- factor improvement, 204
- maximum transfer, 228
- reactive, 203
- station system, 233
- transmission, 52, 108, 223
- true, 203

Pressure gauge, 250
Prestressed concrete beams, 40
Process control system, 272
Programmable logic controller, 273
Projectiles, 64
Pulse-width modulation, 279, 282
Punch, 46
Pyroelectric sensors, 251

Quantisation level, 239
Q-factor, 192, 199

Radian, defined, 66
Radio,
- AM receiver, 236
- communication system, 233

Radius of gyration, 36, 83
Range, 251
Reactance, 181
Reaction, 74
Reactive forces at supports, 15
Rectification, 215
Reinforced concrete beam, 7, 22
Rejector circuit, 198
Relays, 274
Resistance,
- defined, 137
- temperature detector, 248
- to voltage change, 246

Resistors,
- carbon composition, 137
- film, 137
- parallel, 139
- phasors, 182
- series, 138
- series and parallel, 140
- types of, 137
- wire-wound, 137

Resolution,
- of ADC, 241
- of vectors, 63

Resonance,
- mechanical oscillations, 130
- parallel *LCR* circuit, 196
- series *LCR* circuit, 190

Response time, 252
Rise time, 252
Riveted joints, 46
Root-mean-square value, 178
Rubber band, force–extension graph, 5

Scalar quantities, 56
Screwdriver, 45
Second moment of area, 30
Section modulus, 33
Sensitivity, 251
Sensor,
- capacitive, 248
- defined, 247
- elastic, 250
- electromagnetic, 250
- Hall effect, 251
- inductive, 249
- performance terms, 251
- piezoelectric, 251
- pyroelectric, 251
- resistive, 248
- semiconductor, 251
- types of, 248

Shear,
- defined, 45

Shear *(continued)*
force, 17, 20
modulus, 46
strain, 45
stress, 45
Sheet thickness control system, 272
Shot putter, 65
Signal processor, 247
Silicon controlled rectifier, 274
Simple harmonic motion, 116
Simply supported beam,
bending moment diagram, 20, 21, 23
defined, 14
shear force diagram, 20, 21, 23
Ski wax, 78
S/N ratio, 258
Solenoid operated valve, 277
Spanner, 45
Speed,
average, 57
defined, 56
instantaneous, 57
uniform, 58
Spool valve, 277
Spring, oscillations, 122
Stability, 252
Steel, for bridge, 4
Step voltage, 135, 236
Stopping distances, 74
Strain,
defined, 3
energy, 101
gauge, 248
Strawberry jelly, modulus of elasticity, 4
Stress,
bending, 25
defined, 2
Strut, 1, 37
Superposition, principle of, 210, 214
Susceptance, 181
System,
connected, 233
control, 262
block diagram, 232
defined, 232
measurement, 247

Tachogenerator, 250, 266
Tacoma Narrows Bridge, 131
Tension, 1
Thermistor, 248, 252
Thermocouple system, 235, 250
Thyristors, 274
Tie, 1
Time constant, 152, 156, 162, 165, 252
Torque,
and angular motion, 80
defined, 45
internal combustion engine, 52
work done by, 97
Torsion,
circular shafts, 47
defined, 47
Transducer, 247
Transformer,
audio-frequency, 226
construction, 225
copper loss, 226, 227
cores, 225
core-type, 225
eddy current loss, 225
efficiency, 227
harmonics, 218
hysteresis loss, 225
impedance matching, 228
ideal, 222
iron loss, 227
power rating, 226
principles, 222
shell-type, 225
step-down, 223
step-up, 223
symbols, 222
with resistive load, 223
Transients, 135
Transistor, 275
Transmission efficiency, 108
Tuning radios, 191

Universal beam,
defined, 17
properties, 33
torsion, 50

Universal column, properties, 36

Vector quantities, 56
Velocity,
 average, 57
 defined, 56
 instantaneous, 57
 uniform, 58
Voltage division, 142, 246

Washing machine system, 268
Weight, 73
Wheatstone bridge, 246
Wire, force–extension graph, 5
Work, 94